T0329848

FUNDAMENTALS OF HETEROCYCLIC CHEMISTRY

FUNDAMENTALS OF HETEROCYCLIC CHEMISTRY

Importance in Nature and in the Synthesis of Pharmaceuticals

LOUIS D. QUIN
Adjunct Professor, University of North Carolina Wilmington
James B. Duke Professor Emeritus, Duke University
Professor Emeritus, The University of Massachusetts

JOHN A. TYRELL, PH.D.
University of North Carolina Wilmington

A JOHN WILEY & SONS, INC., PUBLICATION

Copyright © 2010 by John Wiley & Sons, Inc. All rights reserved.

Published by John Wiley & Sons, Inc., Hoboken, New Jersey
Published simultaneously in Canada

No part of this publication may be reproduced, stored in a retrieval system, or transmitted in any form or by any means, electronic, mechanical, photocopying, recording, scanning, or otherwise, except as permitted under Section 107 or 108 of the 1976 United States Copyright Act, without either the prior written permission of the Publisher, or authorization through payment of the appropriate per-copy fee to the Copyright Clearance Center, Inc., 222 Rosewood Drive, Danvers, MA 01923, (978) 750-8400, fax (978) 750-4470, or on the web at www.copyright.com. Requests to the Publisher for permission should be addressed to the Permissions Department, John Wiley & Sons, Inc., 111 River Street, Hoboken, NJ 07030, (201) 748-6011, fax (201) 748-6008, or online at http://www.wiley.com/go/permission.

Limit of Liability/Disclaimer of Warranty: While the publisher and author have used their best efforts in preparing this book, they make no representations or warranties with respect to the accuracy or completeness of the contents of this book and specifically disclaim any implied warranties of merchantability or fitness for a particular purpose. No warranty may be created or extended by sales representatives or written sales materials. The advice and strategies contained herein may not be suitable for your situation. You should consult with a professional where appropriate. Neither the publisher nor author shall be liable for any loss of profit or any other commercial damages, including but not limited to special, incidental, consequential, or other damages.

For general information on our other products and services or for technical support, please contact our Customer Care Department within the United States at (800) 762-2974, outside the United States at (317) 572-3993 or fax (317) 572-4002.

Wiley also publishes its books in a variety of electronic formats. Some content that appears in print may not be available in electronic formats. For more information about Wiley products, visit our web site at www.wiley.com.

Library of Congress Cataloging-in-Publication Data:

Quin, Louis D., 1928–
Fundamentals of heterocyclic chemistry : importance in nature and in the synthesis of pharmaceuticals / Louis D. Quin, John A. Tyrell.
p. cm.
ISBN 978-0-470-56669-5 (cloth)
1. Heterocyclic chemistry. 2. Heterocyclic compounds—Synthesis. I. Tyrell, John A. II. Title.
QD400.Q46 2010
547′.59–dc22

2009051019

Printed in the United States of America

10 9 8 7 6 5 4 3

To our wives, Gyöngyi Szakal Quin and Ann Marie Tyrell,
with deep appreciation for their understanding and support
during the preparation of this book

CONTENTS

PREFACE

FOR WHOM THIS BOOK IS WRITTEN

For some 30 years I taught a graduate-level course in heterocyclic chemistry at Duke University and later at the University of Massachusetts. Then in 1997 I was given the opportunity at the University of North Carolina Wilmington to tailor the level of the course so as to be appropriate for undergraduates who had completed only the basic two-semester course in organic chemistry. This new one-semester course was described as a special topics course and met a curriculum requirement. The course was also open to first-year graduate students working toward the M.S. degree.

This book grew out of the lectures in that course. The subject is of course enormous, and the course had to be designed to introduce an appreciation of the vast number of parent heterocyclic systems and the importance of their derivatives (especially in medicine), both in synthetic and in natural structures, without going into excessive detail. Similarly, fundamental aspects of synthesis of representative ring systems and of their special properties as heterocycles were topics given major attention but again without going into the great detail found in more advanced books on this subject. After the first offering of the course, it was apparent that the students would benefit from a brief review of some of the reactions and properties they had encountered in their basic organic course, before these were applied to heterocyclic systems. Such reviews are included in this book.

The emphasis in this book, then, is to *teach* the elements of heterocyclic chemistry; it is not to serve as a broad reference work, and it is not competitive with the numerous more advanced books in this field. It should be noted, however, that chemists at all levels might find it useful to assist them when first entering the field, as for example those headed to research in medicinal chemistry where heterocycles abound.

A subsequent development was the offering of this course on an online basis for chemists working in pharmaceutical and other chemical industries, using the same material given in the lecture course. This course was designed and executed by my colleague Dr. John A. Tyrell and is available through the University of North Carolina Wilmington.

A solutions manual to the end-of-chapter review exercises is available for academic adopters registering through the book's Wiley website: http://www.wiley.com/WileyCDA/WileyTitle/productCd-0470566698.html.

LOUIS D. QUIN
Durham, NC

ACKNOWLEDGMENT

We are indebted to Dr. Kenneth C. Caster for reviewing the entire manuscript and for making numerous valuable comments.

CHAPTER 1

THE SCOPE OF THE FIELD OF HETEROCYCLIC CHEMISTRY

We must start out by examining what is meant by a heterocyclic ring system. To do this, we must use as examples some structures and their names, but we defer discussion of the naming systems for heterocyclic compounds to Chapter 2.

Heterosubstituted rings are those in which one or more carbon atoms in a purely carbon-containing ring (known as a carbocyclic ring) is replaced by some other atom (referred to as a heteroatom). In practice, the most commonly found heteroatom is nitrogen, followed by oxygen and sulfur. However, many other atoms can form the stable covalent bonds necessary for ring construction and can lead to structures of considerable importance in contemporary heterocyclic chemistry. Of note are phosphorus, arsenic, antimony, silicon, selenium, tellurium, boron, and germanium. In rare cases, even elements generally considered to be metallic, such as tin and lead, can be incorporated in ring systems. In a 1983 report, the International Union of Pure and Applied Chemistry (IUPAC) recognized 15 elements coming from Groups II to IV of the Periodic System capable of forming cyclic structures with carbon atoms.[1]

The compound pyridine is an excellent example of a simple heterocycle. Here, one carbon of benzene is replaced by nitrogen, without

Fundamentals of Heterocyclic Chemistry: Importance in Nature and in the Synthesis of Pharmaceuticals,
By Louis D. Quin and John A. Tyrell Copyright © 2010 John Wiley & Sons, Inc.

interrupting the classic unsaturation and aromaticity of benzene. Similarly, replacement of a carbon in cyclohexane by nitrogen produces the saturated heterocycle piperidine. Between these extremes of saturation come several structures with one or two double bonds.

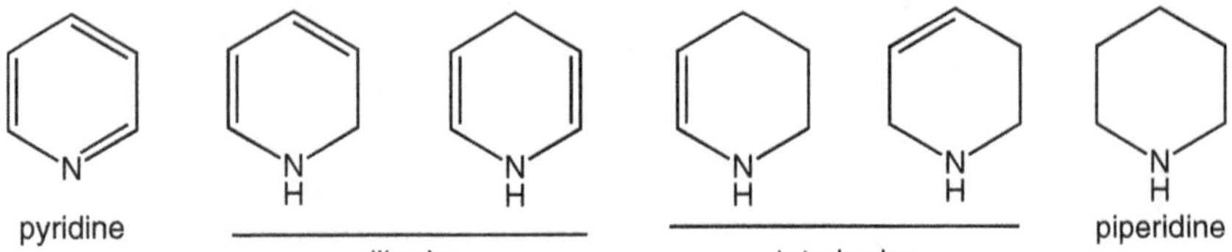

pyridine — dihydro — tetrahydro — piperidine

Rings may have more than one heteroatom, which may be the same or different, as in the examples that follow.

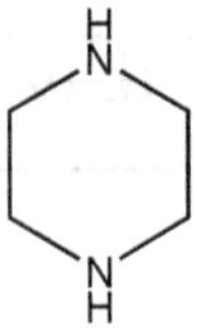

piperazine 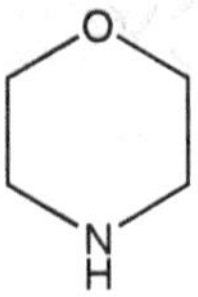 morpholine

To broaden the field, other rings may be fused onto a parent heterocycle. This gives rise to many new ring systems.

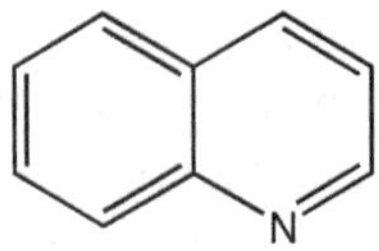

quinoline 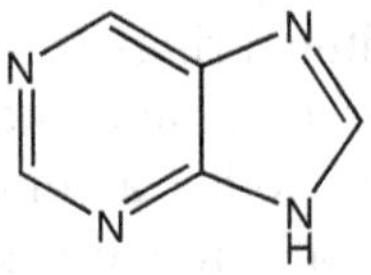 purine

By such bonding arrangements, 133,326 different heterocyclic ring systems had been reported by 1984,[2] and many more have been reported since then. But that is not the whole story; hydrogens on these rings can be replaced by a multitude of substituents, including all the functional groups (and others) common to aliphatic and aromatic compounds. As a result, millions of heterocyclic compounds are known, with more being synthesized every day in search of some with special properties, which we will consider in later chapters. A recent analysis of the organic compounds registered in *Chemical Abstracts* revealed that as of June 2007, there were 24,282,284 compounds containing cyclic structures, with heterocyclic systems making up many of these compounds.[3]

Heterocyclic compounds are far from being just the result of some synthetic research effort. Nature abounds in heterocyclic compounds, many of profound importance in biological processes. We find heterocyclic rings in vitamins, coenzymes, porphyrins (like hemoglobin), DNA, RNA, and so on. The plant kingdom contains thousands of nitrogen heterocyclic compounds, most of which are weakly basic and called alkaloids (alkali like). Complex heterocyclic compounds are elaborated by microorganisms and are useful as antibiotics in medicine. Marine animals and plants are also a source of complex heterocyclic compounds and are receiving much attention in current research efforts. We should even consider that the huge field of carbohydrate chemistry depends on heterocyclic frameworks; all disaccharides and polysaccharides have rings usually of five (called furanose) or six (called pyranose) members that contain an oxygen atom. Similar oxygen-containing ring structures also are important in monosaccharides, where they can be in equilibrium with ring-opened structures, as observed in the case of D-glucose.

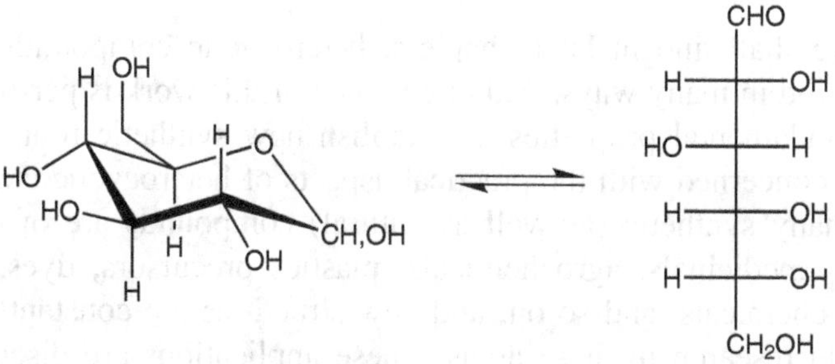

However, in this book we will not give additional attention to carbohydrates, which constitute a field all to themselves.

A low concentration of nitrogen and sulfur heterocycles also can be found in various petroleums. Coal was for years the major source of pyridine-based heterocycles, obtained by pyrolysis in the absence of oxygen (destructive distillation). An intriguing new detection of heterocycles in nature has occurred in the field of chemistry of the solar system. Pyridine carboxylic acids have been detected in a meteorite that landed in Canada (near Tagish Lake).[4] Nicotinic acid and its two isomers were isolated along with 12 methylated and other derivatives.

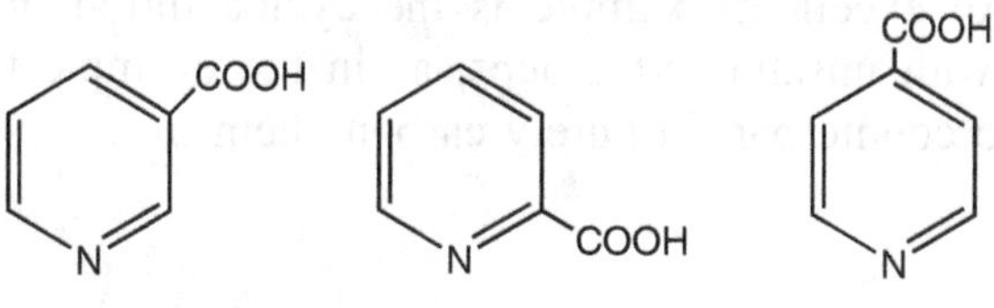

Here, great caution had to be exerted to ensure that contamination by terrestrial compounds had not occurred. One wonders what other heterocycles can be detected (and confirmed) in the current intensive research activity in astrochemistry. In this connection, molecules known as porphyrins that contain the porphin nucleus have been tentatively identified spectroscopically on the moon.

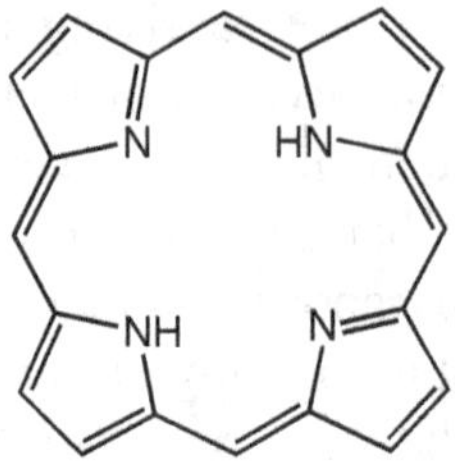

porphin

As we shall find in later chapters, heterocyclic compounds can be synthesized in many ways. Although some of this work is performed to study fundamental properties or establish new synthetic routes, much more is concerned with the practical aspects of heterocyclic chemistry. Thus, many synthetic (as well as natural) compounds are of extreme value as medicinals, agrochemicals, plastics precursors, dyes, photographic chemicals, and so on, and new structures are constantly being sought in research in these areas. These applications are discussed in Chapter 11. Medicinal chemistry especially is associated intimately with heterocyclic compounds, and most of all known chemicals used in medicine are based on heterocyclic frameworks. We shall observe many of the prominent biologically active heterocyclic compounds as this book proceeds to develop the field of heterocyclic chemistry.

Is heterocyclic chemistry somehow different from the much more familiar aliphatic and aromatic chemistry studied in basic organic chemistry courses? Certainly, many reactions used to close rings and to modify ring substituents are common to these fields, and as they are encountered, the reader should review them in a basic organic chemistry textbook. However, some reactions can be found only in heterocyclic chemistry. An excellent example is the cycloaddition of 1,3-dipolar compounds with unsaturated groups, as in the example that follows, which has no counterpart in purely carbon chemistry.

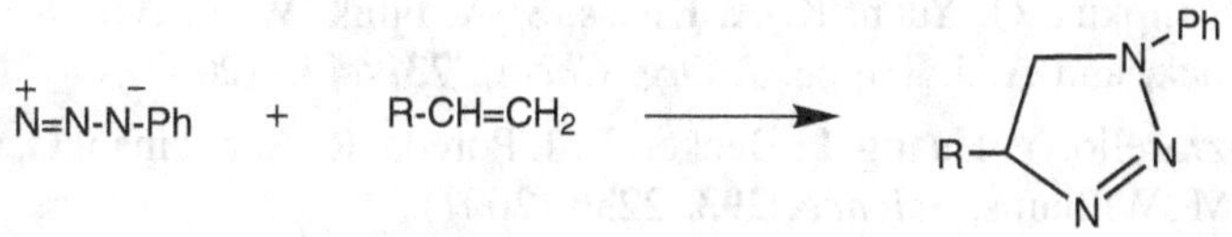

Heterocyclic compounds find use in other synthetic processes. In some cases, heterocyclic ring systems can be opened to give valuable non-cyclic compounds useful in synthetic work. Acting through their lone electron pairs or pi-systems, they can be useful ligands in the construction of coordination complexes. An example of a heterocycle frequently used for this purpose is 2,2′-bipyridyl, which is shown here as complexed to cupric ion.

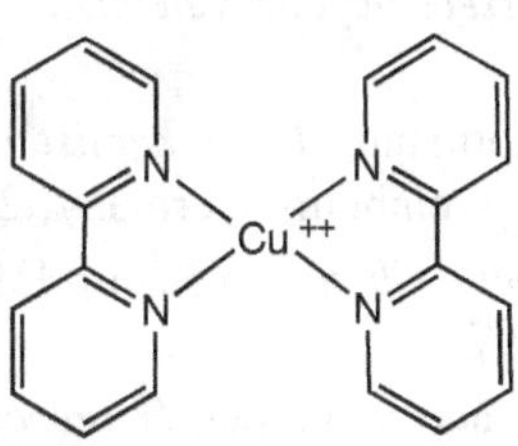

A large amount of literature is available on the subject of heterocyclic chemistry. There are advanced textbooks to help expand the knowledge imparted in this book, and there are expansive collections that cover almost all types of heterocycles and are exhaustive in providing methods of synthesis and treatment of their properties. Information on these books is given in the Appendix of this chapter. Particularly valuable is the series *Comprehensive Heterocyclic Chemistry*,[5] and this is often the first place to go for detailed information on a particular heterocyclic family. The third edition (2008) consists of 15 volumes. Other series cover physical properties or provide detailed reviews of topics or compound families in heterocyclic chemistry. There are also many books on specific topics or types of heterocycles, but these are not listed in the Appendix.

REFERENCES

(1) W. H. Powell, *Pure Appl. Chem*., **55**, 409 (1983).

(2) American Chemical Society, *Ring Systems Handbook*, Chemical Abstracts, Columbus, OH, 1984, p. 2.

(3) A. H. Lipkus, Q. Yuan, K. A. Kucas, S. A. Funk, W. F. Bartelt III, R. J. Schenck, and A. J. Trippe, *J. Org. Chem.*, **73**, 4443 (2008).

(4) S. Pizzarello, Y. Huang, L. Becker, R. J. Poreda, R. A. Nieman, G. Cooper, and M. Williams, *Science*, **293**, 2236 (2001).

(5) A. R. Katritzky, C. A. Ramsden, E. F. V. Screeven, and R. J. K. Taylor, *Comprehensive Heterocyclic Chemistry III*, Elsevier, New York, 2008.

APPENDIX

1. Some textbooks published since 1980 include the following:

   D. I. Davies, *Aromatic Heterocyclic Chemistry*, Oxford University Press, Oxford, UK, 1992.

   T. Eichner and S. Hauptmann, *The Chemistry of Heterocycles*, Second Edition, Wiley-VCH, Weinheim, Germany, 2003.

   T. L. Gilchrist, *Heterocyclic Chemistry*, Third Edition, Prentice Hall, Upper Saddle River, NJ, 1997.

   R. R. Gupta, M. Kumar, and V. Gupta, *Heterocyclic Chemistry*, Vols. 1-2, Springer Verlag, Berlin, Germany, 1998.

   J. A. Joule, *Heterocyclic Chemistry*, Wiley, New York, 2000.

   J. A. Joule and K. Mills, *Heterocyclic Chemistry at a Glance*, Blackwell Publishing, Oxford, UK, 2007.

   A. R. Katritzky, *Handbook of Heterocyclic Chemistry*, Second Edition, Pergamon, Oxford, UK, 2000.

   G. R. Newkome and W.W. Paudler, *Contemporary Heterocyclic Chemistry*, Wiley, New York, 1982.

   A. F. Pozharskii, A. T. Soldatenkov, and A. R. Katritzky, *Heterocycles in Life and Society: An Introduction to Heterocyclic Chemistry and Biochemistry and the Role of Heterocycles in Science, Technology, Medicine and Agriculture*, Wiley, New York, 1997.

2. Reference works

   A. R. Katritzky, C. A. Ramsden, E. F. V. Screeven, and R. J. K. Taylor, *Comprehensive Heterocyclic Chemistry III*, Elsevier, New York, 2008.

   A. R. Katritzky, *Advances in Heterocyclic Chemistry*, Academic Press, New York, 2009.

   A. R. Katritzky, *Physical Methods in Heterocyclic Chemistry*, Academic Press, New York, 1974.

   R. C. Elderfield, *Heterocyclic Compounds*, Wiley, New York, 1950

   A. Weissberger, *The Chemistry of Heterocyclic Compounds*, Wiley, New York, 2008.

D. H. R. Barton and W. D. Ollis, Editors, *Comprehensive Organic Chemistry*, Vol. 4, Pergamon, Oxford, UK, 1979.

American Chemical Society, *Ring Systems Handbook*, Chemical Abstracts, Columbus, OH, 1984.

G. W. Gribble and J. A. Joule, *Progress in Heterocyclic Chemistry*, Elsevier, Oxford, UK, 2009.

R. R. Gupta, *Topics in Heterocyclic Chemistry*, Springer, Berlin, Germany, 2009.

CHAPTER 2

COMMON RING SYSTEMS AND THE NAMING OF HETEROCYCLIC COMPOUNDS

2.1. GENERAL

Heterocyclic compounds are among the earliest organic compounds to be purified and recognized as discrete substances, although their structures remained unknown for a long time. The science of chemistry was advancing rapidly in the first half of the nineteenth century thanks to the studies of some brilliant chemists. The structure of organic compounds remained a mystery, however, until about 1860 when the work of Archibald Couper of Scotland, Friedrich Kekulé of Germany, and Alexander Butlerow of Russia led to the recognition of the tetrahedral nature of the carbon atom and the devising of the first structural formulas. The structural formulas we use today are most closely associated with the name of Kekulé. These were exciting days in chemistry, as the true nature of organic compounds began to unfold and many new compounds were being made. A brief but informative account of the development of organic chemistry has been given by H. W. Salzberg in a monograph published by the American Chemical Society.[1] There was no systematic naming system in use in these early days, and chemists simply assigned what we now call common names to these compounds.

Fundamentals of Heterocyclic Chemistry: Importance in Nature and in the Synthesis of Pharmaceuticals,
By Louis D. Quin and John A. Tyrell Copyright © 2010 John Wiley & Sons, Inc.

By and large, these early heterocyclic compounds were isolated from natural sources; versatile synthetic procedures followed only after many years of research. Some examples of early compounds are as follows:

Uric acid (1776, by Scheele from human bladder stones)
Alloxan (1818, by Brugnatelli on oxidation of uric acid)
Quinoline (1834, by Runge from coal distillates, called coal tar)
Melamine (1834, by Liebig by synthesis)
Pyrrole (1834, by Runge in coal tar, but first purified by Anderson in 1858)
Pyridine (1849, by Anderson by pyrolysis of bones)
Indole (1866, by Baeyer from degradation of indigo)
Furan (1870, from wood and cellulose destructive distillation)

Note that not all of the compounds represent the unmodified parent ring; often these were obtained many years later. The structures are given in Table 2.1.

Natural compounds can be complex, beyond the ability of the early chemists to understand them. An excellent example is the first isolation of what proved many years later to be deoxyribonucleic acid (DNA). This was accomplished by Friedrich Miescher in Germany in 1869, who

Table 2.1. Some Early Heterocyclic Compounds of Natural Origins

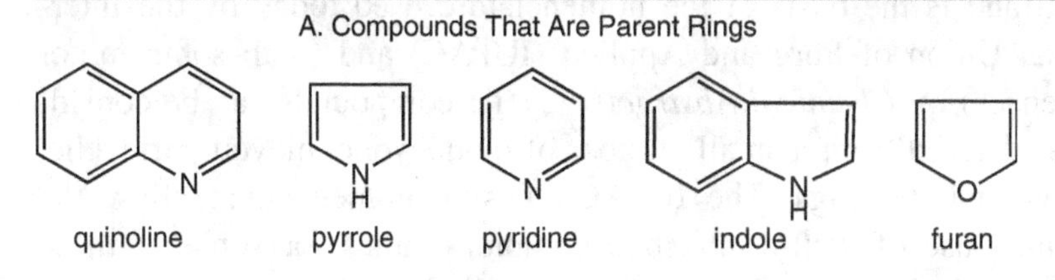

isolated the substance from cell nuclei. He gave it the name nuclein, which was a precursor of our present name nucleic acid. He recognized that it differed from a protein and contained nitrogen and phosphorus, but it could go no further structurally. Many years later, it was recognized that nuclein was rich in several heterocyclic "bases," and from our present viewpoint, we can claim it as a discovery in heterocyclic chemistry (it also is claimed by phosphorus and carbohydrate chemists), and ultimately, its composition and stereo structure as the famous double helix were established. A fascinating discussion of Miescher's truly pioneering work, which is now generally unrecognized, is given by R. Dahm.[2] We will examine its structure in Chapter 9.

It is no different in heterocyclic chemistry than in other branches of organic chemistry; with millions of compounds to deal with, convenient, generally accepted systems must be available for the naming of the compounds, so that they can be classified and their all-important structures can be deduced universally from their names. Many of the early common names are still in use today (e.g., pyridine, quinoline, and nicotinic acid), but as the great proliferation of heterocyclic compounds commenced in the latter half of the nineteenth century, the need for effective nomenclature systems became clear, and a highly versatile system was created by A. Hantzsch in 1887 and independently by O. Widman in 1888 for the naming of 5- and 6-membered rings containing nitrogen. The system was later applied to different ring sizes and to rings with other heteroatoms. It is now known as the Hantzsch–Widman system and is the basis of the nomenclature used today by the International Union of Pure and Applied (IUPAC) and (with some minor differences) by *Chemical Abstracts*. Cyclic compounds can be considered as derived from a small number of monocyclic, bicyclic, tricyclic, or larger parent rings. The IUPAC rules of nomenclature allow the continued use of well-established common names for some of these fundamental ring systems, but as we will find, there are systematic names also in use for them. Parent rings are known where the number of atoms in the rings can be from 3 to 100, but much of heterocyclic chemistry is centered around rings of 5 or 6 members, just as is true of all-carbon (carbocyclic) systems. The naming of complex heterocycles can be a difficult task and is beyond the purpose of the present discussion. Here, we will concentrate on the simpler cases, but more complete discussions can be found in the published IUPAC rules[3] and also in *Chemical Abstracts*.

2.2. NAMING SIMPLE MONOCYCLIC COMPOUNDS

The common, accepted names of most of the important monocyclic parents are given in Table 2.2, along with the systematic names from the Hantzsch–Widman naming system. The latter names are derived from the following four rules:

1. The heteroatom is given a name and is used as a prefix: N, aza-; O, oxa-; S, thia-; P, phospha-; As, arsa-; Si, sila-; Se, selena-, B, bora, and so on. The “a” ending is dropped if the next syllable starts with a vowel. Thus “aza-irine” is properly written “azirine.”
2. Ring size is designated by stems that follow the prefix: 3-atoms,-ir-; 4-atoms, -et-; 5-atoms, -ol-; 6-atoms, -in-; 7-atoms, -ep-; 8-atoms, -oc-; 9-atoms, -on-; and so on.
3. If fully unsaturated, the name is concluded with a suffix for ring size: 3-atoms, -ene (except -ine- for N); 4-, 5-, and 6-atoms, -e; 7-, 8-, and 9- atoms, -ine.
4. If fully saturated, the suffix is -ane for all ring sizes, except for N, which uses -idine for rings of 3-, 4-, or 5-atoms, and for 6-atoms, a prefix of hexahydro- is used. Also, the name oxane, not oxinane, is used for the 6-membered ring with O present. Other exceptions exist for P, As, and B rings, but they will not be given here.

Table 2.2 shows the application of the above rules to several N, O, and S rings. However, it is preferable and acceptable to use the common names in some cases, and these are included in parentheses.

The naming system easily accommodates the case of partial saturation of the double bonds by designating with numbers the positions on the ring where hydrogen has been added. For this purpose, the heteroatom is designated position 1 on the ring, and the numbering proceeds through the site of hydrogenation. If one double bond is removed, the prefix dihydro- is used; with two double bonds removed, it is tetrahydro-. The following examples will make this system clear.

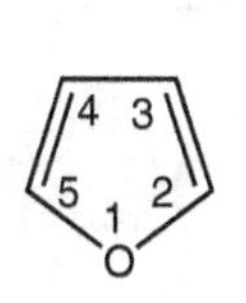

furan

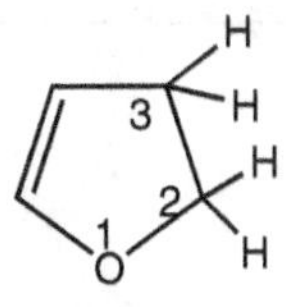

2,3-dihydrofuran

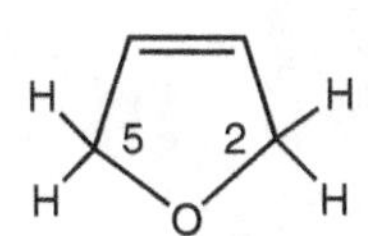

2,5-dihydrofuran

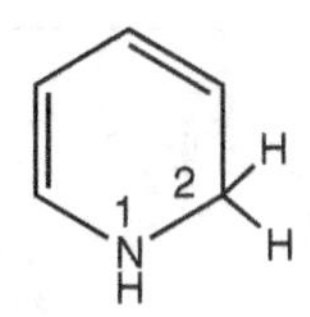

1,2-dihydropyridine

Table 2.2. IUPAC and Common Names for Monocyclic Heterocycles

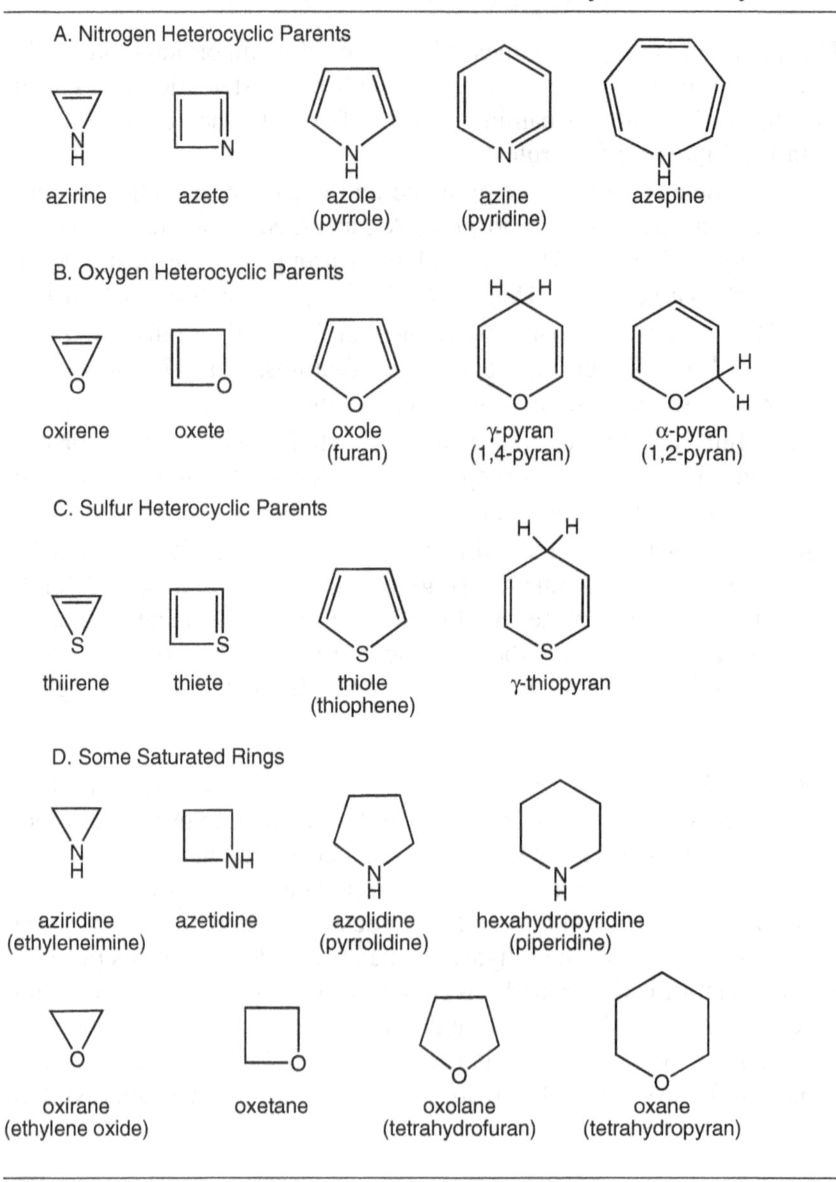

There is an alternative system, sometimes useful in complex structures, where the position of the remaining double bond in a partially hydrogenated compound is indicated by a Greek "delta" with a superscript of the ring positions bearing the double bond. Using the dihydro furans as examples, we have the following:

$\Delta^{3,4}$-dihydrofuran $\Delta^{2,3}$-dihydrofuran

2.3. HANDLING THE "EXTRA HYDROGEN"

There is a special problem resulting from isomerism in certain heterocyclic systems that requires clarification in the name. Consider the case of pyrrole: There are actually two additional isomeric forms that result from apparent 1,3-shifts of hydrogen starting from the familiar structure we have already observed. This is referred to as the "extra-hydrogen" problem, and the naming of the isomers is handled by simply adding a prefix that indicates the number of the ring atom that possesses the hydrogen, thus, 1H, 2H, and 3H. In the case of pyrrole, there is no stability to the isomers when the extra hydrogen is on carbon, although the double-bond structure can be stabilized by proper substitutions of the hydrogens. When these unstable forms are created in a synthesis, they immediately rearrange to the form with H on nitrogen. This is properly known as 1H-pyrrole, but the convention is followed that the 1H designation is dropped if H appears on the heteroatom.

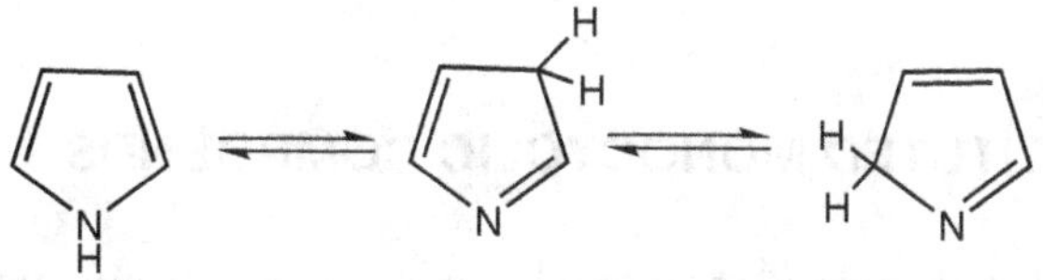

1H-pyrrole 3H-pyrrole 2H-pyrrole

An example of a stabilized 3H-pyrrole is shown as follows:

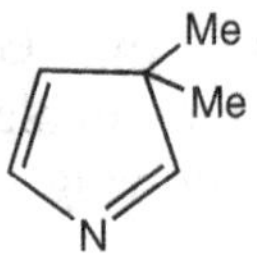

3,3-dimethyl-3H-pyrrole

The extra-hydrogen problem can occur in any ring system of nitrogen containing an odd number of ring atoms but not of course with an even number because there is no H to relocate (as in pyridine). For example, it is known in the azirine system.

1H-azirine

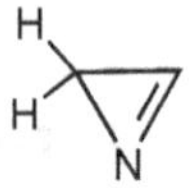

2H-azirine

The extra-hydrogen problem can also appear in odd-numbered rings containing other heteroatoms, phosphorus for example, and in some oxygen cycles, as in the pyrans (and the related thiopyrans). Note here the use of Greek letters to imply the location of the extra hydrogen, using the convention that the carbon next to the heteroatom is designated the alpha position.

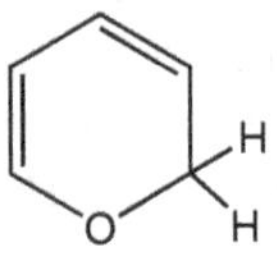

2H-pyran
(α-pyran)

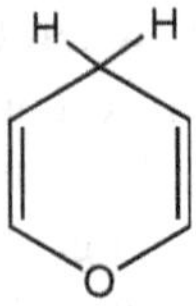

4H-pyran
(γ-pyran)

2.4. SUBSTITUTED MONOCYCLIC COMPOUNDS

With the rules discussed previously, we can name any parent monocyclic heterocycle with a single heteroatom, in any state of unsaturation. Compounds in which ring hydrogen is replaced by one or more of the common functional groups of organic chemistry also are readily named, by assigning numbers to the ring atom(s) bearing the substituents, starting with the heteroatom as number 1. The functional groups are placed alphabetically in the name. Some examples are as follows:

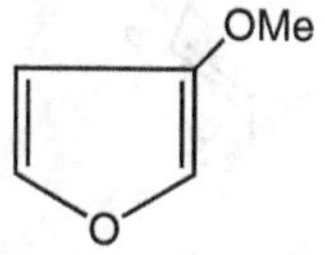

3-methoxyfuran

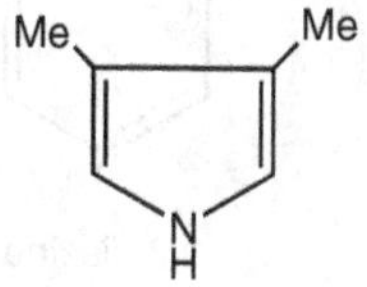

3,4-dimethyl-1H-pyrrole

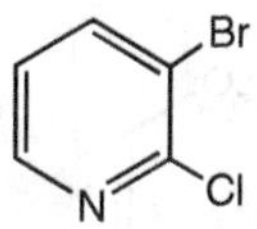

3-bromo-2-chloropyridine

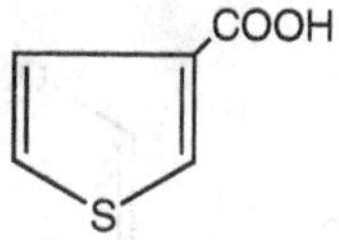

thiophene-3-carboxylic acid

2.5. RINGS WITH MORE THAN ONE HETEROATOM

Now we have to consider the common case where more than one heteroatom is present in the ring. The usual rules for stems to indicate ring size and suffixes for degree of saturation are used, as are the prefixes for the various heteroatoms. They are listed in the following order of priorities, derived from the main groups of the Periodic System, and then within each group by increasing atomic number: Group VI (O > S > Se > Te) > Group V (N > P > As) > Group IV (Si > Ge) > Group III (B). This listing can be simplified greatly by taking out the most commonly found heteroatoms in their order, which gives O > S > N > P. Each heteroatom is then given a number as found in the ring, with that of highest priority given position 1. Some additional points include the following (examples in Table 2.3 will illustrate these points):

- A saturated heteroatom with an extra-hydrogen attached is given priority over an unsaturated form of the same atom, as in 1H-1,3-diazole (see the following discussion).
- The numbers are grouped together in front of the heteroatom listings (thus, 1,3-oxazole, not 1-oxa-3-azole).
- The heteroatom prefixes follow the numbers in the priorities given previously.

Table 2.3. Some Multiheteroatom Systems

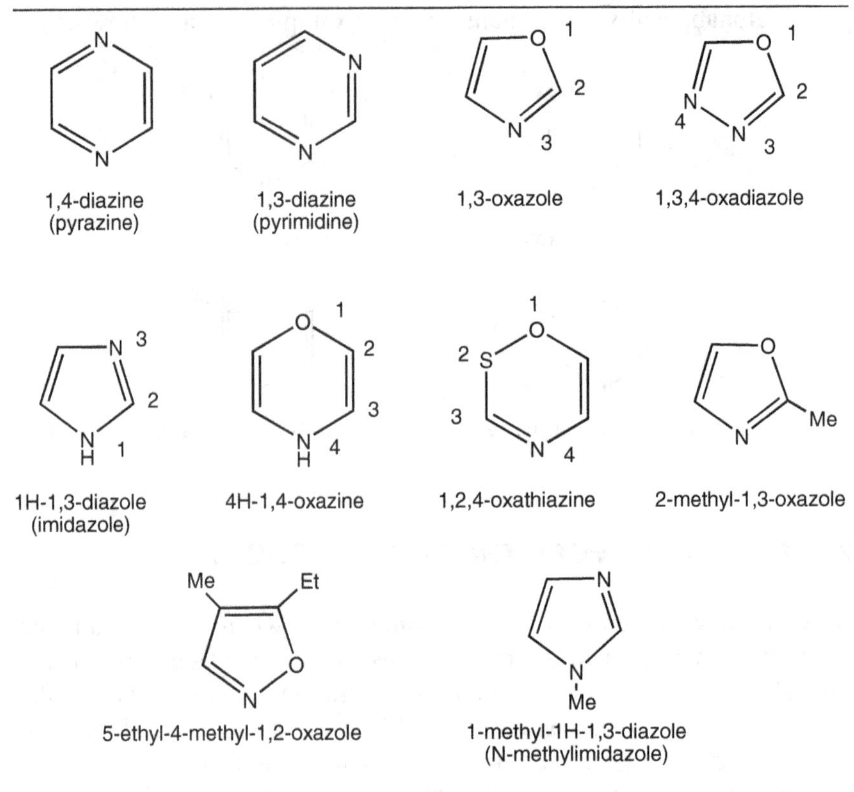

- Punctuation is important; in the examples to follow, a comma separates the numbers and a dash separates the numbers from the heteroatom prefixes.
- A slight modification is used when two vowels adjoin; one is deleted, as in the listing for "oxaaza," which becomes simply "oxaza."
- As for monohetero systems, substituents on the ring are listed alphabetically with a ring atom number for each (not grouped together).

2.6. BICYCLIC COMPOUNDS

We have now seen rules that will allow the naming of *any* monocyclic heterocycle. We next consider systems where two rings share a common single or double bond, which are said to be fused rings. A common case is where a benzene ring is fused to a heterocyclic ring. The name begins with the prefix "benzo." The point of attachment is indicated by a letter that defines the "face" of the heterocycle involved. Thus, the 1,2- position on the heterocyclic ring is always the "a-face," 2,3- is the "b-face," 3,4- is the "c-face," and so on. After the name is established, the ring atoms are given new numbers for the entire bicycle. In Table 2.4 and in subsequent examples, the letters for the faces of the monocycle are placed inside the ring, and the numbers for ring positions of the bicycle taken as a whole are shown on the outside. Note that the final numbering always begins at a position next to the benzo group and that the heteroatoms are given the lowest numbers possible, observing the O > S > N > P rule. The positions of ring fusion bear the number of the preceding ring atom with the letter "a" attached. Brackets are used around the face letter, and the name is put together without spaces, except that a dash separates the bracket from ring numbers if present, as in benzo[d]-1,3-thiazole. A convention frequently followed is to write the structure with the heteroring on the right and with its heteroatom at the bottom.

If two heterocyclic rings are fused, additional rules are required. A parent ring is selected, and the other ring is considered fused on, as was observed for benzene fusion. Some rules are as follows:

- If one ring contains N, it is considered the parent, and its name is placed last in the compound's name.

Table 2.4. Benzo-Fused Systems

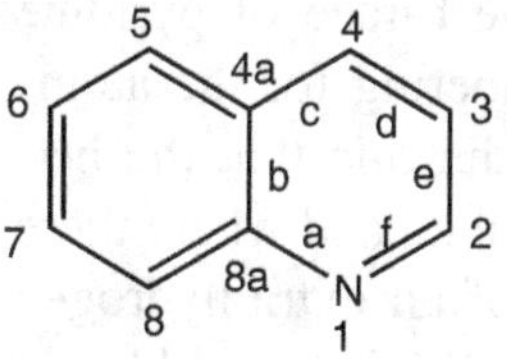

benzo[b]pyridine (quinoline)	benzo[c]pyridine (isoquinoline)	1H-benzo[b]pyrrole (indole)

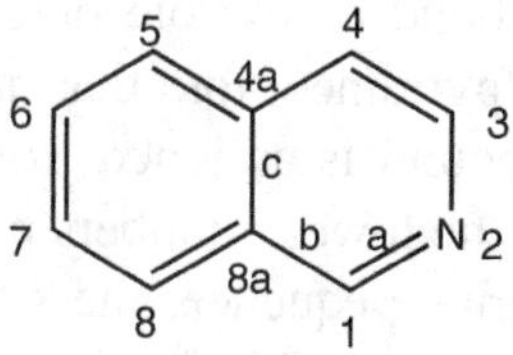

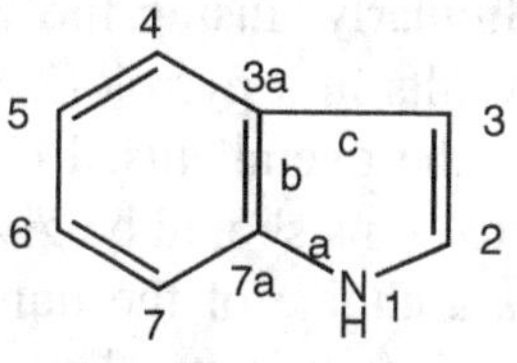

- If both rings contain N, the larger ring is the parent.
- If both rings are of the same size, that with the most N atoms is the parent, or if the same number of N atoms is present, that fusion of the rings that gives the smallest numbers for N when the bicycle is numbered is chosen.
- If no N is present, O has priority over S over P, and then the above rules are applied.
- The ring fused onto the parent has the suffix "o"; common names are used (with modification) where possible to simplify the name. Some examples are pyrido for pyridine, pyrrolo for pyrrole, thieno for thiophene, furo for furan, imidazo for imidazole, pyrimido for pyrimidine, pyrazino for pyrazine, among others.
- The face letter of the parent ring where the fusion occurs is placed in brackets preceding the name of that ring. The position numbers of the fused ring are placed inside the brackets before the face letter of the parent ring, separated by a comma. The proper numbers for the fused ring are those that are encountered as one goes around the ring in the same direction as going alphabetically around the faces of the parent. (One can liken this to the meshing of two gears.) These need not be in numerical order. Some examples will illustrate the two possible situations. Thus, fusing the 2,3-bond of furan onto the b-face of pyrrole, taken as the parent, results in the name 6H-furo[2,3-b]pyrrole.

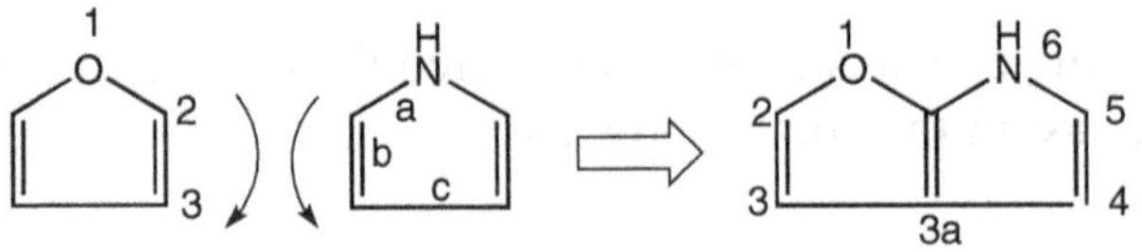

furo-[2,3-b fusion]pyrrole

6H-furo[2,3-b]pyrrole

Similarly, fusing the 2,3-bond of pyrrole onto the b-face of pyridine results in a pyrrolo[2,3-b]pyridine. Note that numbering for the atoms in the overall fused compound is assigned from the rule that the heteroatoms should be given the lowest numbers possible, and where there is a choice of the numbering sequence, the site of an extra-hydrogen is given priority. Thus, in pyrrolo[2,3]pyridine, numbering could begin from either N as position one, because the numbers would be 1,7 starting from either N, but NH has priority.

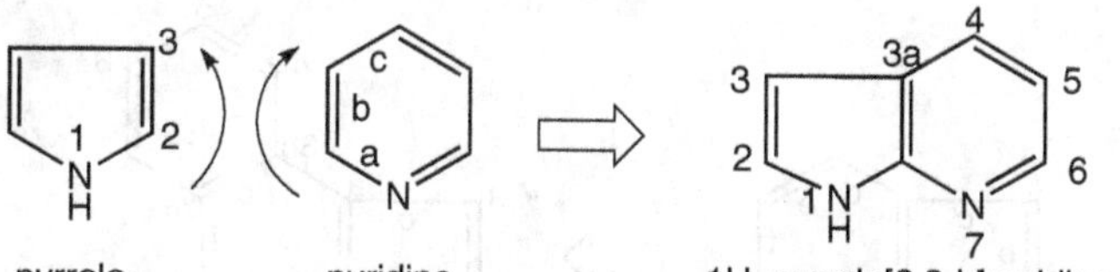

pyrrolo- pyridine 1H -pyrrolo[2,3-b]pyridine

An example where the numbers are in reverse order is pyrrolo[3,2-b]pyrrole. Note that the numbering technique clearly distinguishes between isomers of pyrrolopyrrole.

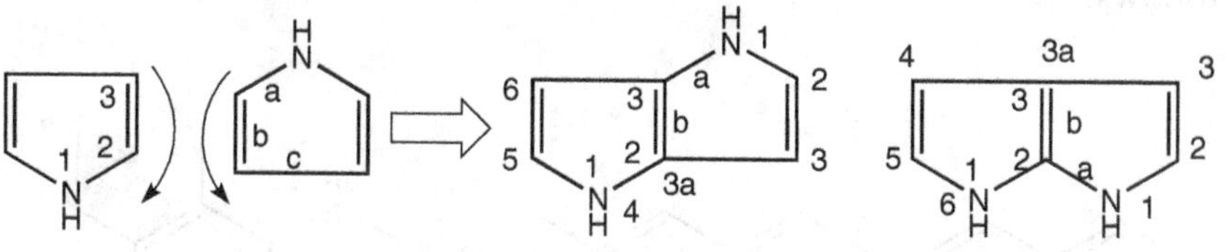

(3,2- not 2,3-) 1H,4H-pyrrolo[3,2-b]pyrrole 1H,6H-pyrrolo[2,3-b]pyrrole

Some other examples are as follows:

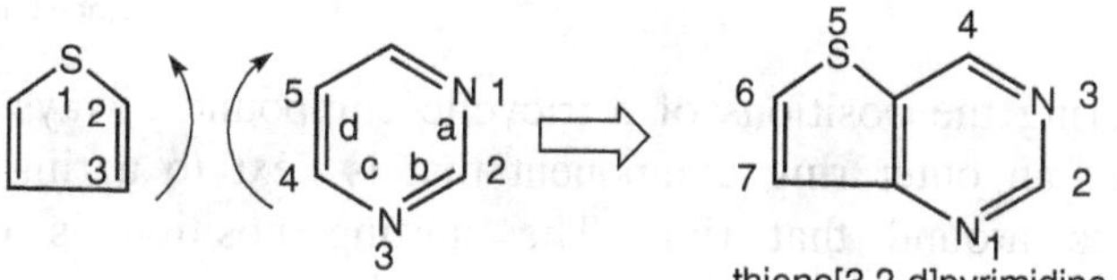

thieno[3,2-d]pyrimidine

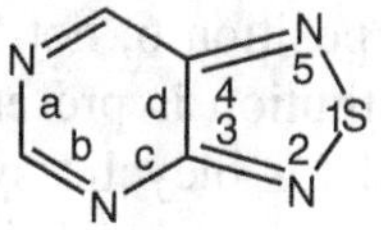

1,2,5-thiadiazolo[3,4-d]pyrimidine 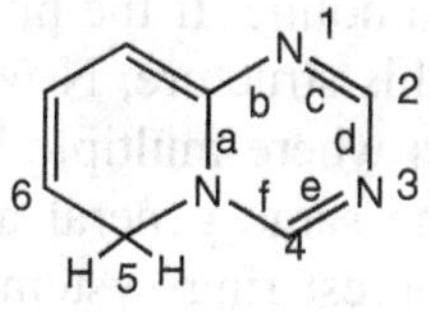 5H-pyrido[1,2-a]1,3,5-triazine

2.7. MULTICYCLIC SYSTEMS

The general approach is similar to that for bicyclic compounds. The parent is taken as the largest multicyclic system with a common name, and then other rings are fused on as observed in the preceding section. The fusion of benzene is illustrated by the compound benzo[e]indole, with indole being the name for the largest heterocycle that can be recognized.

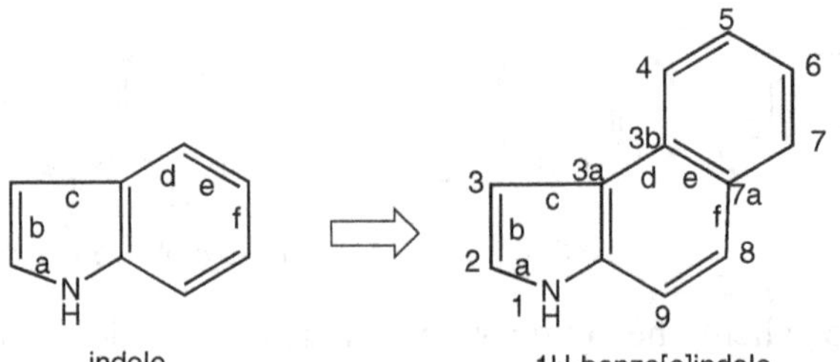

indole 1H-benzo[e]indole

Two of the isomers that can be formed from quinoline are shown as follows:

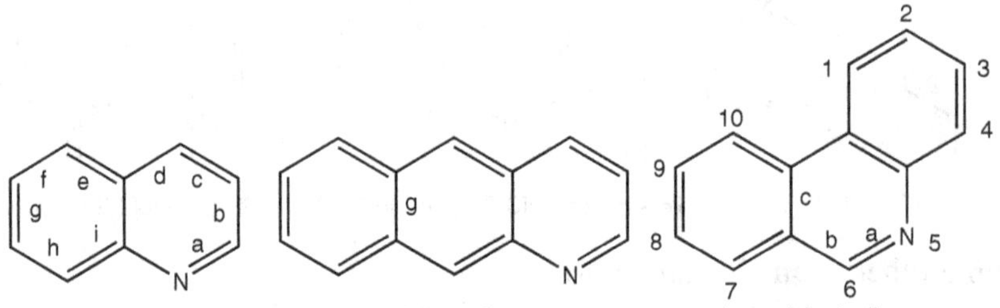

quinoline benzo[g]quinoline benzo[c]quinoline (phenanthridine)

Numbering the positions of a tricyclic compound always starts at an atom of an outer ring component that is next to a ring fusion and proceeds around that ring. The starting position is chosen that gives the heteroatoms the lowest possible numbers, as shown for benzo[c]quinoline. If the numbering had started at the position marked as 10 on this structure, N would have been position 6, not 5.

Systems where multiple heteroatom substitution is present are handled by the same general approach as used for bicyclic systems; we find the largest ring system that has a simple name and then specify the point of attachment of other rings. We observe in one example that follows, a fusion of furan at its 3,4- position with the parent cinnoline. As before, the numbers outside the rings are the final numberings for all members of the compound.

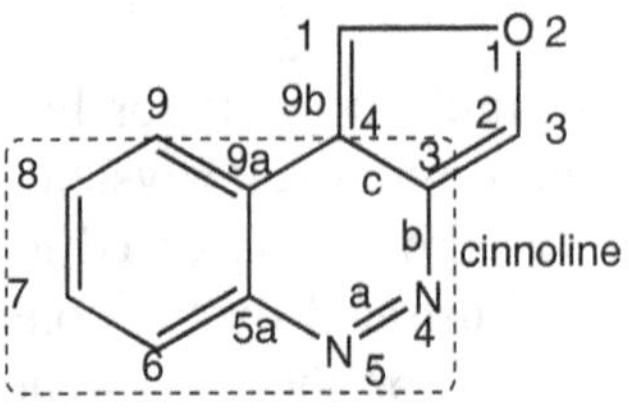

furo[3,4-c]cinnoline

This systematic approach to naming heterocycles can be extended to those containing many rings, and indeed there are a multitude of such structures. Just glancing at Chemical Abstracts *Ring Systems Handbook* will confirm this point. The *Handbook* is of major value in finding correct names for complex multicyclic structures. We will not go into the naming of complex multiring systems, but we will present just one example to illustrate the approach. Again, the naming begins with recognition of a parent with a common name if possible. An attached ring is then specified by the usual approach, and then another ring attached to that is specified. In the example that follows, the parent ring is quinoxaline; there is an imidazole ring attached to that, and then a pyridine ring is attached to the imidazole. The correct name is pyrido[1′,2′:1,2]imidazo[4,5-b]quinoxaline. The designation of the fusion of the third ring component differs slightly; ring numbers are used to indicate a face rather than a letter, and the ring numbers for the attachment of the last ring are given with a prime symbol.

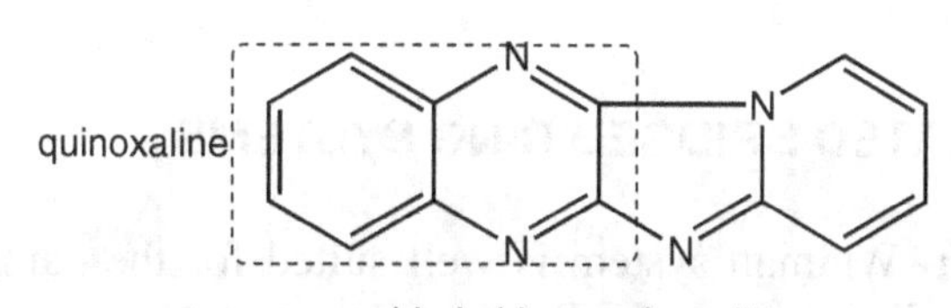

a pyrido-imidazo-quinoxaline

2.8. THE REPLACEMENT NOMENCLATURE SYSTEM

At this point, we can introduce an entirely different system of nomenclature that is nevertheless accepted by IUPAC and is extremely valuable in multicyclic and bridged saturated systems. This is the "replacement system," where the hydrocarbon name that would correspond to the entire ring structure, as if no heteroatom were present, is stated, and then given a Hantzsch–Widman prefix and number for the heteroatom(s). Thus, phenanthridine shown previously has the ring framework of the hydrocarbon phenanthrene, with N at position 5. The replacement name would be 5-azaphenanthrene.

phenanthrene 5-azaphenanthrene

This system also can be used for simpler ring systems. Thus, phosphabenzene was the name first used when this compound initially was synthesized in 1971 by A. Ashe.[4] The Hantzsch–Widman name would be the less informative phosphinine.

2.9. SATURATED BRIDGED RING SYSTEMS

The Hantzsch–Widman system is well suited for the naming of saturated monocyclic compounds. Bridged ring systems, however, present a special challenge, and here the replacement system is preferred. An example is provided by the framework of the bicyclic hydrocarbon norbornane. Heteroatoms may in principle be substituted for any of the carbons. To illustrate, nitrogen substitution at position 7 would give rise to the name 7-azanorbornane.

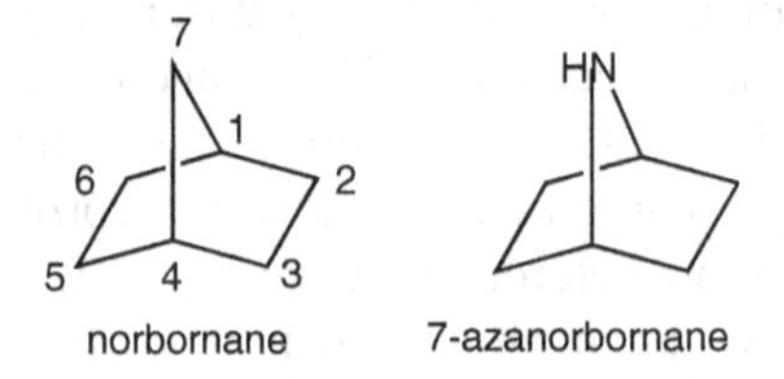

norbornane 7-azanorbornane

The parent bridged hydrocarbons also have systematic IUPAC names, and the replacement system can be applied to these names. The rules for their naming can be found elsewhere.[5] In brief, numbering begins at a bridgehead atom and proceeds through the longest chain connecting the two bridgehead atoms, then goes through the next longest, and finally the shortest. The alkane name, corresponding to the total number of atoms in the rings, is preceded by "bicyclo." The numbers are placed in brackets in decreasing order of the three different chains attached to the bridgehead atoms. Thus, norbornane has the IUPAC name bicyclo[2.2.1]heptane, and 7-azanorbornane would be 7-azabicyclo[2.2.1]heptane.

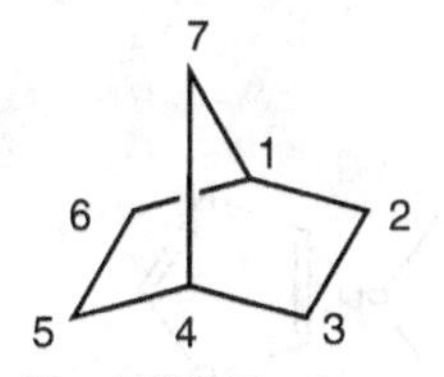

bicyclo[2.2.1]heptane

1-azabicyclo[2.2.1]heptane

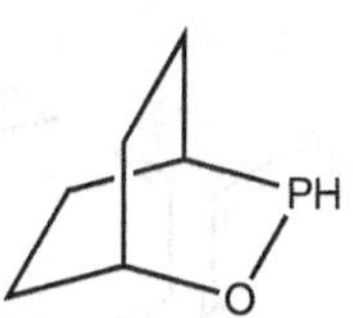

2-oxa-3-phosphabicyclo[2.2.2]octane

REFERENCES

(1) H. W. Salzberg, *From Caveman to Chemist*, American Chemical Society, Washington, D.C., 1991.

(2) R. Dahm, *American Scientist*, **96**, 320 (2008).

(3) IUPAC Commission on Nomenclature of Organic Compounds, *Pure Appl. Chem.*, **55**, 409 (1983).

(4) A. Ashe, III, *J. Am. Chem. Soc.*, **93**, 3293 (1971).

(5) IUPAC, *Nomenclature of Organic Compounds*, Pergamon Press, Oxford, UK, 1979.

REVIEW EXERCISES

2.1. Write the trivial name for each of the following structures:

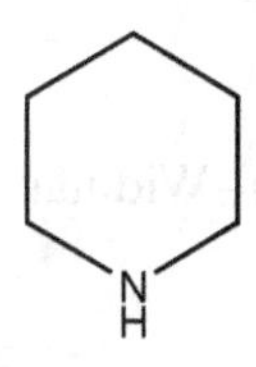

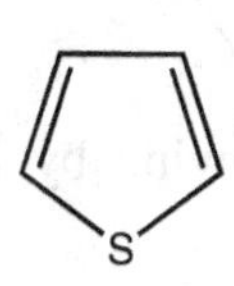

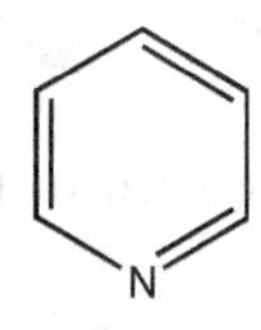

2.2. Write the structure for each of the following trivial names:

indole pyran isoquinoline

2.3. Fill in the Hantzsch–Widman name for each structure:

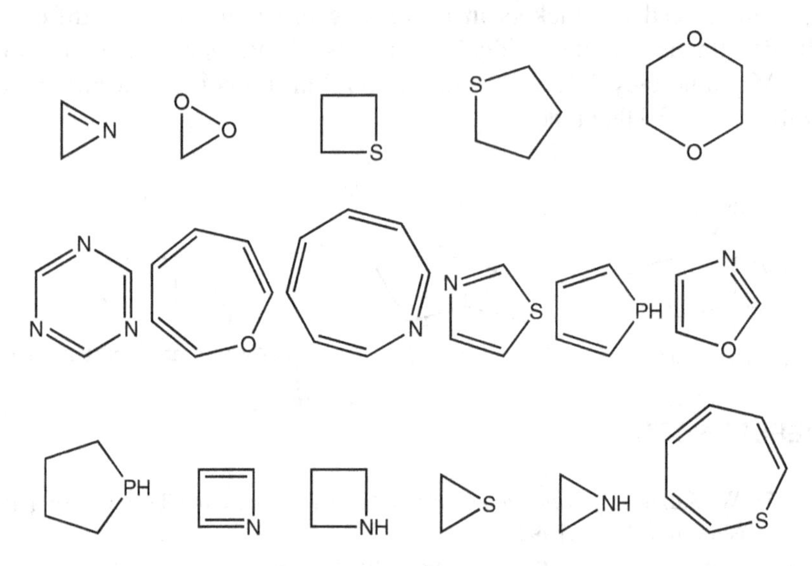

2.4. Write a structure for each name:

2H-azirine dioxirane thietane thiolane 1,4-dioxane
1,3,5-triazine oxepine azocine 1,3-thiazole phosphole
phospholane azete azetidine thiirane aziridine

2.5. Name each of the following by replacement nomenclature:

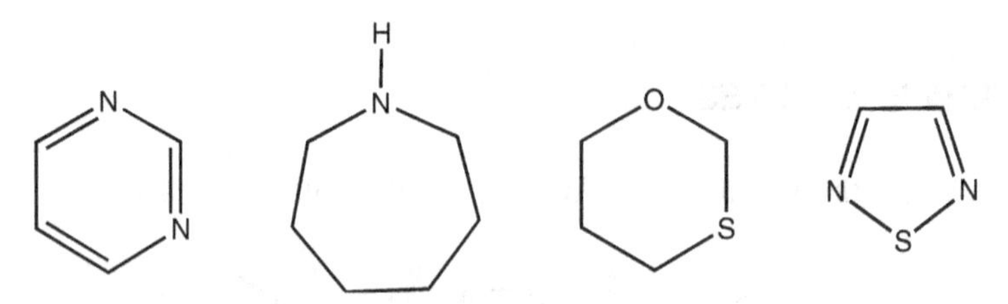

2.6. Name each of the following by Hantzsch–Widman nomenclature:

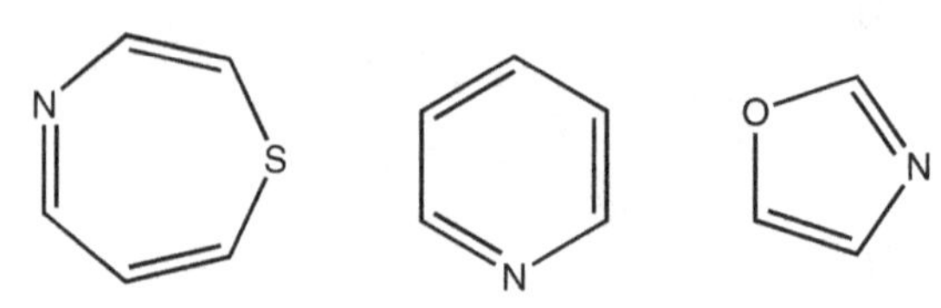

2.7. Pyrimidine has which of the following features?

a. Five-membered ring with two nitrogen atoms
b. Six-membered ring with a single nitrogen atom
c. Six-membered ring with two nitrogen atoms
d. Six-membered ring with one oxygen and one nitrogen atom
e. Two fused 6-membered rings

2.8. Consider the structure

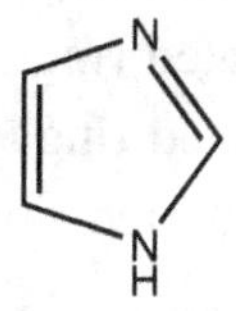

Which is the trivial name?

a. imidazole
b. pyrimidine
c. piperidine
d. piperazine

2.9. Consider the structure

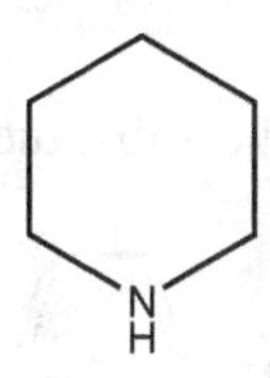

Which is the trivial name?

a. pyrrolidine
b. carbazole
c. chroman
d. morpholine
e. piperidine

2.10. Consider the structure

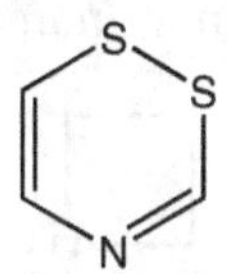

Which is the correct Hantzsch–Widman name?

a. dithiapyridine
b. 1,2,4-dithiazine
c. 1,3,4-dithiazine
d. 1,2,4-azathiazine

2.11. In the Hantzsch–Widman system, the suffix "epane" indicates what structural feature?

a. A 7-membered saturated ring
b. A 6-membered saturated ring
c. A 5-membered saturated ring
d. A 6-membered unsaturated ring

2.12. Sketch the structure for each of the following:

piperidine isoquinoline selenacyclopentane

2.13. Give the trivial name for the following structures:

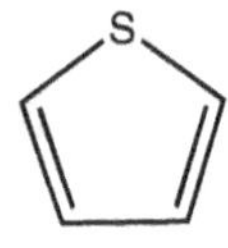

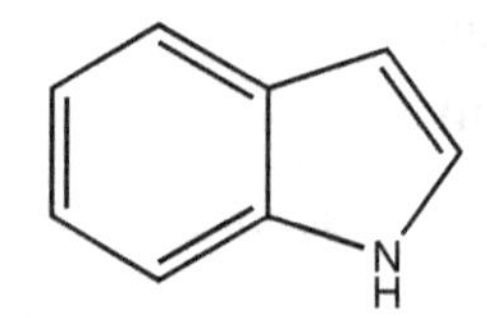

2.14. Select the number for the substituent for the following oxazole:

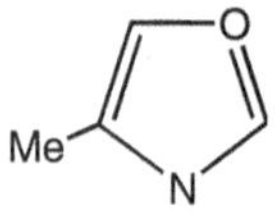

Me O N

a. 1-methyl
b. 2-methyl
c. 3-methyl
d. 4-methyl
e. 5-methyl

2.15. Sketch the structure for thiepine.

2.16. Give the Hantzsch–Widman name for the following:

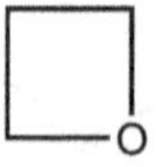

2.17. Write the structure for azete.

2.18. Write the structure for 1-thia-4-azacycloheptane.

2.19. The best name for the following fused compound is

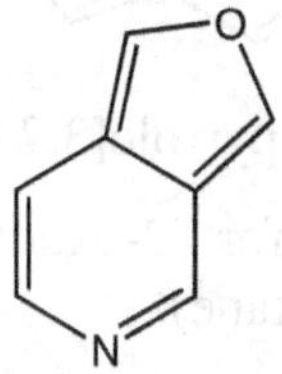

a. furo[3,4-c]pyridine
b. furo[4,3-b]pyridine
c. pyrido[3,4-c]furan
d. pyrido[4,3-b]furan

2.20. The best name for the following fused compound is

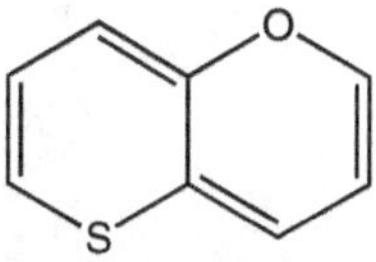

a. thiino[2,3-b]pyran
b. thiino[3,2-b]pyran
c. thiolo[2,3-a]pyran
d. pyrano[2,3-b]thiine

2.21. Draw the structure for benzo[c]quinoline.

2.22. Provide the name for each compound.

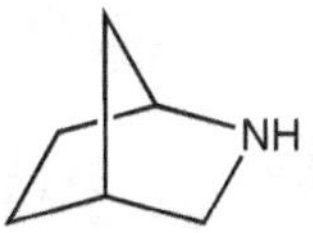

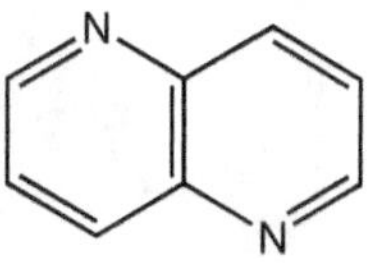

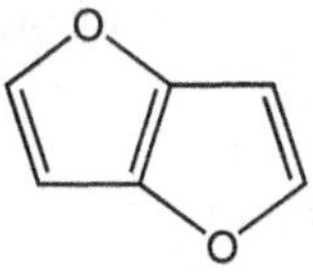

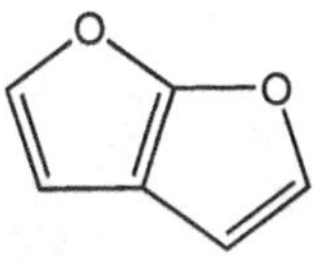

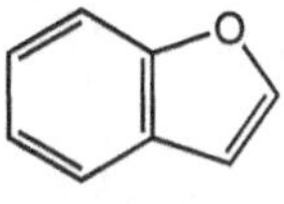

2.23. Draw the structure for 2-ethyl-3H-pyrrole.

2.24. Provide the name for:

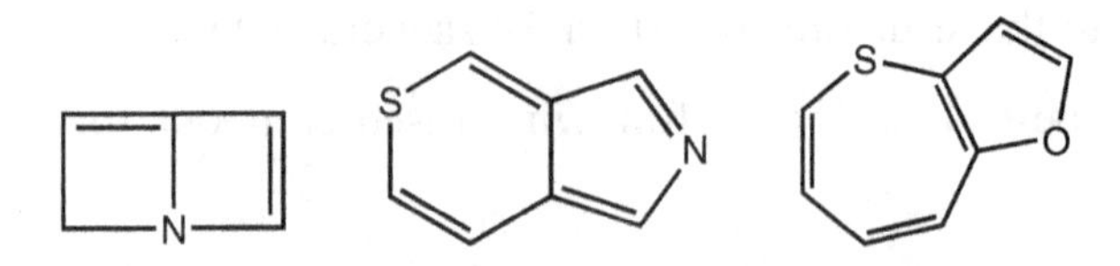

2.25. Draw the structure for pyrrolo[3,2-b]pyridine.

2.26. Draw the structure for 2-oxanorbornane (also known as 2-oxabicyclo[2.2.1]heptane).

2.27. Name the following (use both fused ring nomenclature and replacement nomenclature).

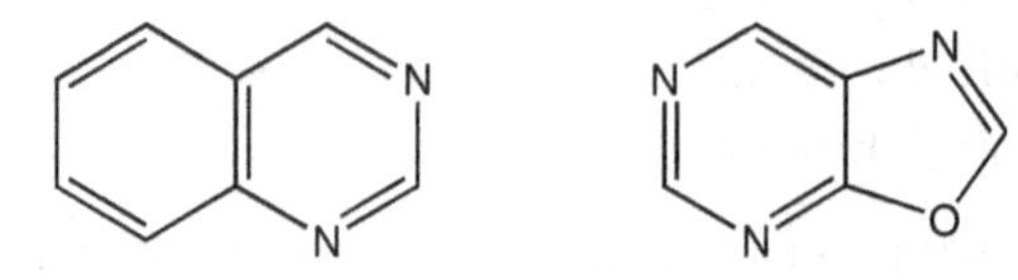

CHAPTER 3

NATURE AS A SOURCE OF HETEROCYCLIC COMPOUNDS

3.1. GENERAL

We have seen just a glimpse of the enormity of the scope of heterocyclic structures in our consideration of nomenclature systems. Now, we consider the question of reality about these structures; drawing them is one thing, but do they have a real existence and importance? There are in fact two main sources of heterocyclic compounds. They abound in nature, many in complex forms, and we will consider some of the most important of these in this chapter. In later chapters, we will learn that efficient procedures have been developed for the laboratory synthesis of heterocyclic compounds. There also can be a melding of these two sources; natural compounds often are the starting materials for laboratory elaboration into other novel structures, many of which have important biological properties or are of value in commercial applications. Natural compounds have intrigued chemists from the earliest times; as noted in Chapter 2, they were among the first to be obtained in pure form. In the formative years of organic chemistry, deducing their structures was a formidable task. This depended heavily on simple methods such as elemental analysis to obtain the stoichiometry of the elements present and, hence, their molecular

Fundamentals of Heterocyclic Chemistry: Importance in Nature and in the Synthesis of Pharmaceuticals, By Louis D. Quin and John A. Tyrell Copyright © 2010 John Wiley & Sons, Inc.

formulas, on their conversion to derivatives by applying known chemical reactions, and on their degradation to simpler structures more easily studied or even of already established structures. There was no chromatography to assist in separation of mixtures; this was largely done by extraction, crystallization, or distillation techniques. There were no spectroscopic methods, which we depend on so heavily today. Indeed, it is remarkable that so many structures could be correctly (but some incorrectly) deduced when these limited resources were applied by skilled chemists with imaginative minds. Until roughly the post-World War II period and the advent of spectroscopic methods, the general approach to proving structures of natural compounds remained unchanged. This early work remains of such great importance to heterocyclic chemistry that an illustration of the thinking used to establish structures is provided in Section 3.2.1.

3.2. NATURALLY OCCURRING NITROGEN HETEROCYCLIC COMPOUNDS

3.2.1. The Alkaloids

3.2.1.1. General. The plant kingdom has an abundance of nitrogen compounds, most being heterocyclic, with some of great complexity. Because they are weakly basic and form salts with mineral acids, the compounds from plants became known long ago as alkaloids, meaning "alkali like," although the term has been extended to include those compounds of animal origin as well, especially those from the marine environment. The discussion in this section will make use only of the plant alkaloids, which will illustrate the great structural diversity of heterocyclic rings to be found in nature. The basic character of alkaloids makes them easily extracted with acids and then regenerated with a base, and they were among the first natural organic compounds to be isolated and studied. It has been stated that more than 8000 alkaloids are known and that more than 100 are discovered annually in current research,[1] many of which have been structurally characterized. They occur in all parts of plants, frequently as families of related compounds. Thus, from tobacco, many minor alkaloids are structurally related to the major one, which is the well-known nicotine. A few are shown in Figure 3.1.

Nomenclature in the alkaloid family is not systematic and does not make use of the International Union of Pure and Applied Chemistry (IUPAC) conventions. It has been the practice for many years for the

nicotine nornicotine anabasine N-methylanabasine

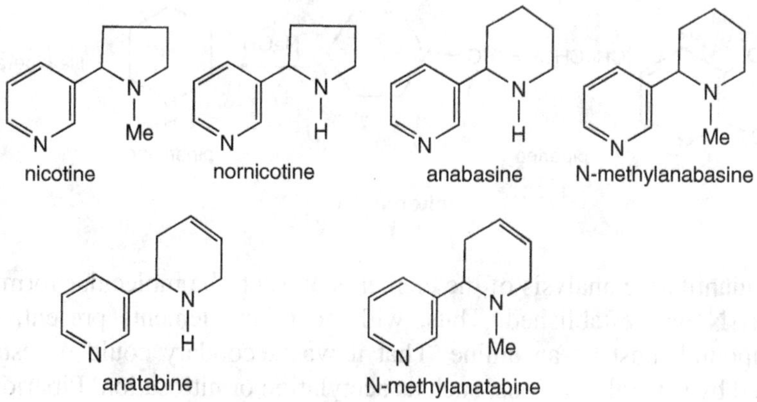

anatabine N-methylanatabine

Figure 3.1. Some tobacco alkaloids.

discoverer to create appropriate common names that always have the ending "-ine." However, a feature of alkaloid nomenclature may be pointed out here. The prefix "nor" is used to indicate the absence of an N-methyl group if one were present in the parent alkaloid.

Alkaloids in mixtures extracted from plants have been known since ancient times. They usually have some form of biological activity, which can range from high mammalian toxicity to valuable therapeutic properties of many different kinds. Mention will be made in Section 3.2.1.3 of some of these biological properties.

There is another important aspect to the (dominating) heterocyclic members of the alkaloid family: They have one or more, sometimes many, asymmetric carbons (with four different groups attached) but are found as single stereoisomers displaying optical activity and never occur as racemic mixtures. Organic chemists have for many years used alkaloids to form diastereoisomeric salts from racemic carboxylic acids for resolution into the enantiomers. The salts can be separated by crystallization or chromatographic techniques, and the optically active acid is then regenerated. Alkaloids are currently finding another use; some form optically active coordination complexes with various metallic species that are useful as catalysts in reactions to generate an excess of one enantiomer where a racemic mixture would otherwise be formed.

The early work on alkaloids remains of such great importance to heterocyclic chemistry that an illustration of the thinking used to establish structures is provided in this section. We take a simple case: proving the structure of piperidine, which was first obtained as in Scheme 3.1 from the basic hydrolysis of piperine (isolated in 1821 from black pepper).

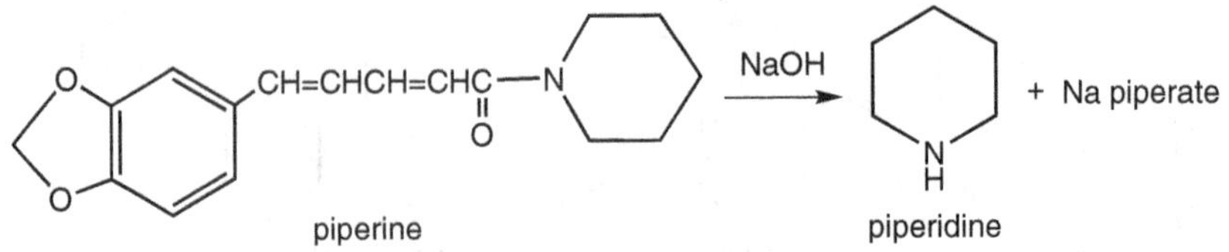

CH=CHCH=CHC(=O)—N + NaOH → piperidine + Na piperate

piperine piperidine

Scheme 3.1

By quantitative analysis of the elements present, the molecular formula $C_5H_{11}N$ was established. Thus, with no other elements present, the compound must be an amine. That it was secondary could be established by several reactions, such as acetylation or nitrosation. Piperidine was found not to have any double bonds present, and from the molecular formula, it was clear that there must be a ring present because a saturated noncyclic compound would have the formula $C_5H_{13}N$. It was not until 1881 that a method was devised by A. Hofmann for the opening of the ring and removal of nitrogen to lead to a hydrocarbon of known structure. Hofmann's method consisted of reaction with methyl iodide first to generate a tertiary amine, then a second methylation to generate the quaternary ammonium iodide structure, followed by replacement of iodide ion by hydroxide ion through reaction with moist silver oxide. The methylation sequence is generally known as Hofmann exhaustive methylation. On heating the dry quaternary ammonium hydroxide, one carbon–nitrogen bond is cleaved with the formation of a double bond, usually the least substituted one if two isomers are possible (the Hofmann rule). The general method is illustrated first by application to a noncyclic amine (Scheme 3.2).

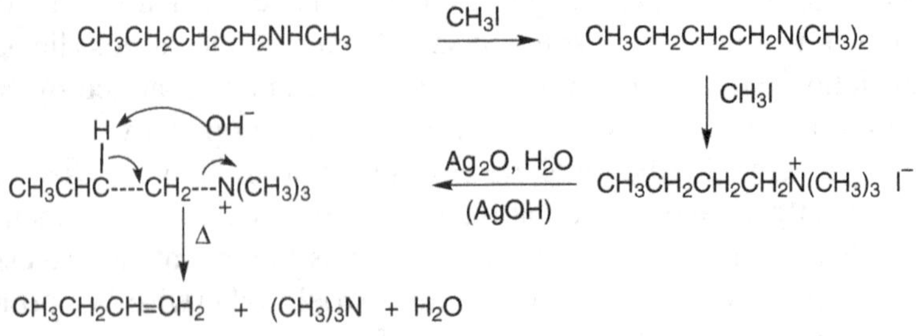

$CH_3CH_2CH_2CH_2NHCH_3 \xrightarrow{CH_3I} CH_3CH_2CH_2CH_2N(CH_3)_2$

$\xrightarrow{CH_3I} CH_3CH_2CH_2CH_2\overset{+}{N}(CH_3)_3\ I^-$

$\xrightarrow{Ag_2O,\ H_2O\ (AgOH)} CH_3CH(H)C\text{---}CH_2\text{---}\overset{+}{N}(CH_3)_3\ \ OH^-$

$\xrightarrow{\Delta} CH_3CH_2CH{=}CH_2 + (CH_3)_3N + H_2O$

Scheme 3.2

If the amine is cyclic, only one C–N bond is broken and the product will be an aminoalkene. To prove its structure, the entire sequence of methylation and pyrolysis of the quaternary ammonium hydroxide is repeated, this time eliminating the amine fragment with the formation of a second double bond. Recognizing the carbon sequence of this diene, usually by conversion to known derivatives, would reveal the structure of the original cyclic amine. Occasionally the initially formed diene rearranges its double bonds so as to achieve the lower energy state of conjugation (as shown in Scheme 3.3). The Hofmann elimination has been much studied from a mechanistic standpoint, which will not be discussed more here. Suffice it to say that the method has been applied to many natural amines, and it was a great structure-determination aid in the early days of heterocyclic chemistry. In Scheme 3.3, the method as originally used to deduce the structure of piperidine is given. Another example is shown in Scheme 3.4. Of course, modern spectroscopic methods can solve such structures with great ease, and the degradation approach is largely of historical importance.

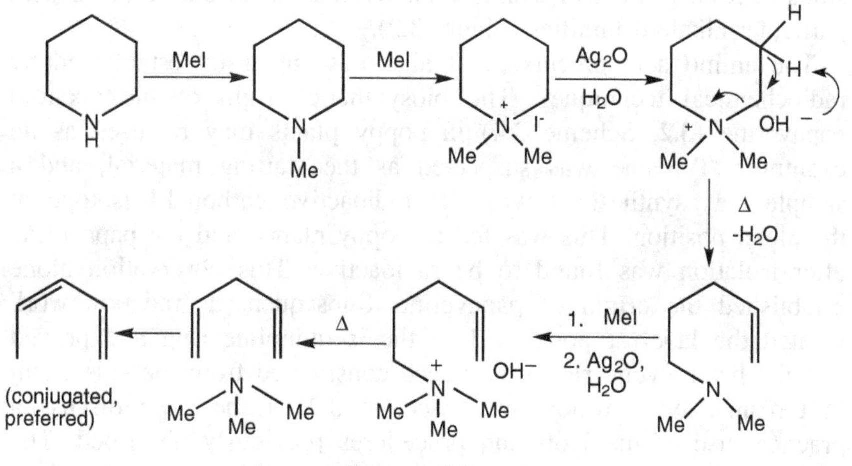

Scheme 3.3

Many important alkaloids have been synthesized in the laboratory, some by laborious, yet ingenious, multistep procedures. We will not discuss these syntheses in this book, leaving the topic to more advanced treatises.

3.2.1.2. Alkaloid Biosynthesis. Alkaloids seem to have no biological functions in plants and are referred to as secondary metabolites derived from major chemicals of the metabolism of the life processes.

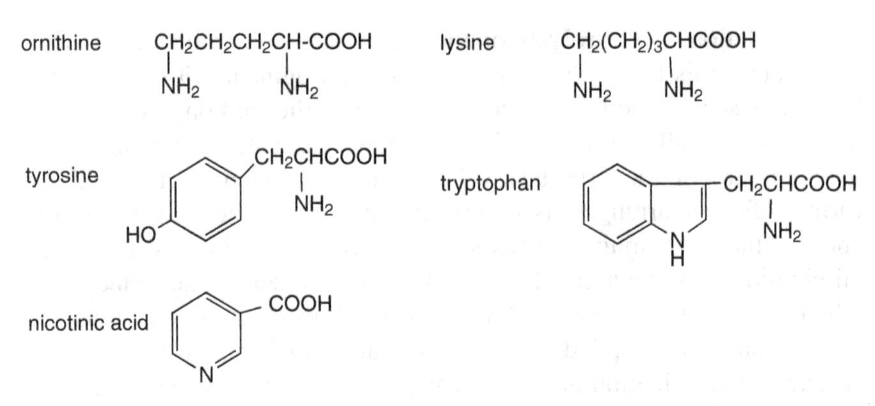

Figure 3.2. Amino acid precursors of some alkaloids.

Most alkaloids can be traced back to a particular primary metabolite, commonly an amino acid, as its precursor. Some amino acids of great importance as alkaloid precursors are shown in Scheme 3.3. In discussions to follow, we will examine the association of amino acids with particular alkaloid families (Figure 3.2).

The amino acid precursors of alkaloids are easily established by radiochemical techniques. The biosynthetic origin of the alkaloid papaverine (**3.2**, Scheme 3.4) in poppy plants may be used as an example.[2] Tyrosine was suspected as the starting material, and a sample was synthesized with the radioactive carbon-14 isotope at the alpha position. This was fed to poppy plants, and the papaverine after isolation was found to be radioactive. This observation alone established the origin of papaverine. Subsequent degradation work located the label at position 3 of the isoquinoline ring and proved that the heterocyclic ring was indeed constructed from the side chain in tyrosine. As a follow-up to section 3.2.1.1, here we discuss a practical use of the Hofmann procedures previously described. The pyridine ring in the radioactive papaverine (**3.2**) was reduced to give a saturated amine (**3.3**), which was subjected to exhaustive methylation and thermal degradation of the quaternary ammonium hydroxide to open the ring and form compound **3.4**. A second Hofmann reaction cycle produced nitrogen-free compound **3.5**. This was subjected to additional degradation, not relevant to our discussion here, to remove the radioactive carbon. The value of the Hofmann procedures in alkaloid structural studies is well illustrated by this example.

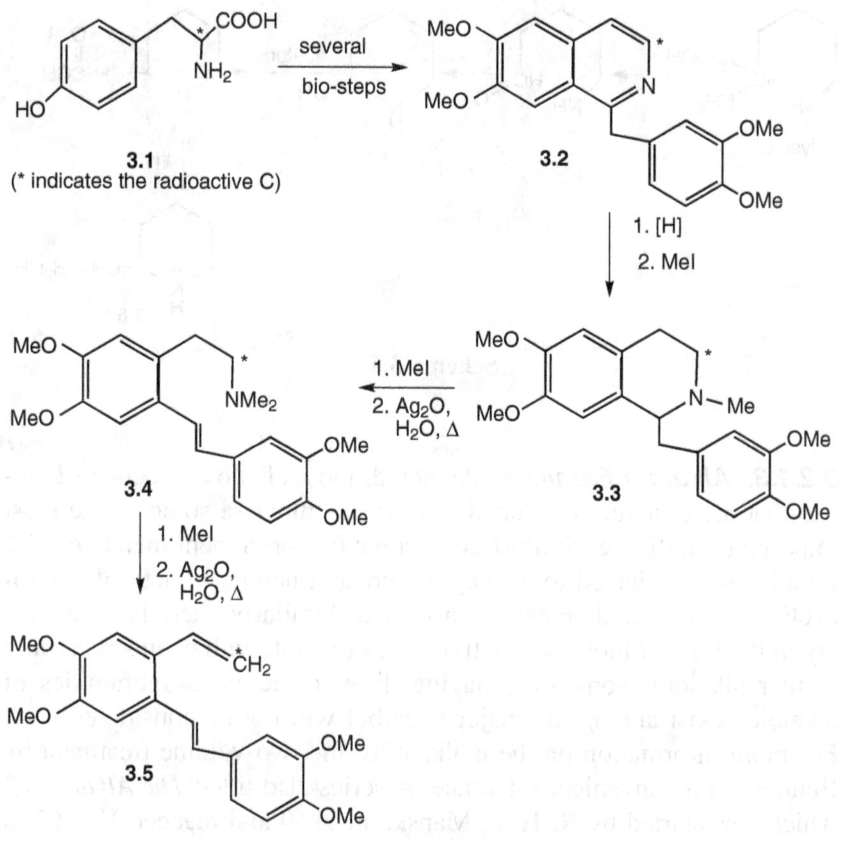

3.1
(* indicates the radioactive C)

several bio-steps

3.2

1. [H]
2. MeI

3.3

1. MeI
2. Ag_2O, H_2O, Δ

3.4

1. MeI
2. Ag_2O, H_2O, Δ

3.5

Scheme 3.4

Reactions in plants are of course catalyzed by enzymes, but the reactions are simply those allowed by organic chemical principles. Familiar processes such as methylation, oxidation, reduction, decarboxylation, aldol condensations, and so on, are frequently involved. In many cases it is difficult to determine the exact sequence of events, but the broad outlines of the biosynthesis of many alkaloids have been elucidated. To take a simple case for illustration of a biosynthetic pathway, the synthesis of coniine in the hemlock tree will be presented. Coniine (**3.6**) is the poison in hemlock that was used to kill Socrates. As found in Scheme 3.5, the amino acid lysine is the precursor of coniine.[3] The use of radioactive isotopic labels is invaluable in proving such pathways.

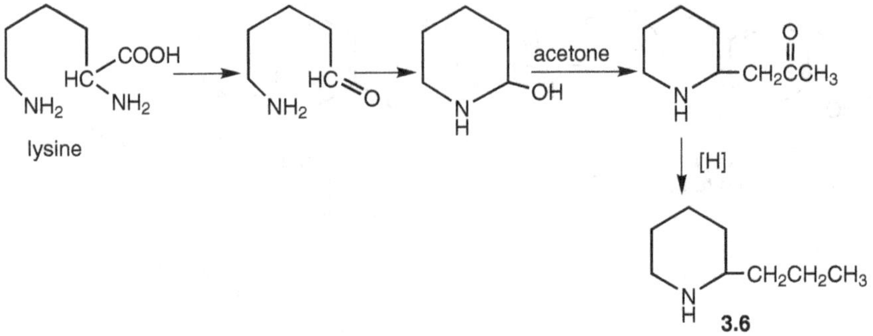

Scheme 3.5

3.2.1.3. Alkaloid Families. As noted, most alkaloids occur in families that are structurally related. We will summarize some of the most important families, each illustrated with a few prominent members. The families were selected to display the great structural variety of heterocyclic systems found among the alkaloids. Similarly, there is great variety in the type of biological activity they exhibit, with even structurally similar alkaloids sometimes having different activity. Subfamilies of alkaloids exist among the major ones but will not be considered here. For more information on these alkaloids, the two-volume treatment by Bentley[3] is a convenient reference. A series also titled *The Alkaloids*,[4] which was started by R. H. F. Manske in 1950 and reached Vol. 63 in 2006, contains reviews on particular alkaloids that are a major source of information in this field.

3.2.1.3.1. The Pyrrolidine Family. The alkaloid hygrine (**3.7**) isolated from leaves of plants from the Coca group is an example of a simple pyrrolidine derivative. It and other members of this family have been found to originate from the amino acid ornithine (Scheme 3.6).

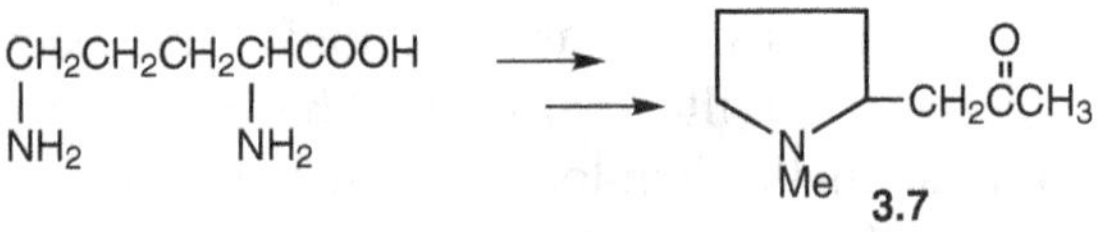

Scheme 3.6

The pyrrolidine ring also can be embodied in the bicyclic structure of bicyclo[3.2.1]octane. The parent N-methyl compound is known as tropane (**3.8**), the 3-keto derivative is tropinone (**3.9**), and the 3-hydroxy derivative, with the OH group on the opposite side of the nitrogen bridge, is tropine (**3.10**). In such bridged structures, a substituent may be on either side with respect to the bridging group. If on the same side, it is the *syn* isomer; if on the opposite side, it is the *anti* isomer.

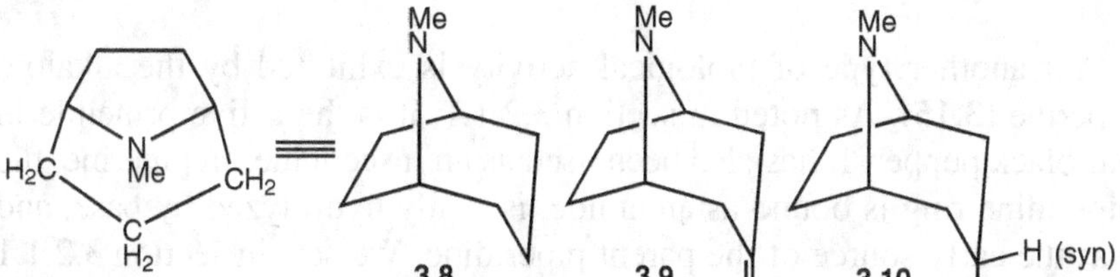

3.8 tropane **3.9** tropinone **3.10** tropine

Two well-known alkaloids, cocaine (**3.11**) and atropine (**3.12**), are ester derivatives of the 8-azabicyclo[3.2.1]octane ring system. Cocaine, isolated from a variety of the poppy plant, has been used as a topical anesthetic, but it is highly addictive if it enters the bloodstream and is now a controlled substance. Atropine, however, is highly useful in medicine with anticholinergic properties. It is isolated from the Belladonna plant and has been used for many years to dilate the pupil of the eye. It is also an effective antidote to poisoning by anticholinesterase chemicals, when these are used as insecticides or in extremely toxic form as chemical warfare agents.

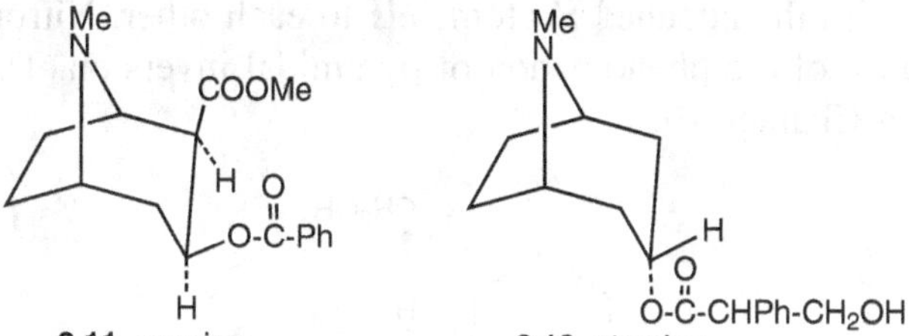

3.11, cocaine **3.12,** atropine

3.2.1.3.2. The Piperidine Family. It was noted in section 3.2.1.2 that the piperidine ring is biosynthesized from the amino acid lysine, and the example of coniine biosynthesis was given in that section. On the pathway to coniine (**3.14**), a highly toxic compound isolated from the hemlock tree is the ketone derivative **3.13**; this happens to be the well-known alkaloid pelletierine, which is isolated from the pomegranate tree (Scheme 3.7). It has found use in the treatment of parasitic infections (called an anthelmintic agent).

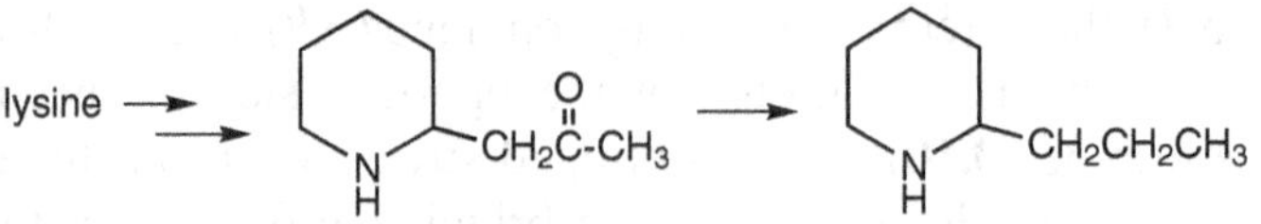

lysine → → 3.13, pelletierine → 3.14, coniine

Scheme 3.7

Yet another type of biological activity is exhibited by the alkaloid piperine (**3.15**). As noted in section 3.2.1.1, it is the active principle in our black pepper. It has also been used as an insecticide. In piperine, the piperidine ring is bound as an amide, is easily hydrolyzed by base, and was the early source of the parent piperidine. We saw in section 3.2.1.1 the importance of piperidine in the historical development of heterocyclic chemistry.

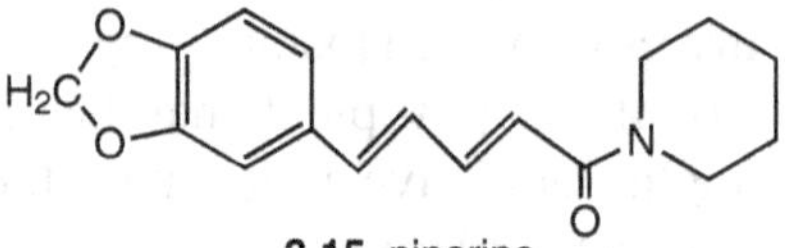

3.15, piperine

We observe another structural variation for a piperidine alkaloid in the compound lupinine (**3.16**), which is isolated from seeds of Lupinus and Anabasis plants. Here, nitrogen is found at the junction of two fused rings. There are two chiral carbon atoms, so there will be four possible stereoisomers. Natural lupinine has only one of these isomeric structures, with the attached H atoms *cis* to each other. Nitrogen is not chiral because of the phenomenon of pyramidal inversion. This will be discussed in Chapter 10.

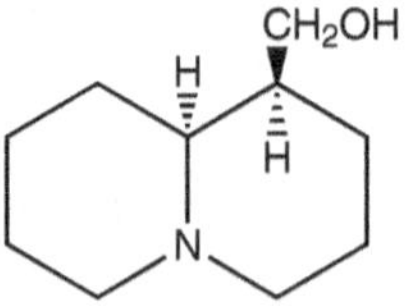

3.16, lupinine

3.2.1.3.3. The Pyridine Family. The tobacco alkaloids are the best representatives of this family. We have already viewed the structures of several of these alkaloids in Figure 3.1. They may have a pyrrrolidine ring as in nicotine or a piperidine ring as in anabasine attached to the 3-position of pyridine. They are biosynthesized from nicotinic acid (Scheme 3.8).

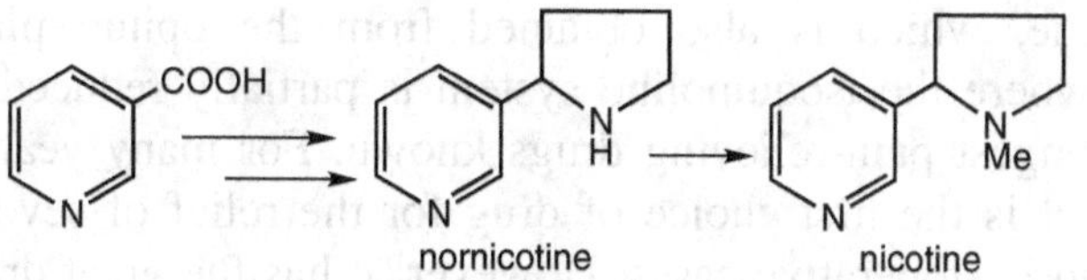

nornicotine nicotine

Scheme 3.8

Anabasine is also the chief alkaloid in another plant, the Asiatic *Anabasis aphylla*.

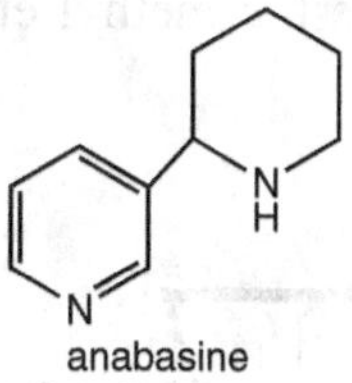

anabasine

3.2.1.3.4. The Isoquinoline Group. Many biologically important alkaloids contain the isoquinoline ring system, or a reduced form of this system. An entire book has been written about them.[2] We will only consider three of them here that illustrate some of the molecular complexity associated with this family.

We first want to note that the isoquinoline ring is biosynthesized from the amino acid tyrosine.

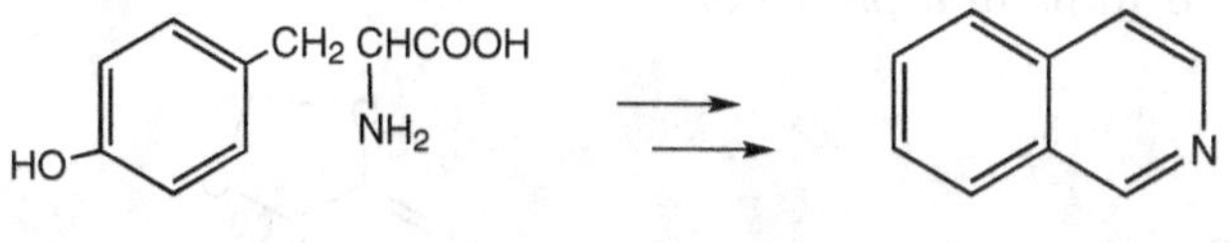

Scheme 3.9

The isoquinoline ring system is observed in the alkaloid papaverine, which is found in the opium plant. It has been used in medicine for many years, and it is still employed as a muscle relaxant and vasodilator. It is a member of a subclass of isoquinoline alkaloids called the benzylisoquinoline alkaloids because of the presence of a form of a benzyl substituent ($C_6H_5CH_2$-) at the 1-position.

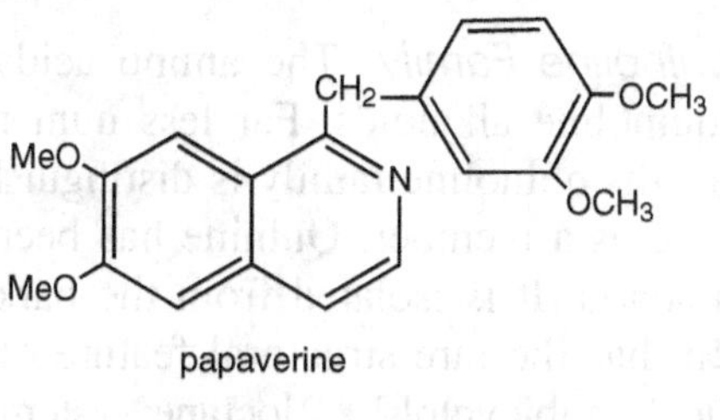

papaverine

Morphine, which is also obtained from the opium plant, is an example where the isoquinoline system is partially reduced. It is one of the strongest pain-relieving drugs known. For many years and still currently, it is the first choice of drug for the relief of severe pain in trauma and postoperative cases. However, it has the great drawback of being highly addictive, so its clinical use must be controlled carefully. The structure of morphine is frequently expressed by **3.17**, where the reduced rings of isoquinoline are labeled A (the former benzene ring) and B (the former pyridine ring). Codeine, which is a milder pain reliever but still addictive, is the methyl ether **3.18**.

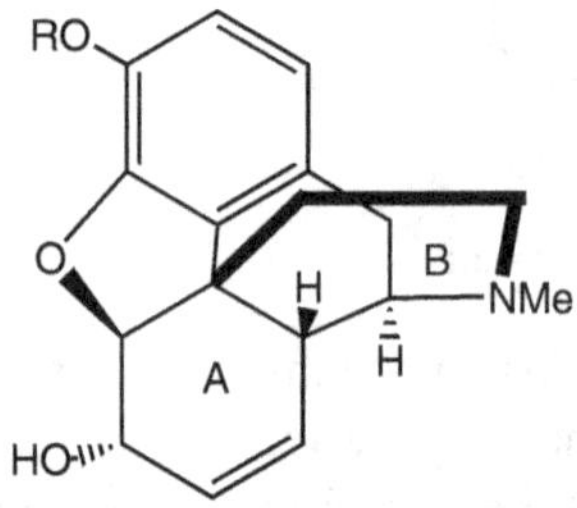

3.17, R=H, morphine
3.18, R=Me, codeine

Another structural feature is found in the alkaloid berberine (**3.19**), which is yellow and widely distributed in plants. Here, the isoquinoline component has nitrogen in the quaternary salt condition. There are five fused rings in this structure. The 1,3-dioxolane ring is frequently encountered in natural products.

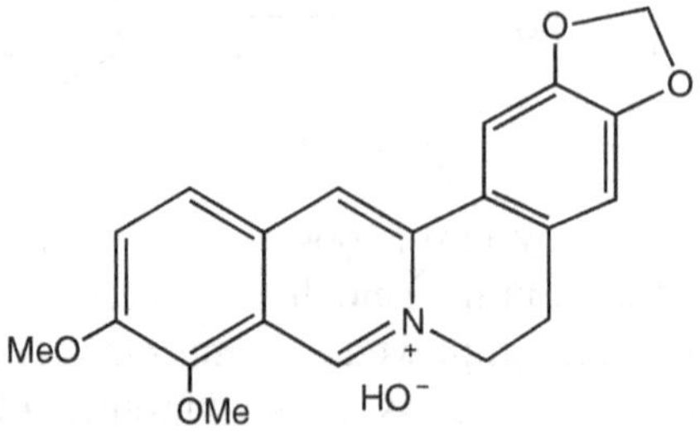

3.19, berberine

3.2.1.3.5. The Quinoline Family. The amino acid tryptophan is the precursor of the quinoline alkaloids. Far less numerous than the isoquinoline alkaloids, the quinoline family is distinguished by having the famous drug quinine as a member. Quinine has been used for ages in the treatment of malaria. It is isolated from the bark of the Cinchona tree. Quinine (**3.20**) has the rare structural feature of a bicyclic amine (a derivative of the 1-azabicyclo[2.2.2]octane system) as a substituent.

Quinoline has a long history in the development of heterocyclic chemistry, as has been detailed in an account titled "The Quest for Quinine" published in 2005.[5]

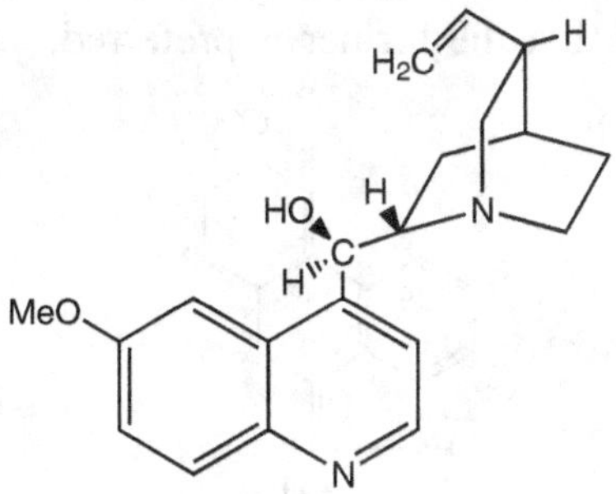

3.20, quinine

Another feature found in quinoline alkaloids is the presence of a furan ring fused to the quinoline system, as in dictamine (**3.21**).

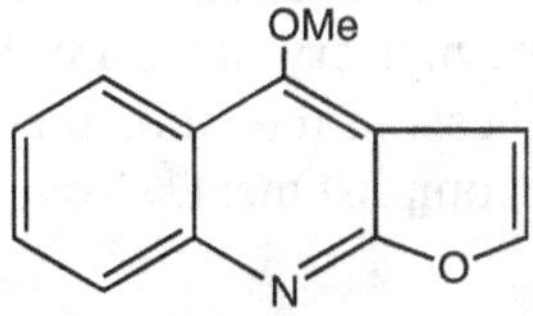

3.21, dictamine

3.2.1.3.6. The Indole Family. Tryptophan is also the precursor of another large alkaloid family based on the indole system. Usually, indole is found fused to other rings, as in physostigmine (**3.22**). Here, a pyrrolidine ring is fused to the b-face of indole. This alkaloid occurs in calabar beans. It is a potent inhibitor of cholinesterase, and like other such chemicals, it causes contraction of the pupil of the eye. In harmine (**3.23**), a pyridine ring is fused to the b-face.

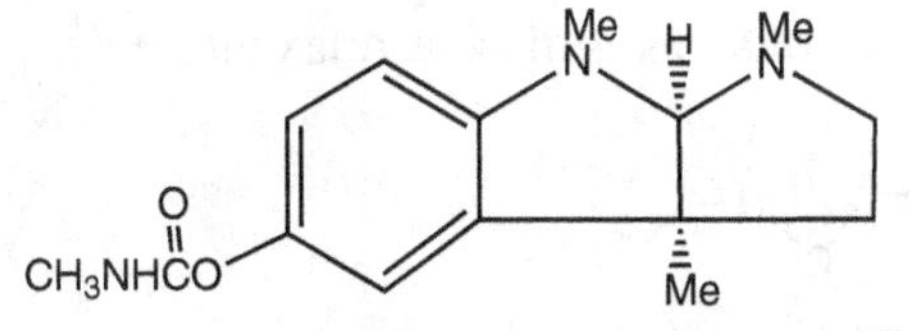

3.22, physostigmine

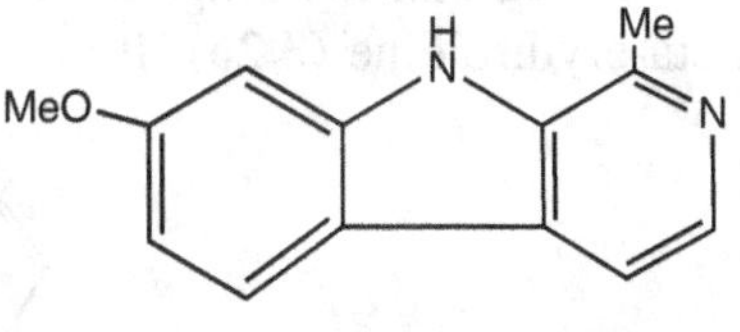

3.23, harmine

Alkaloids from the fungus *Claviceps purpurea* are collectively extracted as ergot, and this mixture has been used for years because of its action on the vasomotor system and its ability to induce labor. On hydrolysis, lysergic acid (**3.24**) is formed from them. The diethylamide made synthetically is the infamous hallucinogen lysergic acid diethylamide (LSD). Lysergic acid presents yet another unique bonding pattern with its tetracyclic structure. It can be viewed as either an indole alkaloid or a quinoline alkaloid, but the former is preferred.

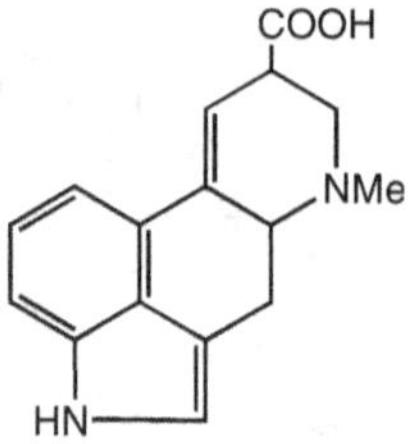

3.24

Finally, we come to another complex indole-derived structure, the alkaloid strychnine (**3.25**). This is another infamous structure; it is highly toxic, and in earlier years, it was used as a rat poison and in tiny doses in various treatments. It has also been much used to murder humans, at least in the mystery literature. Strychnine has six chiral centers, and its total synthesis in the laboratory with the correct stereochemistry was a major accomplishment.[6] Novel also is the 7-membered oxygen-containing ring.

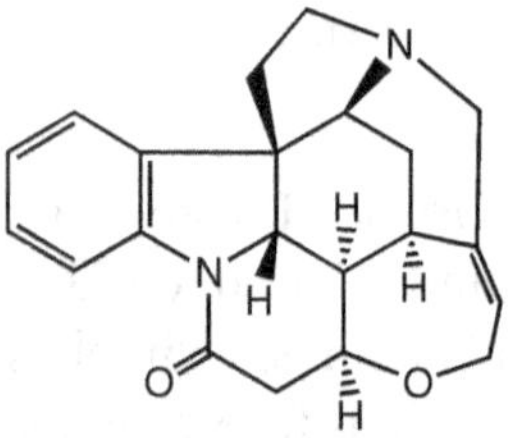

3.25, strychnine

3.2.1.3.7. The Erythrina Family. The final alkaloid family we will consider presents another novel feature, a spiro carbon (meaning from a single carbon two rings are formed). The best example is the alkaloid beta-erythroidine (**3.26**). It has found use as a muscle relaxant.

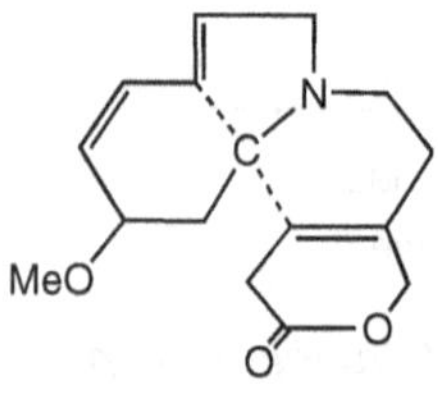

3.26, beta-erythroidine

3.2.1.3.8. Other Plant Alkaloids. It is not possible to describe all types of alkaloids in this brief discussion. Other ring systems acting as bases for significant alkaloid families include imidazole, quinazoline, and pyrrolizidine.

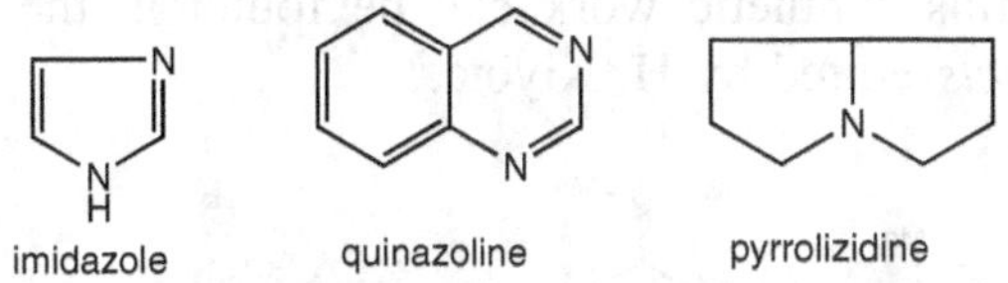

imidazole quinazoline pyrrolizidine

In the last group alone, there are more than 160 members, some of which are toxic. Many other alkaloids, some of great complexity, can be found in plants and frequently are referred to by the plant name. Thus, the Amaryllidaceae (exemplified by the common narcissus plant) alkaloids are a rich collection of complex structures. From the moss family Lycopodiaceae are obtained a group known as the Lycopodium alkaloids. Many miscellaneous alkaloids also are known. The book by Aniszewski[1] is an excellent source of information on other types of alkaloids, their botanical distribution, and their biological and other features.

3.2.1.4. Marine Alkaloids. In recent years, the marine environment has been recognized as a rich source of novel structures, some of which have valuable biological properties. Intensive research is being conducted in this area. Here, the animal source is more common than the plant source, with invertebrate animals being of special importance. A concise survey of some of the many compounds isolated has been given by Faulkner.[7] There is little consistency in the structural types; many structures are extremely complex and are not easily collated. There is an exception to this statement; the backbone shown as follows, which contains a reduced pyrrolo[3,4-c]pyridine moiety, does show up in several compounds.

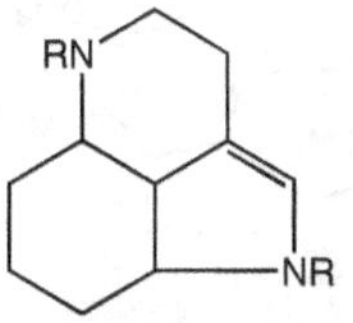

Just to illustrate the complexity of the structures formed in the marine environment, we observe here the examples of diazonamide A (from a Phillipines invertebrate animal), securamine A (from North Sea bryozoans), ecteinascidin-743 (from a Caribbean sea-squirt, used in refractory cancer), and norzoanthamine (from a sea anemone). To elucidate complex structures such as these is a major undertaking and a tribute to those who can accomplish this task. Original references to these and many other structure proofs can be found in the reference by Faulkner.[7] Equally challenging is the devising of multistep syntheses for the complex marine natural products after their structures have been established. Examples of this synthetic work can be found in the book *Marine Natural Products* edited by H. Kiyota.[8]

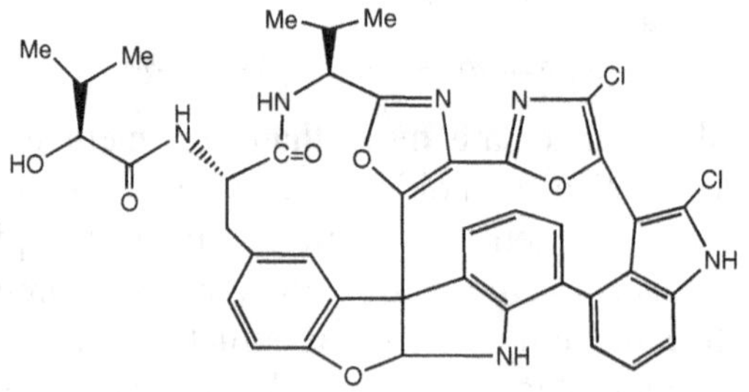

diazonamide A

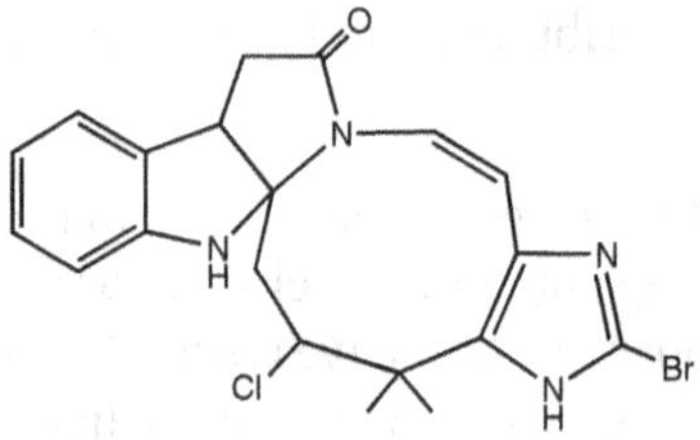

securamine A

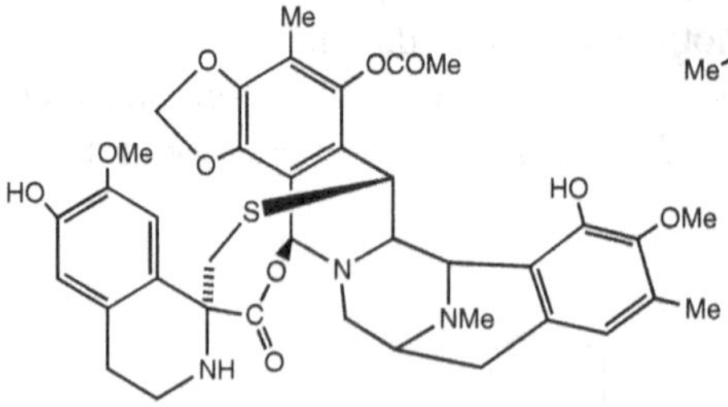

ecteinascidin-743

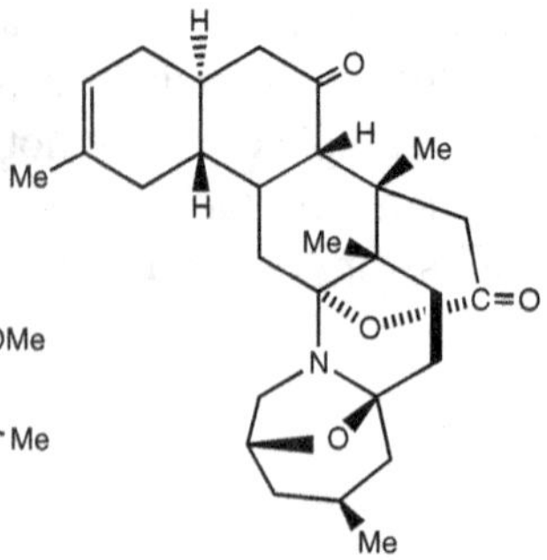

norzoanthamine

3.2.2. Porphyrins

One of the most important families of natural heterocyclic compounds is based on the porphin (**3.27**) ring system. Here, there are four pyrrole units linked in a cycle by four sp^2 carbons. The derivatives of this system are known as porphyrins. Porphyrins usually have substituents at all beta-positions of the pyrrole units. Generally, these substituents are small groups such as methyl, ethyl, vinyl, CH_2COOH, CH_2CH_2COOH, CHO, and CH_2OH. For a given set of substituents, various isomers are possible and are distinguished by such prefixes as etio, proto, deutero, copro, and so on, but we need not go into the structural significance of these terms. The overall ring is planar and of great stability, with the four N atoms in good position to complex with metals, in which form they are found in nature. As will be discussed in Chapter 7, the system can be viewed as having conjugation involving 9 double bonds (18 pi-electrons), and thus, it is aromatic according to the Hückel 4n+2 rule, where n = 4. It does indeed display characteristics of aromaticity. Note that two double bonds indicated by parentheses on structure **3.27** are not included in the cyclic delocalization. Porphyrins are common in living systems and are involved in various forms of biological activity. They are, for example, involved in the respiratory systems of both animals and plants, and they are fundamental to life as we know it. The two most well-known porphyrins are the iron(II) complex heme (**3.28**), which when associated with the protein globin is found in blood as the oxygen-carrying hemoglobin, and chlorophyll (**3.29**), which is a magnesium complex involved in photosynthesis. One of the vitamins (B_{12}, cyanocobalamine) also contains a modified porphin ring, which is here complexed to cobalt.

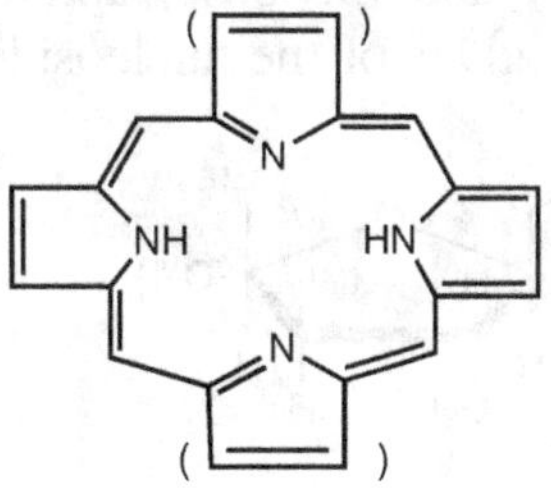

3.27, porphin

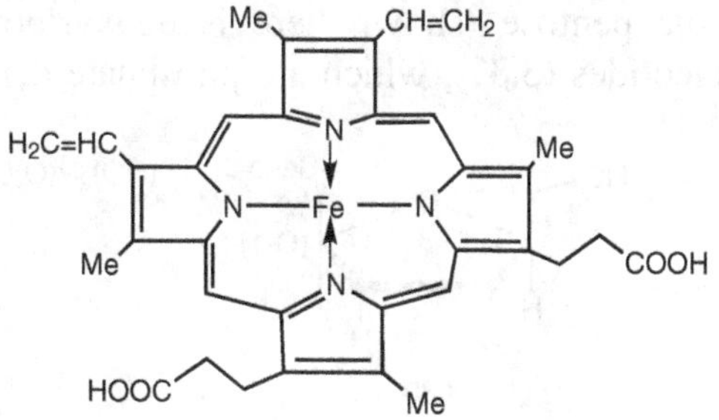

3.28, heme

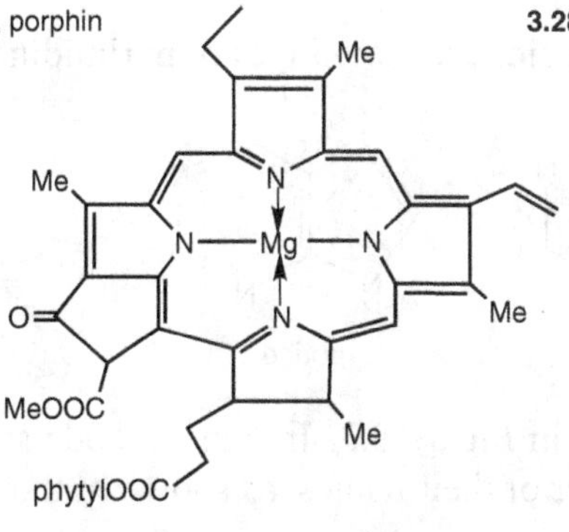

3.29, chlorophyll a, phytyl=$C_{20}H_{39}$

Porphyrins have a long history in the development of organic chemistry. As early as 1853, a crystalline derivative was obtained from the hydrochloric acid hydrolysis of hemoglobin. This proved to be the iron (III) chloride complex of the porphyrin in heme and was named hemin. It was the subject of extensive research over the years, which culminated in the total synthesis by H. Fischer and K. Zeiler in 1929.[9] This is a classic piece of work in the annals of organic chemistry; Fischer received the Nobel Prize in Chemistry in 1930 for this and other work.

3.2.3. The Bases of Nucleic Acids, Nucleosides, and Nucleotides

Deoxyribonucleic acid (DNA) and ribonucleic acid (RNA) are composed of long chains (polymers) of repeating pentose and phosphate groups. Attached to the 1-position of each pentose is a heterocyclic compound known as a base, albeit a rather weak one. It is the sequence of the bases on the polymer chain that gives rise to the all-important genetic code. Here, we will only mention the structure of the heterocyclic bases, which is consistent with the goal of this chapter to reveal the wide distribution and biological significance of heterocycles, and defer subsequent consideration of the nucleic acids to Chapter 9. Monomeric building blocks for the nucleic acids also are found in living systems; these are the nucleosides (**3.30**), which are composed of the pentose (shown here is deoxyribose) and the base, and the nucleotides (**3.31**), which are phosphate derivatives of the nucleosides.

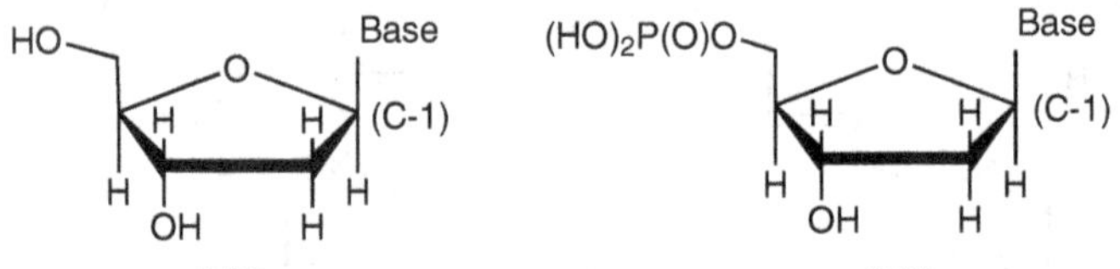

HO Base O H H (C-1) H H OH H

3.30

$(HO)_2P(O)O$ Base O H H (C-1) H H OH H

3.31

The bases of the nucleic acids are derivatives of pyrimidine or purine.

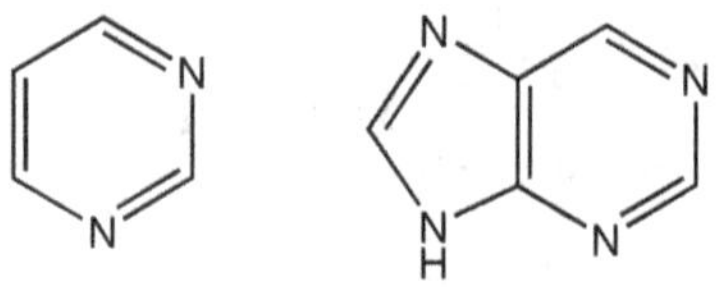

N N

pyrimidine

N N N N H

purine

Their structures are shown in Figure 3.3. In genetic code studies, they are denoted by the first letter of their names, as shown in the figure. The

cytosine (C) thymine (T) uracil (U) adenine (A) guanine (G)

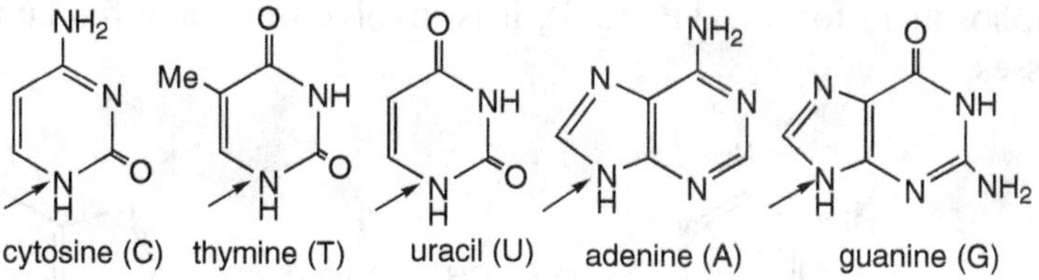

Figure 3.3. The bases of nucleic acids.

bases are attached at a ring nitrogen atom to carbon-1 of the pentoses. The nitrogen of attachment is indicated by an arrow on the structures. DNA and RNA differ in the bases incorporated on the polymeric chain: DNA makes use of A, G, C, and T, whereas RNA uses A, G, C, and U.

Four of the bases possess carbonyl groups. In early studies, the oxygen was written in the tautomeric hydroxyl form, and this confused the structural assignment of their bonding to the pentose. The tautomerism is expressed in Scheme 3.10 for cytosine as an example, where the enol form can be viewed as a hydroxypyrimidine. In the solid state, only the keto form is observed, but in an aqueous solution, a small amount of the enol form is present in the tautomeric equilibrium (e.g., for cytosine, 0.2% enol and 99.8% keto). Recognition of the keto form as dominant played a significant role in the unraveling of the structure of the nucleic acids and led J. D. Watson and F. Crick to propose the famous double helix held together by H-bonds to the carbonyl oxygens. We will review details of this bonding in Chapter 9. An interesting account of the discovery of the double helix and clarification of the structure of the pyrimidine bases has been given by Watson.[10]

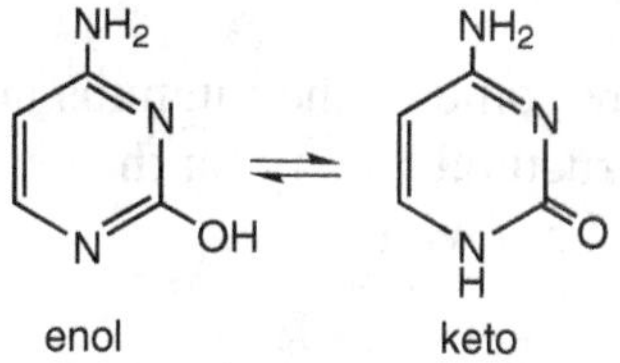

enol keto

Scheme 3.10

Some of the nucleosides and nucleotides that are found in nature are generally derivatives of the same bases that are present in the nucleic acids. Of great biological importance is the nucleoside adenosine (**3.32**), which is formed from adenine. As found with the phosphorus group in the triphosphate form (ATP, **3.33**), it is involved in many fundamental processes.

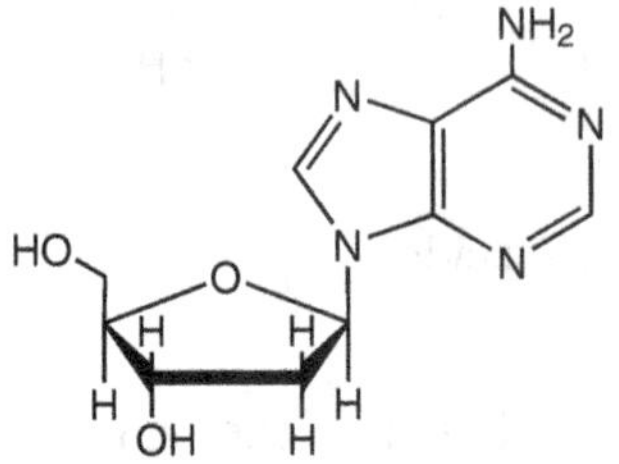

3.32, adenosine

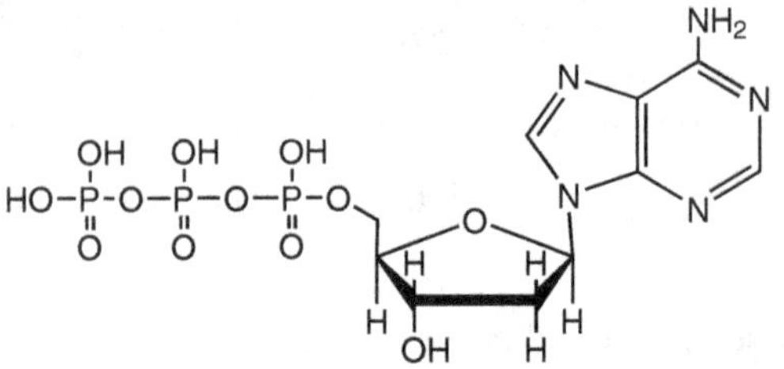

3.33, adenosine triphosphate (ATP)

Adenine is also found in the coenzyme nicotinamide adenine dinucleotide (NAD) (**3.34**); here, dinucleotide refers to the presence of the diphosphate group to distinguish it from the monophosphate of the nucleotide. Another feature of NAD is the presence of a quaternized pyridine ring.

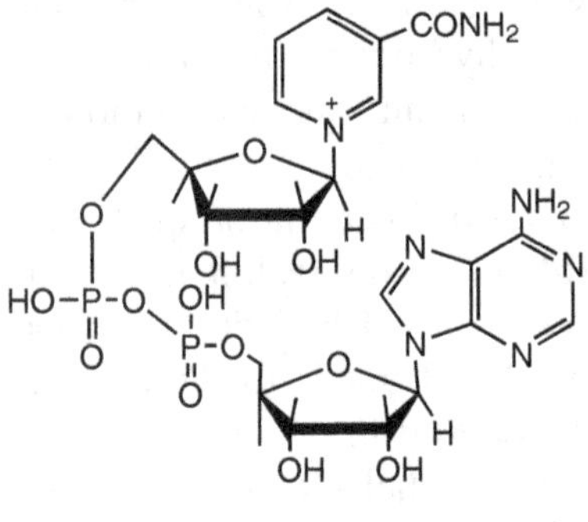

3.2.4. Vitamins

Shown in Figure 3.4 are some of the vitamins and chemicals essential for growth. The great structural variety for the heterocycles incorporated in these structures will be obvious.

3.2.5. Antibiotics

The name antibiotic was coined for substances isolated from microorganism growth media that had potent antibacterial, antifungal, antineoplastic, and so on, activity. The well-known antibacterial penicillin G was the first to be discovered. Its activity was noted by Fleming in England in 1929, but the work of Florey and Chain in Oxford during

riboflavin, vitamin B_2 thiamine, vitamin B_1

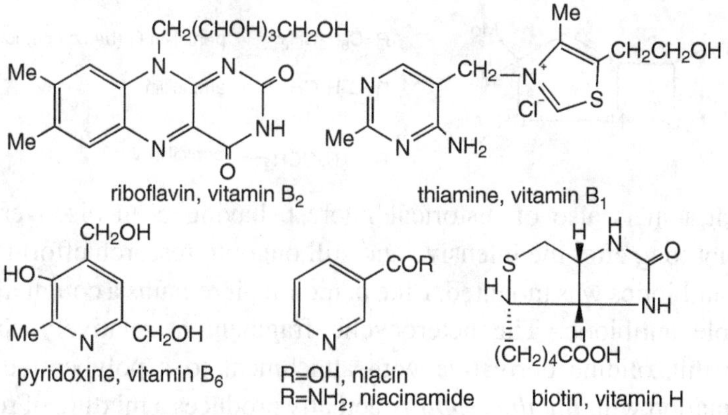

pyridoxine, vitamin B_6 R=OH, niacin
R=NH_2, niacinamide biotin, vitamin H

Figure 3.4. Structures of some vitamins.

the next 10 years led to the first clinical trials in 1941, with astounding results. Penicillin went on to become one of the most famous drugs ever discovered; it revolutionized the treatment of bacterial infections (gram-positive staining only), and along with the synthetic sulfa drugs (Chapter 8), saved countless lives during World War II. It and some derivatives remain of major use even today, although many antibiotic substances have since been discovered that have a broader spectrum of activity or are more potent. Numerous antibiotics are on the market now, many of which are based on heterocyclic frameworks or substituents. There is little consistency in the types of structures involved, and cataloging them is difficult. We will show here only four antibiotics to stress the range of heterocyclic structures encountered in the research for new drugs.

The structure of penicillin was worked out only after an intensive war-time effort. It was found to have the unstable beta-lactam ring (a keto azetidine or azetidinone derivative), which is a structural type not well known before this work. This ring was fused onto a thiazolidine ring. Penicillin was produced by large-scale fermentation of *Penicillium notatum* or *Penicillin chrysogenum*; the search for synthetic methods to avoid the fermentation procedures was not successful until 1957, when this was accomplished by J. Sheehan and K. R. Henery-Logan at M.I.T.[11] Now, other beta-lactams have become known, and this ring system is no longer a chemical rarity. Numerous drugs are available with the beta-lactam structure. The Sheehan synthesis will be described in Chapter 9.

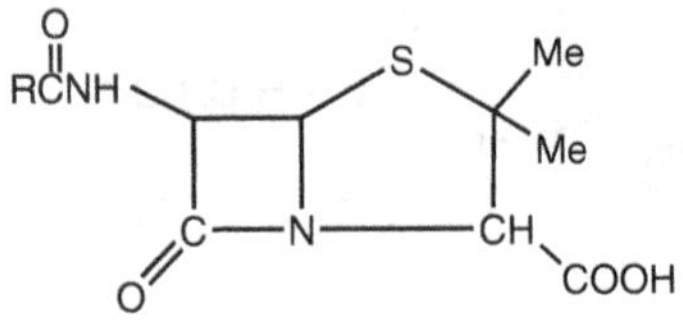

$R=C_6H_5CH_2-$ penicillin G (benzyl penicillin)

$R=C_6H_5CH(NH_2)-$ ampicillin

$R=C_6H_5OCH_2-$ penicillin V

Bacitracin is also of historical interest, having been discovered in 1943 not long after the intensive and still ongoing research effort to find other antibiotics was mounted. Like penicillin, it remains a commercially available antibiotic. The heterocyclic fragment is relatively simple; it is a thiazolidine derivative with attachment to a polypeptide. The fermentation with *Bacillus subtilis* actually produces a mixture of related compounds, the major one being bacitracin A with structure **3.35**.

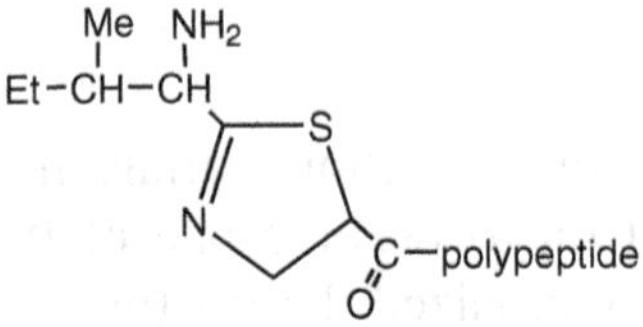

3.35, bacitracin A, $C_{66}H_{103}N_{17}O_{16}S$

Anthramycin is an antibiotic with antitumor activity. It is isolated from the growth of *Streptomyces refuineus*. Anthramycin (**3.36**) is unique in having the 7-membered azepine ring (partially reduced) present, fused to a pyrroline ring (a dihydropyrrole).

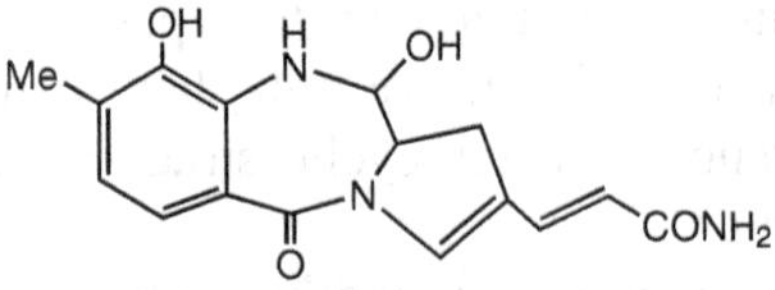

3.36, anthramycin

Mitomycin C (**3.37**) is an antitumor antibiotic isolated from the growth of *Streptomyces caespitosus*. Its unique feature is an aziridine ring fused onto a pyrrolidine ring. Other related aziridines are also present in the fermentation broth.

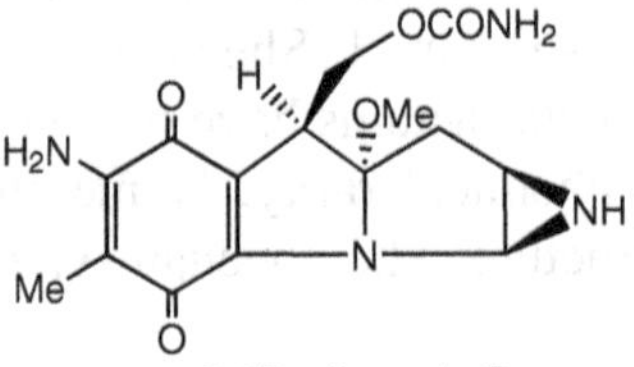

3.37, mitomycin C

3.2.6. Amino Acids

We have already discussed the importance of heterocyclic systems in many biological compounds. The final group to consider is the amino acid family from which all proteins are constructed. Three heterocyclic compounds, of a simple but profoundly important structure, are found in this family: histidine, proline, and tryptophan. Histidine on biological decarboxylation is the precursor of the notorious histamine, which is the cause of much human discomfort and of the creation of a vast family of pharmaceuticals called antihistamines.

histidine proline tryptophan histamine

3.3. OXYGEN COMPOUNDS

3.3.1. General

We have reviewed several naturally occurring cyclic structures containing oxygen, either in a ring containing nitrogen (as in oxazoles) or as a substituent on a nitrogen heterocycle. Also, as mentioned in Chapter 1, cyclic ethers (furan and pyran derivatives) are the base for monosaccharides and monosaccharide units of disaccharides and polysaccharides, as well as for the nucleoside building blocks (section 3.2.3). In this section, we will describe other compounds based on oxygen heterocyclic systems. Such structures are not as numerous as nitrogen-based heterocycles but nevertheless have their own level of importance. It is convenient to catalog these compounds as derivatives of furan, benzofuran, pyran, benzopyran, and complex multiring systems.

3.3.2. Furan Derivatives

The furan ring is found in furoic acid (**3.38**). Some alkyl furan derivatives have value as odorants. Thus, rose furan (**3.39**) gives the characteristic odor of rose oil. Other derivatives have potent biological activity, such as beta-furyl isoamyl ketone (perilla ketone, **3.40**), which causes pulmonary edema in animals grazing on plants containing this substance (such as *Perilla Frutescens* Britton). Ascorbic acid (the well-known Vitamin C), which when written as the enolic form (**3.42**) is found to be a trihydroxyfuran, is properly expressed as the tautomeric keto form (**3.41**).

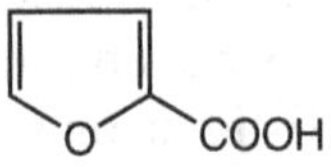

3.38, furoic acid 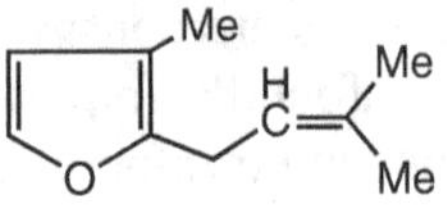 **3.39**, rose furan 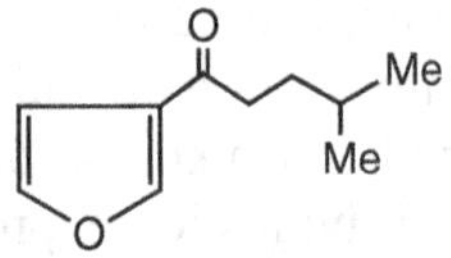 **3.40**, perilla ketone

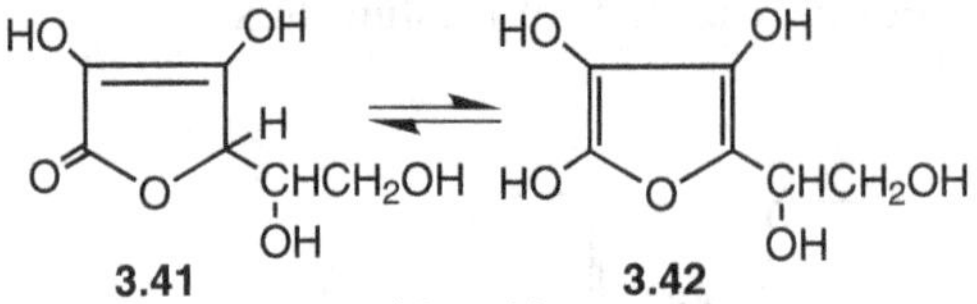

3.41 ascorbic acid **3.42**

3.3.3. Benzofuran Derivatives

Simple benzofuran derivatives can be found in the essential oils extracted from plants. An example is 2-methyl-5-methoxybenzofuran (**3.43**), which is found in the oils of a plant (*Pimpinella cumbrae* Link) from the Canary Islands. The benzofuran without the 2-methyl group has antibacterial properties. A more important compound is Griseofulvin (**3.44**), which has a partially reduced furan ring with a spiro attachment. Griseofulvin, which is isolated from the growth media of *Penicillium griseofulvum* Dierckx, is classed as an antibiotic and is in clinical use against fungal infections. Dibenzofurans are also found in natural sources; compound **3.45**, for example, is present in lichens.

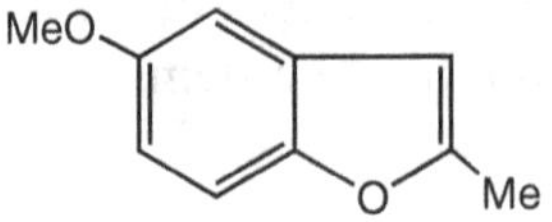

3.43 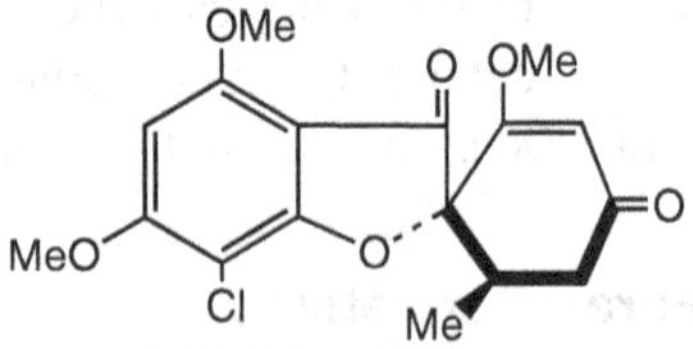 **3.44**, Griseofulvin

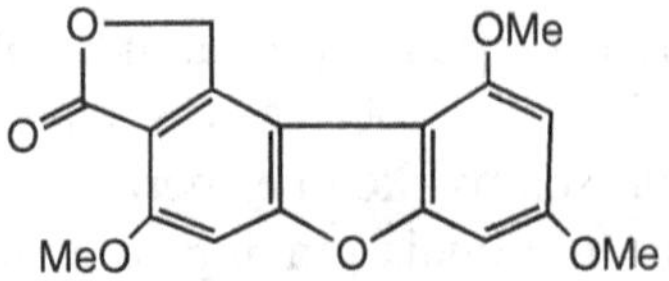

3.45

3.3.4. Pyran Derivatives

Whereas the 6-membered pyran ring usually occurs in natural products as a benzo derivative, some monocyclic structures are also known, most commonly as 4-keto derivatives called gamma-pyrones. Examples are maltol (**3.46**), which is found in the bark and needles of conifers such as pine; chelidonic acid (**3.47**), which is derived from poppy and other plants; and kojic acid (**3.48**), which is formed in molds such as the *Aspergilli*.

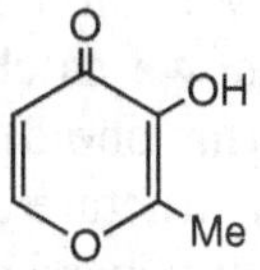

3.46, maltol 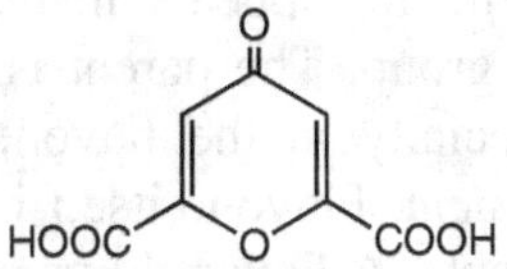 **3.47**, chelidonic acid 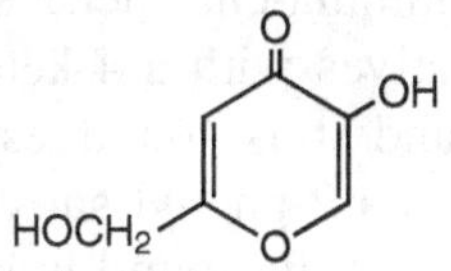 **3.48**, kojic acid

3.3.5. Benzopyran Derivatives

Most of the numerous benzopyrans found in plants can be considered as based on one of the parent structures chroman (**3.49**), alpha-chromene (**3.50**), gamma-chromene (**3.51**), or the benzopyrylium ion (**3.52**). Three-coordinate (tricovalent) oxygen is not common in organic chemistry, but in this ion type, the structure is stabilized by resonance with the benzene ring. We will return to such structures in Chapter 6.

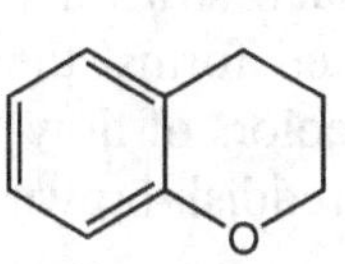

3.49, chroman 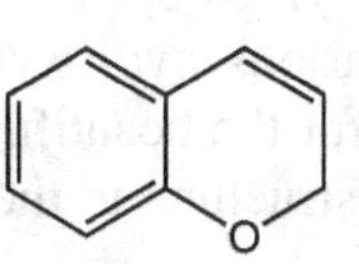 **3.50**, alpha-chromene 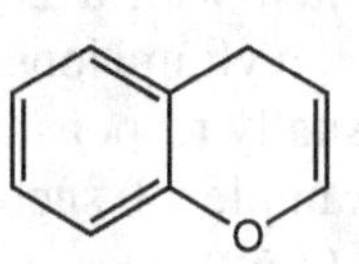 **3.51**, gamma-chromene 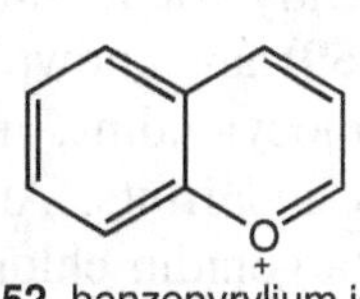 **3.52**, benzopyrylium ion

Alpha-tocopherol (**3.53**) is an example of a chroman derivative. It is found in vegetable oils and is one of several related structures constituting the human growth factor Vitamin E.

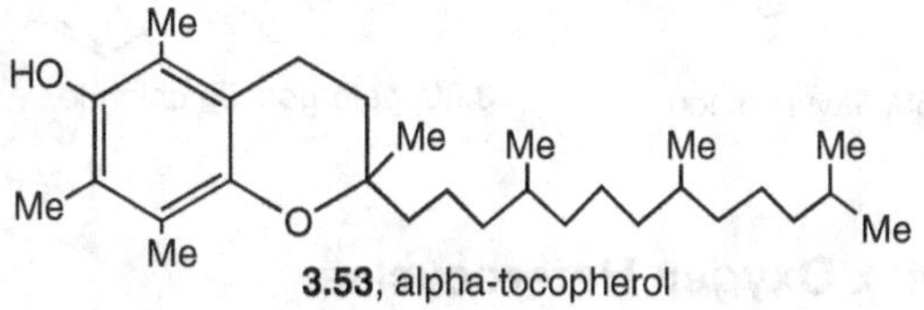

3.53, alpha-tocopherol

Many derivatives of alpha-chromene occur in plants. Typical examples include ageratochromene (**3.54**) from the garden flower ageratum and coumarin (**3.55**), which was once used as a flavoring agent but is now considered too toxic. Coumarin is found in various plants such as the tonka bean, lavender oil, and sweet clover.

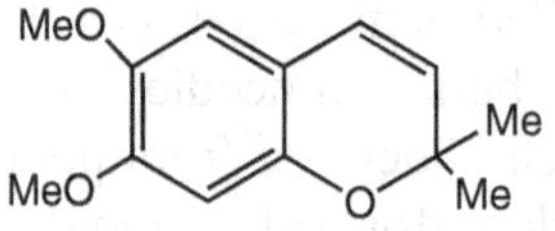

MeO MeO Me Me O O O

3.54, ageratochromen 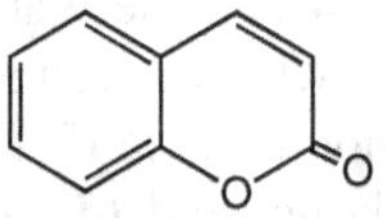 **3.55**, coumarin

The gamma-chromene structure appears in many natural products as derivatives with a 4-keto group. The parent is known as chromone (**3.56**), and it is found especially in the flavone (flavonoid) family, which has a 2-phenyl substituent. Flavone itself has structure **3.57** and is a yellow solid found in plants. A flavonoid from strawberries, fisetin (**3.58**), is said to have weak memory-enhancing ability.[12]

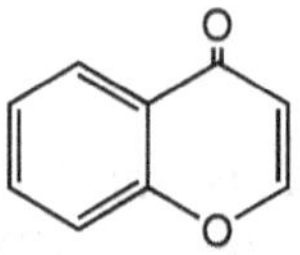

O O O HO OH O O Ph O OH OH

3.56, chromone 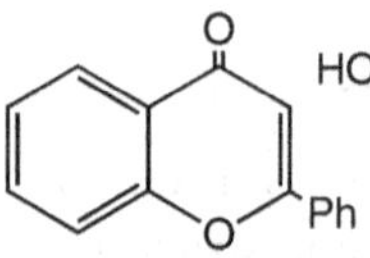 **3.57**, flavone 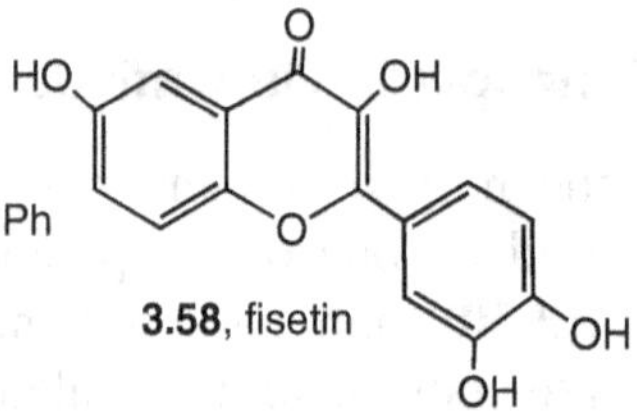 **3.58**, fisetin

The benzopyrylium ion structure is widespread in nature. It is frequently found substituted with a 2-phenyl group, and such structures (**3.59**) are known as flavylium ions. Various hydroxylated forms, the anthocyanidins, are usually responsible for the beautiful colors of flowers and fruits. An example of such a structure is the reddish-brown pelargonidin chloride (**3.60**).

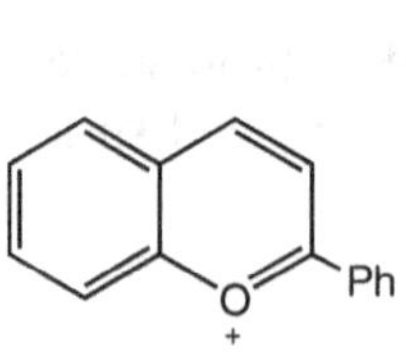

OH OH HO O + Ph O + Cl⁻ OH

3.59, flavylium ion 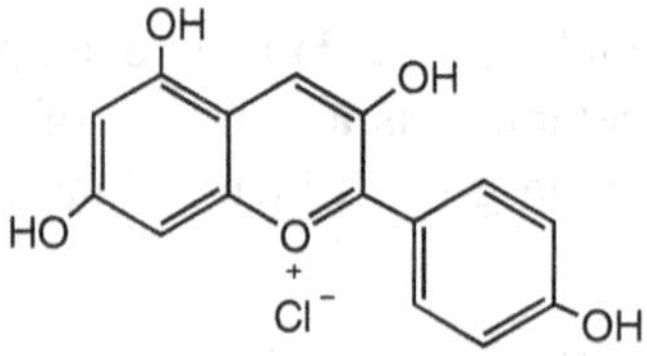 **3.60**, pelargonidin chloride

3.3.6. Complex Oxygen Heterocycles

Found also in the plant kingdom are some complex compounds containing oxygen heterocyclic rings. Coriamyrtin (**3.61**) is an example of such a structure. It is a toxic compound found in *Coraria myrtifolia*, and it is unique in having two oxirane groups. There are seven chiral carbons in this structure. Columbin (**3.62**), which found in roots of *Jatrorrhiza palmate*, is a furan derivative with a beta-substituent containing two alpha-ketopyran (delta-lactone) rings.

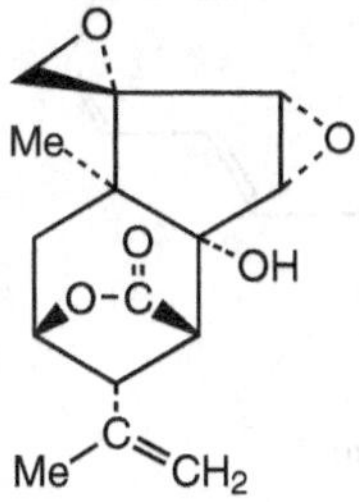

3.61, coriamyrtin

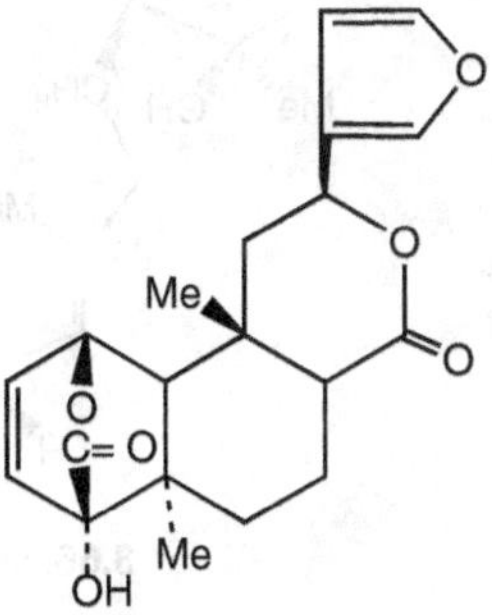

3.62, columbin

3.4. SULFUR AND PHOSPHORUS HETEROCYCLIC COMPOUNDS IN NATURE

It has already been noted that sulfur is found in association with nitrogen in the form of the thiazole ring (e.g., as in thiamine, which is shown in Figure 3.4). Few heterocycles are known in nature where sulfur is the sole heteroatom in the ring. Examples are the thiophene derivatives **3.63**, isolated from the roots of the Tagetes plant and found to have nematocidal properties, and **3.64** from the marigold plant. The vitamin biotin, whose structure is shown in section 3.2.4, has a reduced thiophene (a thiolane) group.

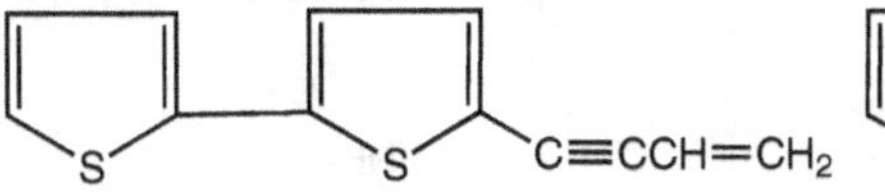

3.63

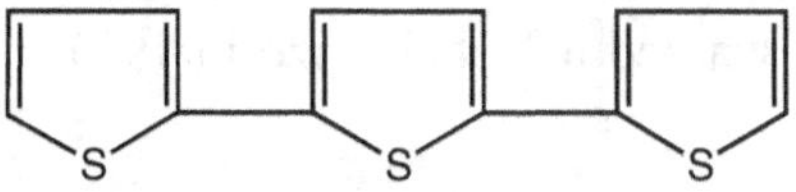

3.64

The antibiotic leinamycin isolated in 1989[13] from a culture broth of a Streptomyces variety has a disulfide group in a 5-membered ring and thus is a 1,2-dithiole derivative. As found in structure **3.65**, one sulfur is in the sulfide state and the other in the sulfoxide state. It has another novel feature, an 18-membered nitrogen heterocycle (in a lactam structure), as well as a thiazole ring. It is of considerable interest as a potent antitumor agent, and it also shows activity against both gram-positive and gram-negative bacteria.

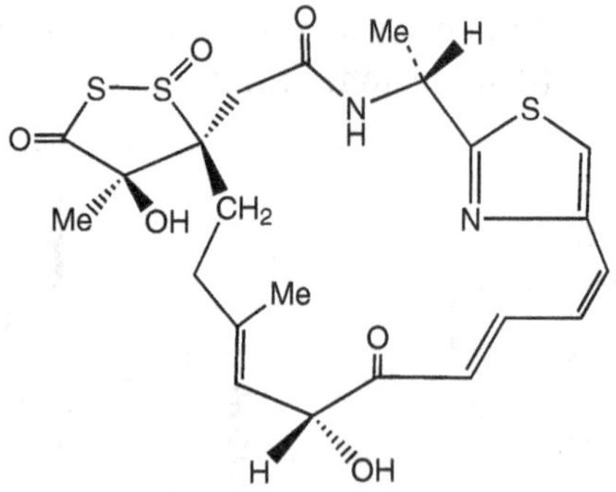

3.65, leinamycin

Until 2009, no cyclic structures containing only phosphorus and carbon have been found in nature, even though this is a large and important class of the huge field of organophosphorus chemistry.[14,15] The only phosphorus-containing ring system found in nature is in fact a cyclic derivative of phosphoric acid, which occurs in a cyclized version of the nucleotide structure. The structure contains the 1,3,2-dioxaphosphorinane ring system (**3.66**, also known as 1,3,2-dioxaphosphinane). This structure was first observed in the form of cyclic adenosine monophosphate (**3.67**, cyclic-AMP or cAMP).[16] It was a major discovery in biochemistry; cAMP was recognized as a second messenger in hormone activity, and in humans, it is involved in the regulation of glycogen, sugar, and lipid metabolism. Adenosine triphosphate is the source of this important compound; it is formed by the action of the enzyme adenylate cyclase on ATP. Earl Sutherland, who was the discoverer of the structure and importance of cAMP, was awarded a Nobel Prize in 1971.

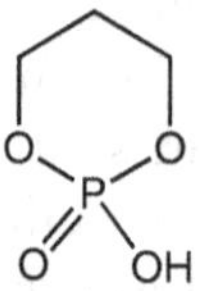

3.66

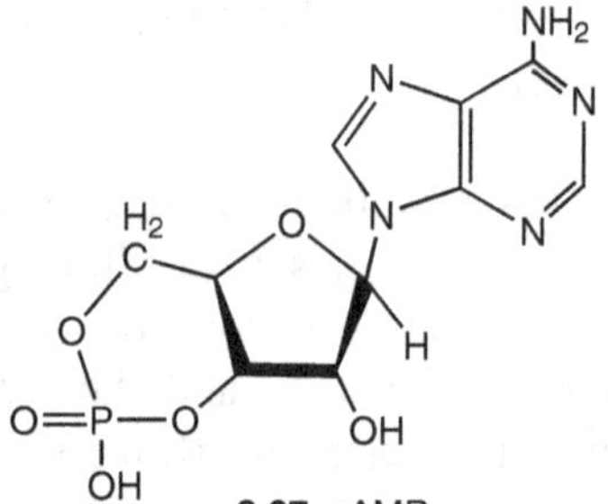

3.67, cAMP

REFERENCES

(1) T. Aniszewski, *Alkaloids-Secrets of Life*, Elsevier, Amsterdam, The Netherlands, 2007, p. 5.

(2) K. W. Bentley, *The Isoquinoline Alkaloids*, Pergamon Press, Oxford, UK, 1965, pp. 238–240.

(3) K. W. Bentley, *The Alkaloids*, Wiley, New York, 1957.

(4) R. H. F. Manske, *The Alkaloids*, Academic Press, New York, 1950.

(5) T. S. Kaufman and E. A. Ruveda, *Angew. Chem., Int. Ed. Engl*., **44**, 854 (2005).

(6) R. B. Woodward, M. P. Cava, A. Hunger, H. V. Daeniker, and K. Schenker, *Tetrahedron*, **19**, 247 (1963).

(7) D. J. Faulkner, *Natural Products Reports*, **17**, 5 (2000).

(8) H. Kiyota, *Marine Natural Products,* in *Topics in Heterocyclic Chemistry*, Springer, Berlin, Germany, Vol. 5, 2006.

(9) H. Fischer and K. Zeiler, *Ann*., **468**, 98 (1929).

(10) J. D. Watson, *The Double Helix*, Atheneum Press, New York, 1968.

(11) J. C. Sheehan and K. R. Henery-Logan, *J. Am. Chem*. Soc., **79**, 1262, (1957); *ibid*., **81**, 3089 (1959).

(12) P. Maher, T. Akaishi, and K. Abe, *Proc. Nat. Acad. Sci*. USA, **103,** 16568 (2006).

(13) M. Hara and I. Takahashi, *J. Antibiot*., **42**, 333 (1989).

(14) L. D. Quin, *The Heterocyclic Chemistry of Phosphorus*, Wiley, New York, 1981.

(15) F. Mathey, Ed., *Phosphorus-Carbon Heterocyclic Chemistry: The Rise of a New Domain*, Elsevier, London, UK, 2001.

(16) E. W. Sutherland and T. W. Rall, *J. Biol. Chem*., **232**, 1077 (1958).

CHAPTER 4

PRINCIPLES OF SYNTHESIS OF AROMATIC HETEROCYCLES BY INTRAMOLECULAR CYCLIZATION

4.1. GENERAL

The synthesis of heterocyclic systems is a major activity of academic research and of the chemical industry, especially in pharmaceutical and agrochemical research and development. The organic chemists of the nineteenth century devised some of the synthetic methods that we still use today, but of course many new methods were discovered later, and in the current chemical literature there are continual reports of new methods. The body of literature on the synthesis of heterocycles is enormous and well covered in the comprehensive reference works listed in the Appendix of Chapter 1. Here, we will only describe some of the more versatile common methods that produce unsaturated rings of five and six members, for such rings constitute major backbones on which more complicated structures can be constructed.

The early chemists recognized that fully unsaturated rings of five and six members were generally stable and possessed some properties

Fundamentals of Heterocyclic Chemistry: Importance in Nature and in the Synthesis of Pharmaceuticals, By Louis D. Quin and John A. Tyrell Copyright © 2010 John Wiley & Sons, Inc.

of aromaticity. We will examine these properties in Chapters 6 and 7. These rings were formed with surprising ease, which led to the still-useful concept that the formation of aromatic rings can be looked at as a driving force in chemical reactions. We now know that there is little angle or steric strain in 5- and 6-membered rings, which is also a factor in making them of common occurrence.

In this chapter, we will examine first some of the classic methods of synthesis and show that they depend on the common reactions of functional groups studied in general organic chemistry. There are many new approaches to these heterocycles, and a selected few will then be described. As we go into these methods, the following generalizations are useful:

- Elimination reactions generally go in the direction that will give an aromatic system.
- Double-bond rearrangements that can result in an aromatic system are common in 5-membered rings.
- Tautomerism (1,3-hydrogen shifts) where possible can convert a structure into the most stable aromatic form.
- Dihydro derivatives are easily oxidized, even on exposure to air, into aromatic systems.

Most reactions that form heterocycles are of the following two general types: (1) intramolecular reactions, discussed in this chapter, between two functional groups at the ends of a chain; and (2) cycloaddition reactions, discussed in Chapter 6, where two different molecules with the proper unsaturation interact with the formation of two new sigma bonds tying the two molecules together. The latter type is of more recent invention. We will present most of these reactions with a logical mechanism, but being precise about some details of complex reactions, where intermediate structures must be created and further react, is not always easy. Some cycle-forming reactions have been studied extensively for many years. The advent of spectroscopic methods to detect the formation and disappearance of intermediates has added greatly to our knowledge of the mechanisms of formation of heterocyclic compounds. We will review an example of this in the first synthetic method to be examined, the Paal–Knorr synthesis of pyrroles (Section 4.2.1). For more detail than can be offered here, an excellent paper on cyclization mechanisms by Katritzky et al.[1] is recommended.

4.2. SOME OF THE CLASSIC SYNTHETIC METHODS

4.2.1. The Use of Familiar Reactions of Aldehydes and Ketones

4.2.1.1. Review. The carbonyl group of aldehydes and ketones is electrophilic and receptive to addition of common nucleophiles, generally those that have "active" (i.e., not bonded to carbon) hydrogen atoms. This includes ammonia, primary and secondary amines, water, alcohols, and thiols (mercaptans). It is common after the addition for a water molecule to be eliminated, either forming a carbon-heteroatom double bond or a carbon-carbon double bond, as directed by the structures. Examples of each are shown in Scheme 4.1. Note that the addition product always undergoes charge neutralization by proton loss from nitrogen and protonation of oxygen that creates the OH group. This reaction is one of the easiest ways to form a carbon–heteroatom bond, and we will observe this sort of chemistry in many syntheses to follow.

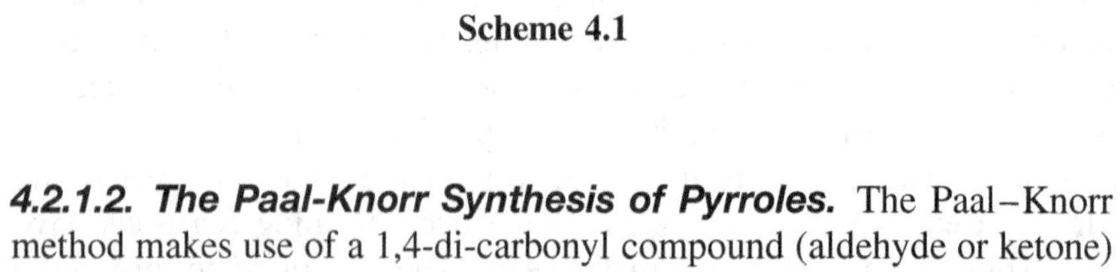

R-C(=O)-R + :NH_2R → R-C(R)(-O^-)-NH^+HR →($-H^+$) R-C(R)(-O^-)-NHR →($+H^+$) R-C(R)(OH)-NHR →($-H_2O$) R-C(R)=NR

R'-CH_2C(=O)-R + :SHR → R'-CH_2C(R)(O^-)-S^+HR → R'-CH_2C(R)(OH)-S-R →($-H_2O$) R'-CH=C(R)-SR

Scheme 4.1

4.2.1.2. The Paal-Knorr Synthesis of Pyrroles. The Paal–Knorr method makes use of a 1,4-di-carbonyl compound (aldehyde or ketone) in reaction with primary amines or ammonia. Many pyrroles have been made by this general process. Alkyl and some other substituents are allowed on the dicarbonyl chain. Diketones, dialdehydes, and ketoaldehydes all serve as reactants. Primary amines give rise to 1-alkylpyrroles. Examples of the overall process are shown in Scheme 4.2.

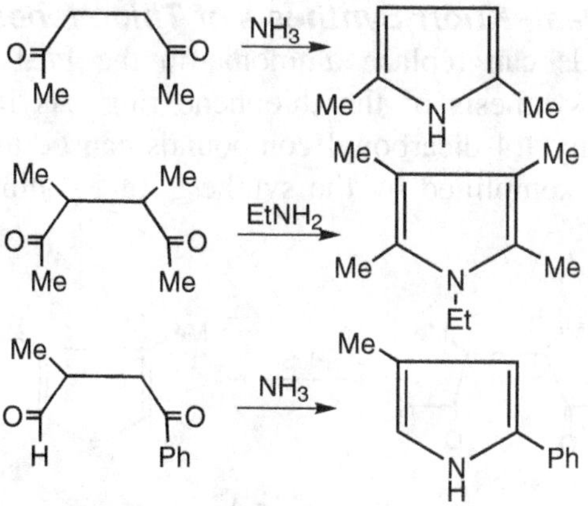

Scheme 4.2

The reaction is known to begin with the N-H group adding to one carbonyl group in the usual manner to form an imine (R–CH=NR). Several mechanisms have been proposed to explain subsequent reactions, but we will show here the one developed by Katritzky et al.[1], which depends on the detection of intermediates by ^{13}C and ^{15}N nuclear magnetic resonance (NMR) spectroscopic measurements during the course of the synthesis. The starting materials were acetonylacetone (**4.1**) and neopentylamine. The process is illustrated in Scheme 4.3. An important feature to note is that imine **4.2** rearranges to the enamine **4.3**, which is the species supplying an N-H group for attack on the second carbonyl group.

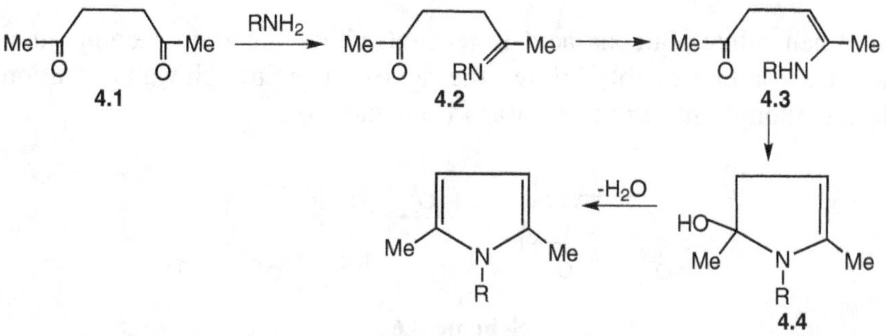

Scheme 4.3

4.2.1.3. The Paal–Knorr Synthesis of Thiophenes and Furans. Hydrogen sulfide can replace ammonia in the Paal–Knorr process and provide a synthesis of the thiophene ring. As in the synthesis of pyrroles, many 1,4-dicarbonyl compounds can be used in this process, which is exemplified by the synthesis of 3,4-dimethylthiophene (Scheme 4.4).

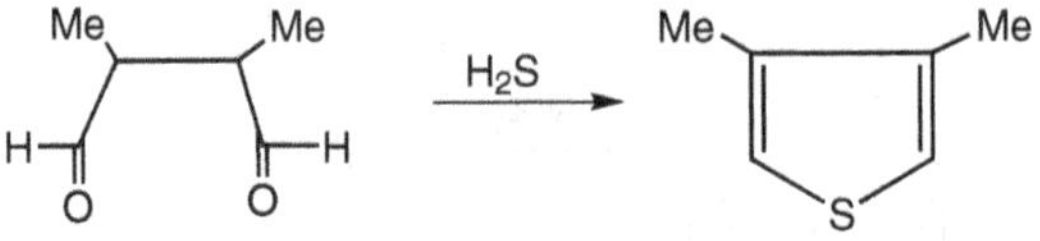

Scheme 4.4

The pyrrole-forming mechanism must be modified because the counterpart of an imine (as in **4.2**) cannot be prepared. The mechanism of thiophene formation can be considered to be that shown in Scheme 4.5.

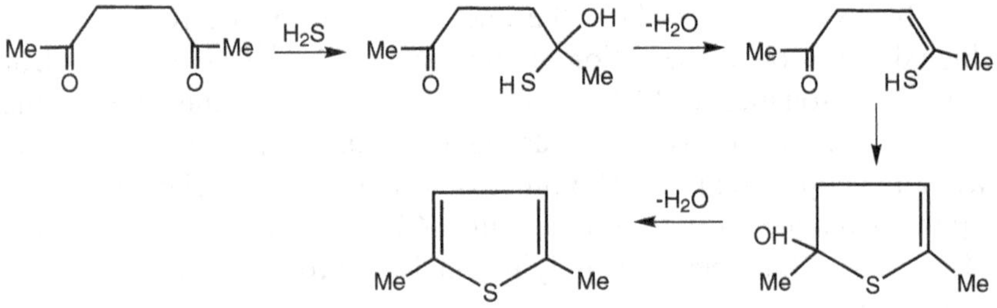

Scheme 4.5

When dilute aqueous acid is reacted with a dicarbonyl compound, the reaction presumably follows the same course through water addition to a carbonyl and provides furans (Scheme 4.6).

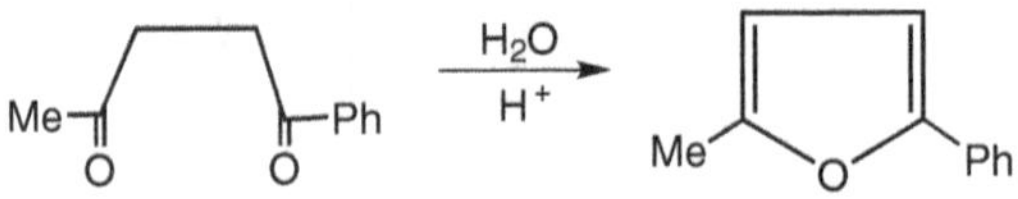

Scheme 4.6

4.2.1.4. Extension of the Paal-Knorr Synthesis to Include 1,3-Thiazoles. In another extension (Scheme 4.7), we review a method to produce the thiazole ring based on cyclization of a gamma-iminoketone. The N-H of the imino group ($HN{=}CR_2$) can add to the carbonyl group in the usual way of an amine to close the ring. The reaction of a bromoketone and a thioamide (**4.6**) is used to construct the carbonyl intermediate with a gamma-imino group (**4.7**). The thioamide is in tautomeric equilibrium with enolic form **4.5**, which is the species displacing bromine. One can conceive of other heterosubstituted dicarbonyl (or equivalent) species that can possibly give similar ring closure.

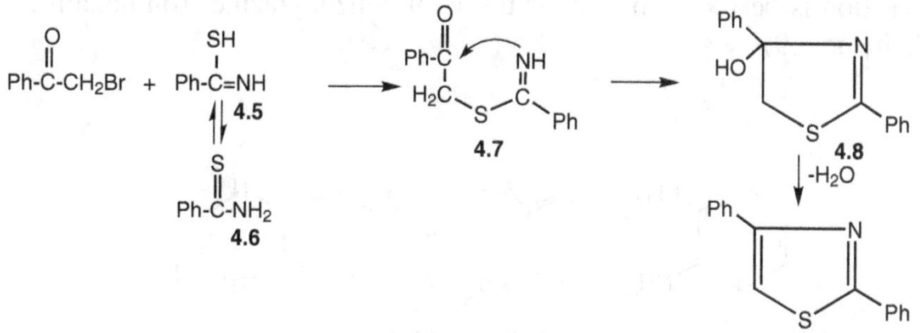

Scheme 4.7

4.2.1.5. Synthesis of Pyrazines from 1,2-Diamines. The condensation reaction of amines and carbonyl compounds can be applied to the synthesis of pyrazines (1,4-diazines). Here, a 1,2-diamine is reacted with an alpha-dicarbonyl compound. The usual reaction takes place to form two C=N bonds thus tying the two reactants together as viewed in Scheme 4.8. The product is a dihydropyrazine (**4.9**). This is a good illustration of the preference for cyclization to form 6-membered rings over the competitive process that would form polymeric linear compounds with C=N bonds (e.g., where C=N formation occurs from the dicarbonyl compound reacting with amino groups in two different molecules of the diamine, rather than with the amino groups in the same molecule). Mild oxidation converts the dihydropyrazine to the aromatic pyrazine (**4.10**). Indeed, dihydropyrazines are so readily oxidized that their isolation can be difficult.

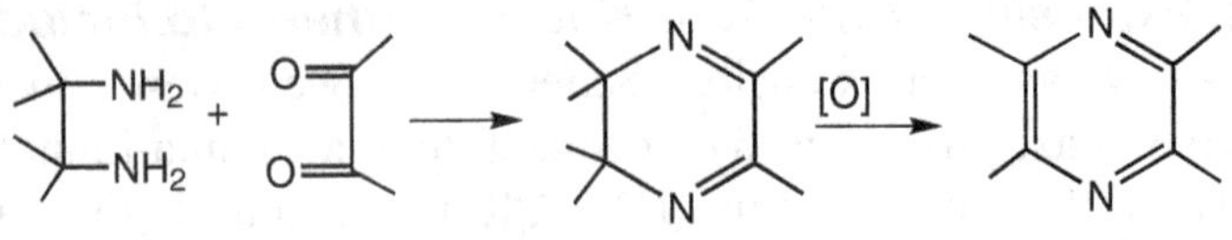

Scheme 4.8

In practice, aliphatic diamines are less useful than ortho-diaminobenzenes (also called ortho-phenylenediamines), and the reaction is best known as a synthesis of benzopyrazines (quinoxalines, Scheme 4.9).

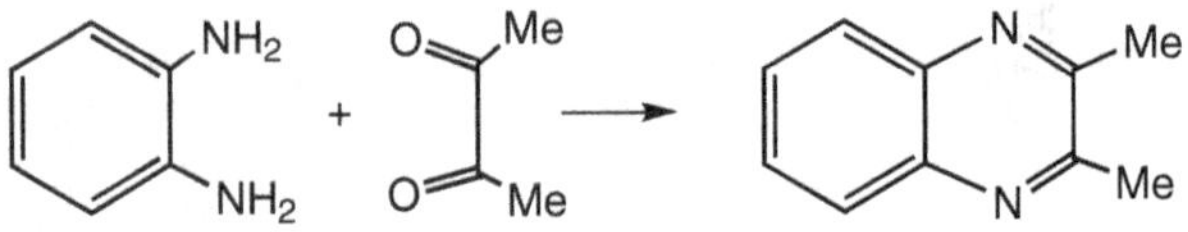

Scheme 4.9

A valuable extension of this process is to react a heterocyclic diamine with a diketone. In Scheme 4.10, a diaminopyrimidine is employed.

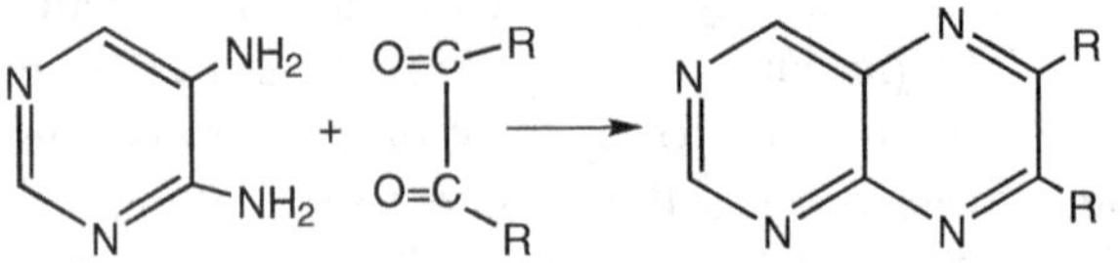

Scheme 4.10

A practical application of this reaction is in the synthesis of folic acid (**4.11**) presented in Scheme 4.11. With an unsymmetrically substituted dicarbonyl compound, two isomeric products are possible, and a minor amount of an isomer of **4.11** is formed in this process.

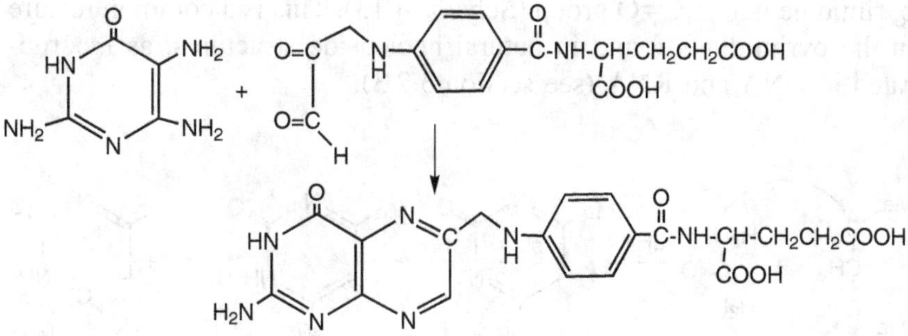

4.11 folic acid

Scheme 4.11

4.2.1.6. Formation of Pyrimidines from beta-Dicarbonyl and 1,1-Diamino Compounds. Yet another application of the amine–carbonyl reaction is viewed in the construction of the pyrimidine ring. Here a 1,1-diamino compound reacts with a beta-dicarbonyl compound, as is outlined in Scheme 4.12.

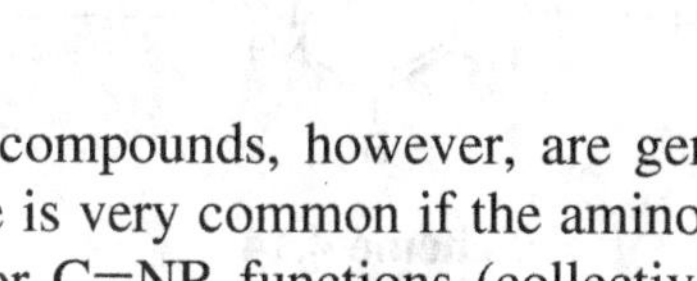

Scheme 4.12

Simple 1,1-diamino compounds, however, are generally unstable, but this structural feature is very common if the amino groups are attached to the C=O, C=S, or C=NR functions (collectively C=X), as in the structures below.

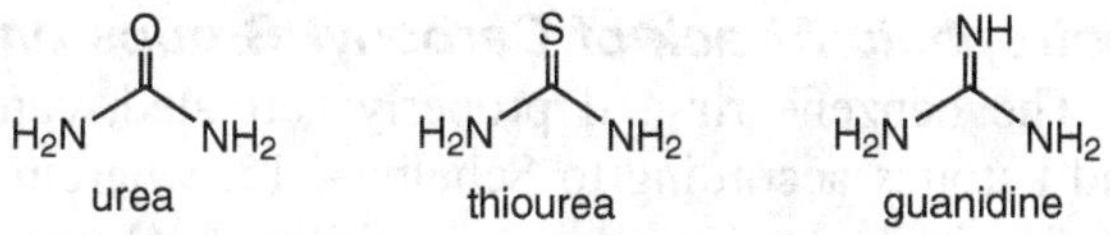

urea thiourea guanidine

When these are used, the pyrimidine formed has the particular advantage of having the extra functionality of the C=X group. We observe this in the reaction of urea with acetylacetone (**4.12**), which gives a pyrimidine with a C=O group (Scheme 4.13). This is a common feature in the pyrimidines found in natural nucleoside structures, as incorporated in DNA and RNA (see section 3.2.3).

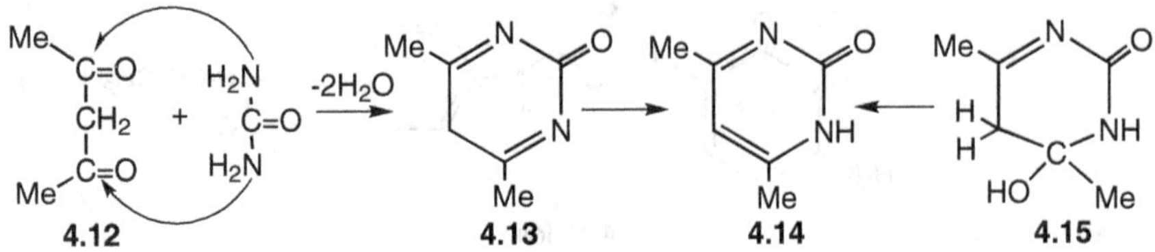

Scheme 4.13

This formulation shows the first product as the expected di-imino structure **4.13**, which then undergoes double-bond rearrangement to the true structure formed, **4.14**. However, another view of the formation of **4.14** is that the water eliminated from one of the initial carbonyl-NH addition products shown as **4.15** involves a proton from carbon rather than nitrogen.

When guanidine is used in the reaction, the product contains an amino group as a result of a tautomeric shift (Scheme 4.14). We will return to the synthesis and properties of the natural pyrimidines found in nucleosides in Chapter 9.

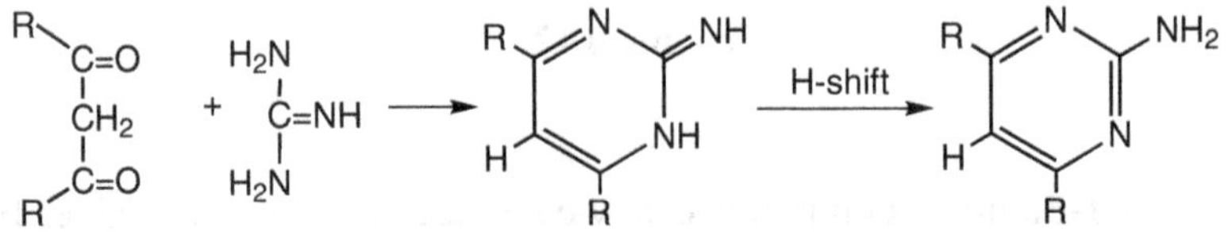

Scheme 4.14

4.2.1.7. Electrophilic Attack of Carbonyl Groups on the Benzene Ring. The benzene ring, if properly activated, can react with aldehydes and ketones according to Scheme 4.15, wherein an electron pair from the ring is donated to the electrophilic C=O group to form a

C–C bond. The intermediate carbocation (**4.16**) is stabilized by resonance and then loses a proton. The formation of an aromatic ring drives the reaction forward.

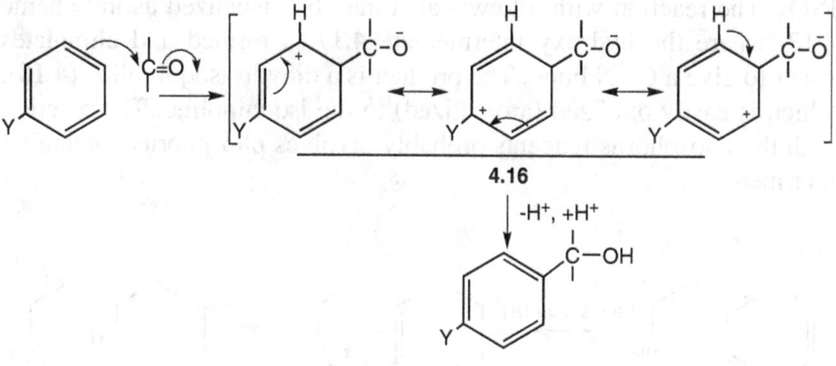

4.16

-H+, +H+

Scheme 4.15

This reaction is employed in the Combes synthesis of quinolines (Scheme 4.16). A side-chain with a terminal carbonyl group is first attached to the benzene ring. Acid catalysis then promotes the attack of C=O on the ring by the adding of a proton to the carbonyl oxygen to increase its electrophilicity; acid also catalyzes the elimination of water from the intermediate to generate the aromatic pyridine component of quinoline. Many quinolines have been prepared by this and related methods in the search for new antimalarial drugs.

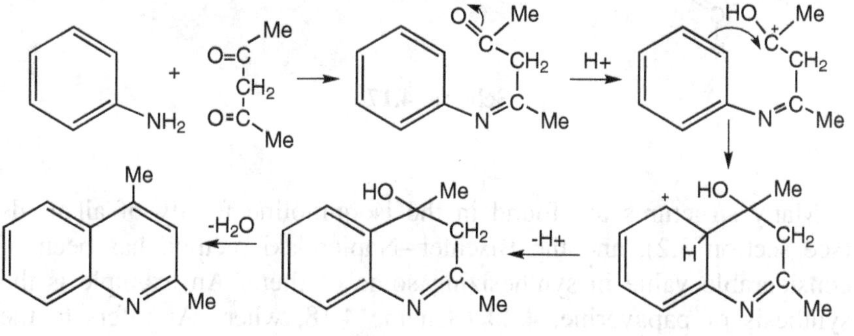

Scheme 4.16

Similar chemistry is employed in the important Bischler–Napieralski synthesis of isoquinolines. Here the carbonyl is in the form of an amide group, at the end of a chain on the benzene ring. The carbonyl attack on the ring is promoted with such reagents as Lewis acids, $POCl_3$, and P_2O_5. The reaction with a Lewis acid may be visualized as in Scheme 4.17, where the hydroxy intermediate **4.17** is formed and eliminates water to give a C=N bond. The product is a dihydroisoquinoline (**4.18**), which is easily oxidized (aromatized) to the isoquinoline. The reaction with the phosphorus reagents probably involves phosphorus-containing intermediates.[2]

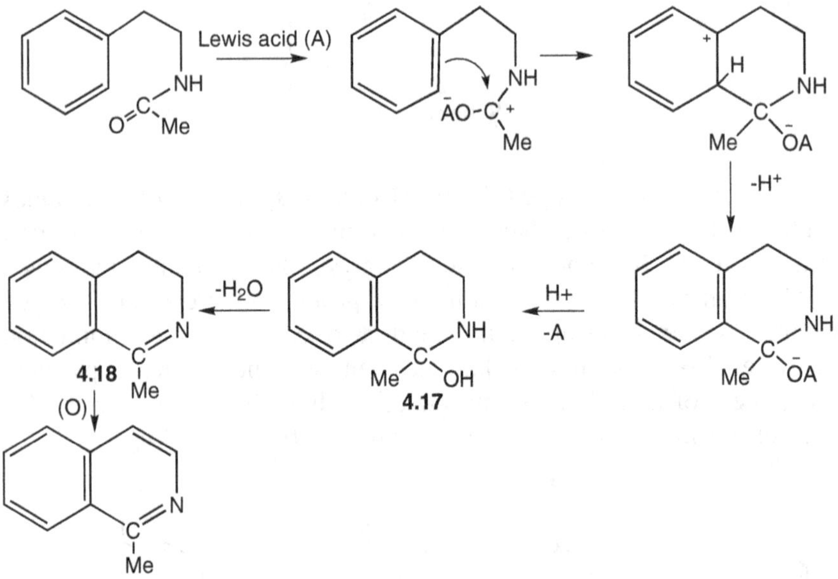

Scheme 4.17

Many structures are found in the isoquinoline family of alkaloids (see section 3.2), and the Bischler–Napieralski method has been of considerable value in synthesizing some of them. An example is the synthesis of papaverine, **4.20** (Scheme 4.18, where Ar refers to the dimethoxyphenyl substituent).

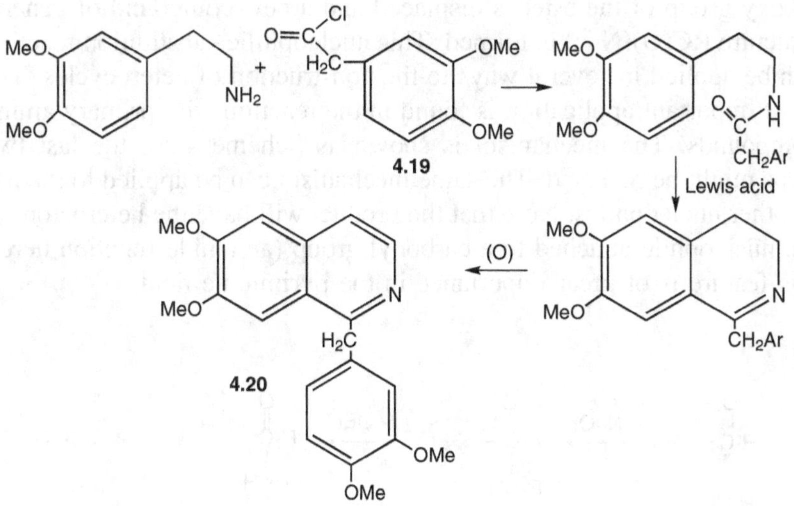

Scheme 4.18

Multicyclic nitrogen heterocycles can also be prepared by extension of the Bischler–Napieralski synthesis. Thus, the azaphenanthrene system (**4.21**) can be prepared as shown in Scheme 4.19.

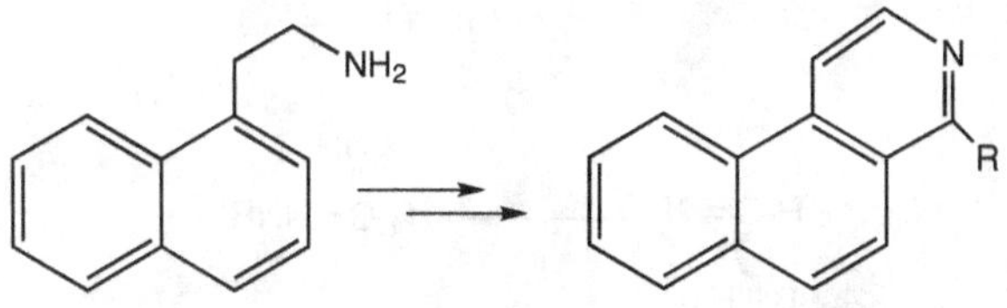

Scheme 4.19

4.2.2. Attack of Nucleophiles on Acid Derivatives

4.2.2.1. Review. When esters are allowed to react with the common nucleophiles (such as alcohols or alkoxides (RO^-M^+), primary and secondary amino compounds, and thiols or thiolates (RS^-M^+)), the alkoxy group of the ester is displaced and a new compound of general structure RC(O)(Nuc) is formed. This nucleophilic substitution reaction can be applied in several ways to the construction of heterocycles. The most important application is found in the reaction with primary amino compounds. The mechanism is shown as Scheme 4.20; the last two steps might be reversed. The same mechanism can be applied to the use of other nucleophiles. Note that the product will have the heteroatom of the nucleophile attached to a carbonyl group (an amide function here); this feature is of great importance in the pyrimidine field.

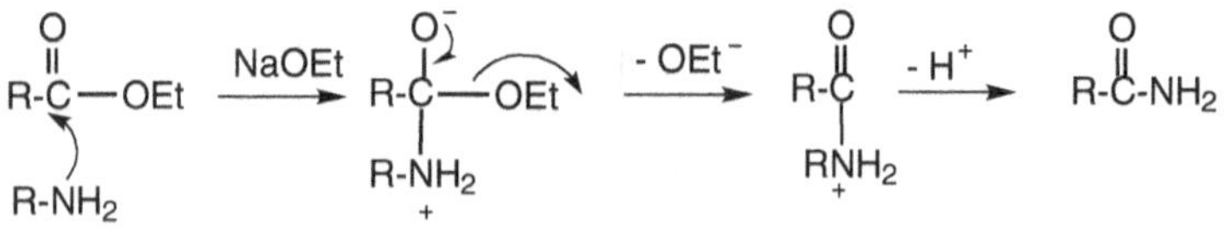

$$R\text{-}C(=O)\text{-}OEt + R\text{-}NH_2 \xrightarrow{NaOEt} R\text{-}C(O^-)(OEt)\text{-}NH_2^+\text{-}R \xrightarrow{-OEt^-} R\text{-}C(=O)\text{-}N^+H_2R \xrightarrow{-H^+} R\text{-}C(=O)\text{-}NH_2$$

Scheme 4.20

Not only esters but also acids, amides, and nitriles can react with nucleophiles, the latter leading to an imino group rather than a carbonyl (Scheme 4.21).

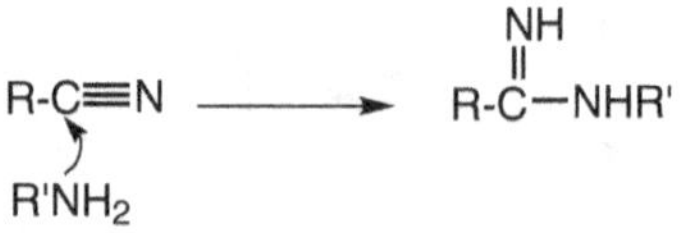

$$R\text{-}C{\equiv}N + R'NH_2 \longrightarrow R\text{-}C(=NH)\text{-}NHR'$$

Scheme 4.21

4.2.2.2. Construction of Pyrimidines. To make use of this reaction for pyrimidine synthesis, a diamino compound, usually urea, thiourea, or guanidine, can be reacted with the diacid malonic acid ($HOOC{-}CH_2{-}COOH$) (Scheme 4.22) or preferably its diethyl ester. In its simplest form, the reaction has been used to make barbituric acid (**4.21**; see also Chapter 8, section 8.2). Salts can be made from barbituric acid with aqueous base; this reaction can be viewed as removing a C–H proton from C-5, or from the corresponding enol (found in structure **4.23**).

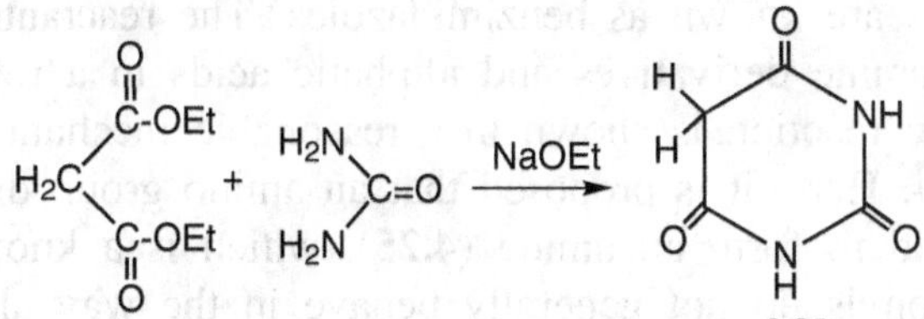

4.22

Scheme 4.22

This structure may not look like it is in the pyrimidine state, but this relation can be observed if we draw barbituric acid in its tri-enol form, **4.23**.

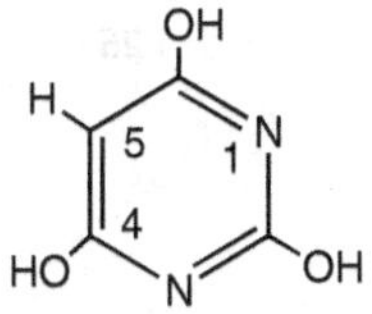

4.23

Many derivatives of barbituric acid with certain substituents at C-5, which are made by using C-substituted malonates, are very useful as hypnotics. The most famous is phenobarbital (**4.24**), which has been used clinically for many years. This compound can be made from diethyl ethylphenylmalonate (Scheme 4.23).

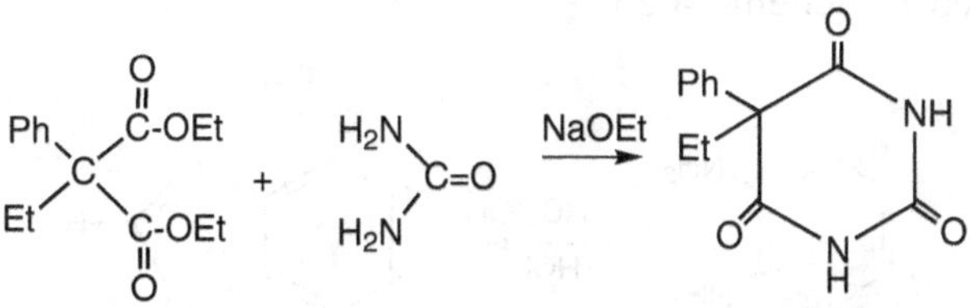

4.24

Scheme 4.23

Although carboxylic acids are less reactive to nucleophilic substitution than are esters, they do participate in a special synthesis (the Phillips synthesis) of the imidazole ring in benzo derivatives.

The products are known as benzimidazoles. The reactants are ortho-phenylenediamine derivatives and aliphatic acids in a refluxing HCl solution. The reaction is shown in a reasonable mechanistic form in Scheme 4.24. Here, it is proposed that an amino group displaces OH from the acid to form an amide (**4.25**), which is a known process. Amide carbonyls do not generally behave in the way aldehyde and ketone carbonyls do, but nevertheless it has been suggested that the second amino group does add to the amide carbonyl, with dehydration of the adduct, to form the double bond in the imidazole ring (**4.26**).

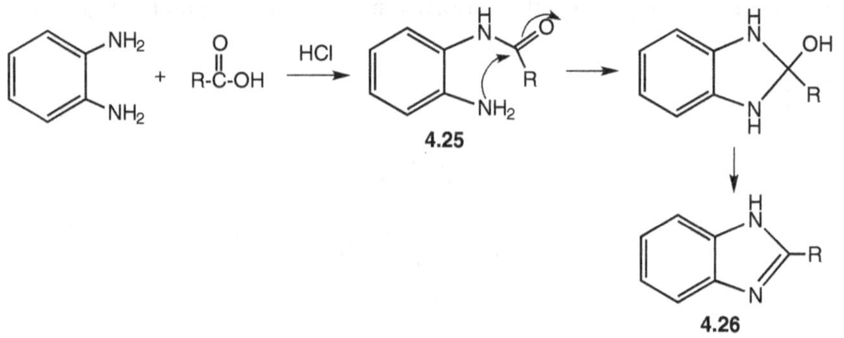

Scheme 4.24

The chemistry of the Phillips synthesis can be extended to prepare the purine ring system. It is simply necessary to use a diaminopyrimidine in the reaction with a carboxylic acid. This is known as the Traube purine synthesis, and it has been of great significance in purine chemistry. It is illustrated in Scheme 4.25.

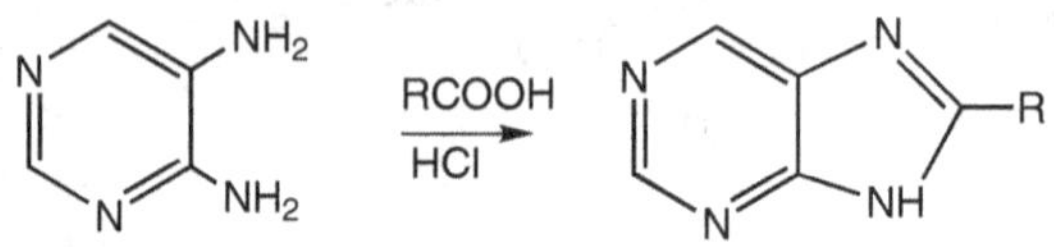

Scheme 4.25

This Traube process has been used to synthesize the important purine "base" guanine (G of the genetic code), as shown in Scheme 4.26. Ethyl formate has also been used instead of formic acid.

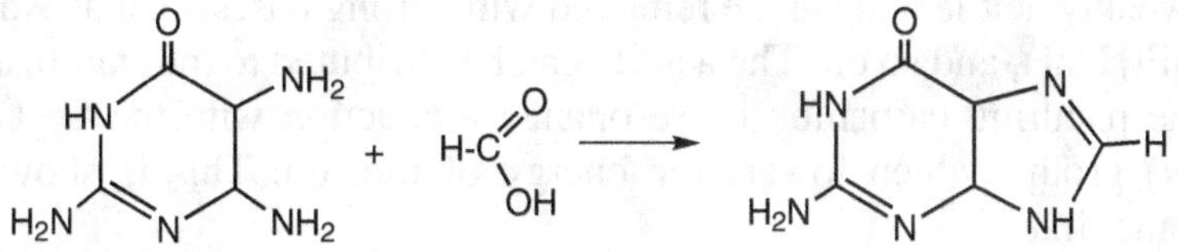

Scheme 4.26

As already shown in Scheme 4.21, an amino group can add to a nitrile group to produce an imine. This reaction can be used to advantage in the construction of pyrimidines. This requires the use of an aldehyde or ketone containing the nitrile group on the alpha-carbon. With a urea derivative, one can consider the reaction proceeding with an amino group adding to C=O in the usual way, then the other amino group adding to the nitrile group. Scheme 4.27 illustrates this process. Unlike the carbonyl group on the pyrimidine ring, the initially formed double bond in the imino group in **4.27** is not retained. Nitrogen accepts a proton in a tautomeric shift, thus becoming an amino group (**4.28**). This constitutes an excellent way to make aminopyrimidines. Compound **4.28** is in fact cytosine, another "base" (C) of the genetic code.

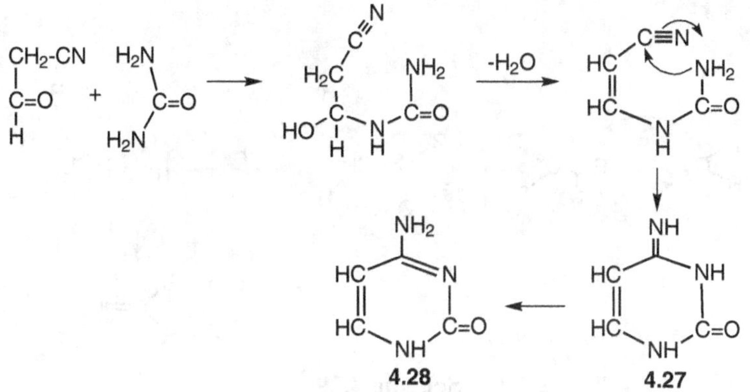

Scheme 4.27

4.2.3. Closing Rings with the Aldol Condensation

4.2.3.1. Review. Hydrogen atoms on the carbon alpha- to a carbonyl or imino group are "activated" and are unlike normal H-C bonds. They are weakly acidic and can be removed with strong bases such as NaOH, NaOEt, NaH, and so on. The acidity can be attributed to the stabilization of the resulting carbanion by resonance interaction with the C=O (or C=N) group, which lowers the energy of the ion. This is shown in Scheme 4.28.

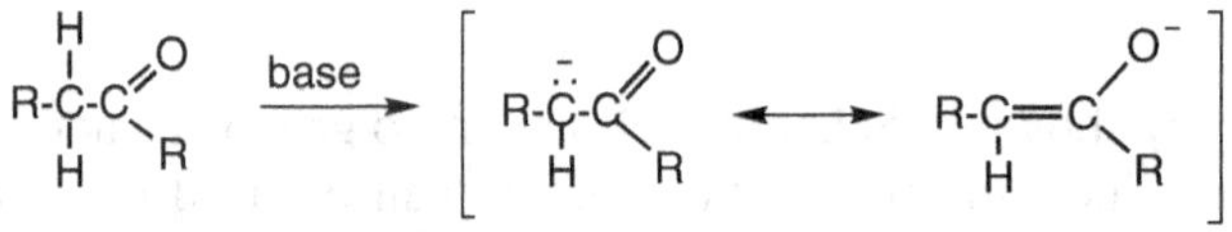

Scheme 4.28

The resonant anion can act as a typical carbanion and add to C=O or C=N bonds, just as a Grignard reagent does. In a true aldol condensation, the anion after formation proceeds to interact with another molecule of the starting aldehyde or ketone, giving a product that is a beta-hydroxy aldehyde or ketone known as an aldol (Scheme 4.29). Frequently, water is eliminated from this compound to produce an α, β-unsaturated carbonyl compound; this process is made easy by the resonance stabilization of such conjugated systems.

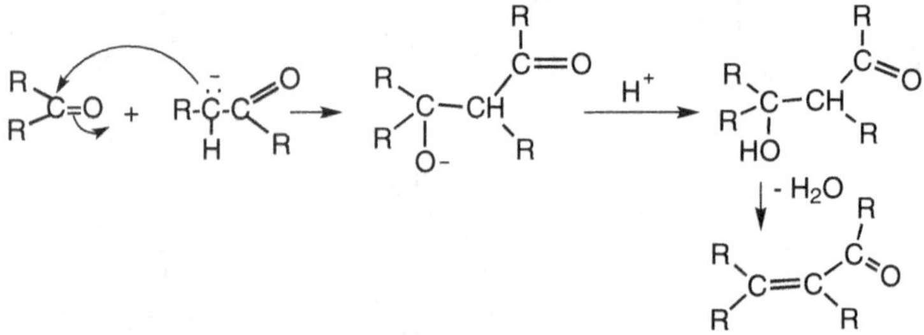

Scheme 4.29

The classic aldol condensation is that between two moles of an aldehyde or a ketone, as illustrated with acetaldehyde in Scheme 4.30, but the process can be broadened by allowing the intermediate carbanion to react with a carbonyl group of a different aldehyde or ketone. This is the process of importance in heterocyclic chemistry, for if both the carbanion and the carbonyl group are in the same molecule, then a ring will be formed. The carbanion can be derived not only from aldehydes and ketones but also from esters and nitriles. We will review this process first as applied originally by Knorr (below) in the construction of pyrroles, but several other well-known syntheses are available depending on carbanion intermediates in condensation reactions.

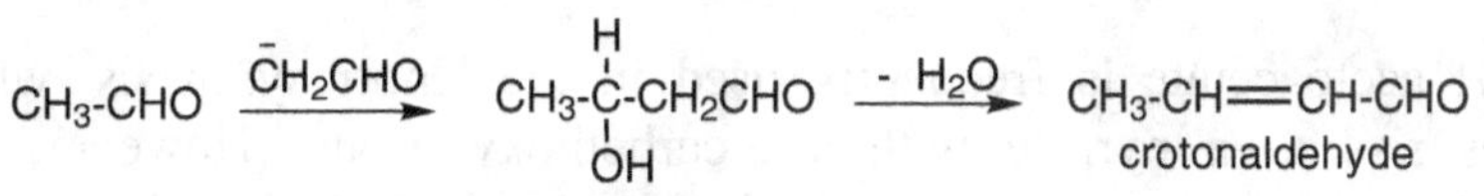

$CH_3\text{-}CHO \xrightarrow{\bar{C}H_2CHO} CH_3\text{-}CH(OH)\text{-}CH_2CHO \xrightarrow{-H_2O} CH_3\text{-}CH{=}CH\text{-}CHO$ (crotonaldehyde)

Scheme 4.30

4.2.3.2. The Knorr Pyrrole Synthesis. In this process, the starting materials are an alpha-amino ketone, or less commonly an alpha-amino aldehyde, and a beta-ketoester. In the latter, the protons on the alpha carbon are activated both by the keto and the ester carbonyls and are especially easily removed. The Knorr process starts with the condensation of the amino group with the keto group in the usual way to tie the two molecules together as in **4.29** of the example shown in Scheme 4.31. This species then undergoes intramolecular aldol-type condensation to form a reduced pyrrole derivative **4.30**.

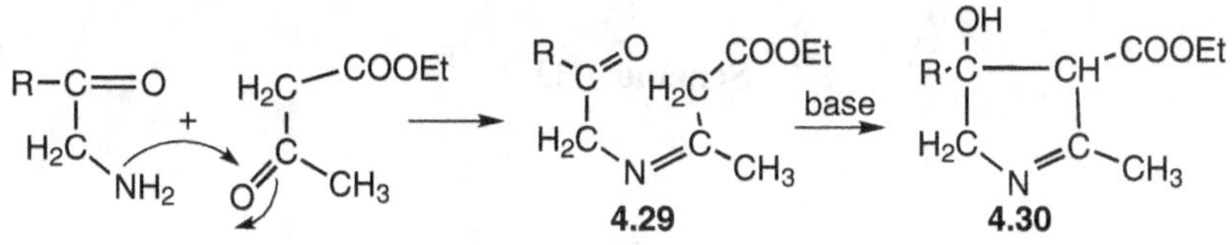

R–C=O, H_2C–NH_2 + H_2C(COOEt)–C(=O)–CH_3 → **4.29** →(base) **4.30**

Scheme 4.31

The final product is formed by elimination of water from **4.30** to form first a 3H-pyrrole (**4.31**), which then undergoes a proton shift to form the 1H-pyrrole as the final product. Both of these processes, easy dehydration and proton shifts to form the most stable structure, are common in heterocyclic syntheses.

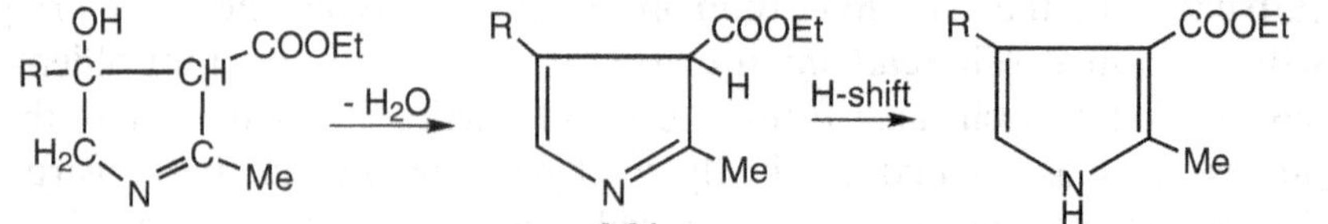

Scheme 4.32

Ethyl acetoacetate is frequently used in the Knorr synthesis, which gives rise to a pyrrole with a 3-carbethoxy group. However, this can be removed easily if desired, first by hydrolysis of the ester to the carboxylic acid and then decarboxylation of the acid group. 2,4-Dimethypyrrole is a well-known compound easily made by the Knorr process and elimination of the carbethoxy group (Scheme 4.33).

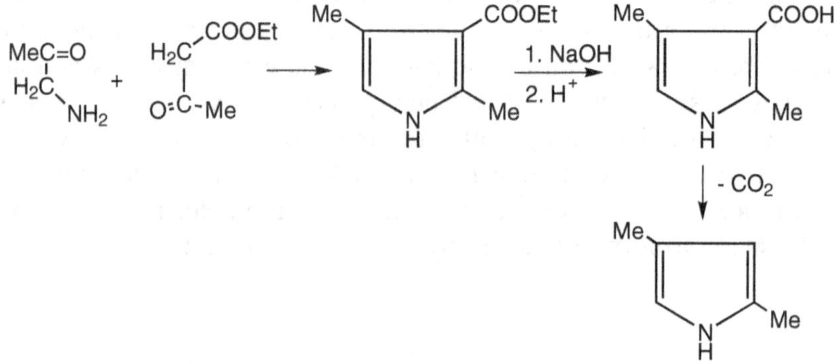

Scheme 4.33

4.2.3.3. The Feist–Benary Furan Synthesis. In this synthesis, an aldol-type carbonyl-carbanion condensation and a halogen displacement by an enol are employed to bring two molecules together in a ring structure, and in this sense it resembles the Knorr synthesis. The reactants are an alpha-haloketone and a ketoester (like ethyl acetoacetate). The mechanistic details remain unclear, but a logical mechanism is shown in Scheme 4.34. This assumes the carbonyl condensation occurs first before the halogen displacement, but the reverse may be true.

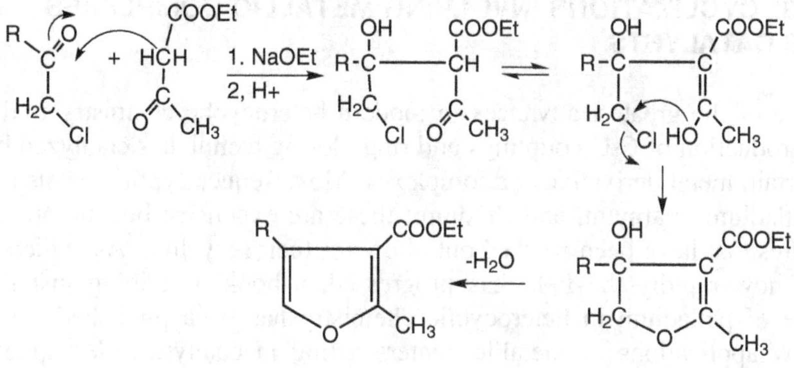

Scheme 4.34

A final example of the use of the carbonyl-carbanion condensation reaction is observed in a synthesis of the pyridine ring. The reactants are a 1,3-diketone such as acetylacetone (**4.32**) and a malonic acid derivative, malonic amide nitrile (**4.33**). The CH_2 group of the malonic derivative easily can form an anion with a base, which can condense with one C=O of the ketone. The amino group condenses with the other C=O group, tying the two reactants together in a 6-membered ring (Scheme 4.35). Dehydration give the pyridine (**4.34**); as with pyrimidines, the C=O is the preferred tautomeric form, but examining its enol form (**4.35**) reveals that it is indeed a pyridine derivative.

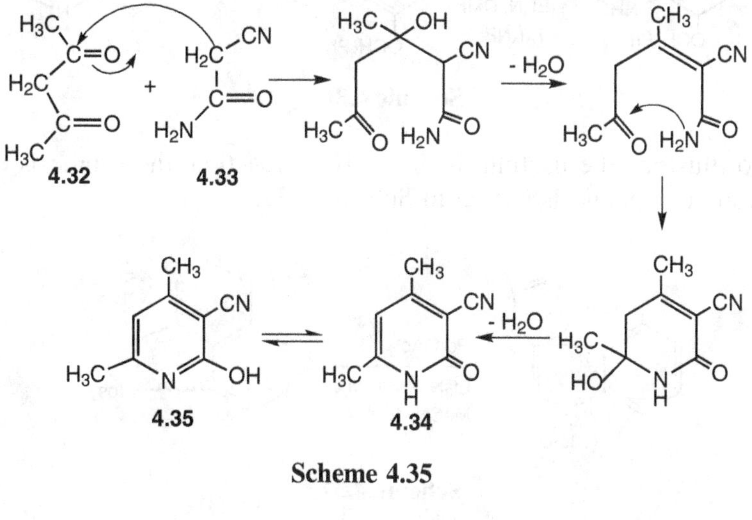

Scheme 4.35

4.3. CYCLIZATIONS INVOLVING METALLIC COMPLEXES AS CATALYSTS

One of the greatest advances in modern heterocyclic chemistry is the introduction of C-C couplings and ring-closing techniques catalyzed by certain metal derivatives or complexes. Most frequently, the metals are palladium, platinum, and rhodium; these are expensive but the procedures that have been worked out allow for their recycling. As evidence of how rapidly the field has progressed, a book describing just the use of palladium in heterocyclic chemistry has been published,[3] and new applications of metallic centers acting in catalytic roles appear regularly in the chemical literature. Of interest in this book are those reactions that cause cyclizations to form some of the common ring systems. One highly useful process is that based on the Heck coupling reaction, which was first introduced independently in the 1970s by the groups of T. Mizoroki in Japan[4] and Richard Heck at the University of Delaware.[5] These researchers effected the coupling of olefinic groups with halides, introducing the palladium in the form of its di-acetate; Pd(0) and $Pd(Ph_3P)_4$ are also used in such reactions. The Heck reaction can be extended to include intramolecular coupling, and with a heteroatom in the chain, heterocyclic systems can be generated. An example of this process is found in an indole synthesis reported by Sundberg[6] to proceed at room temperature in 96% yield (Scheme 4.36).

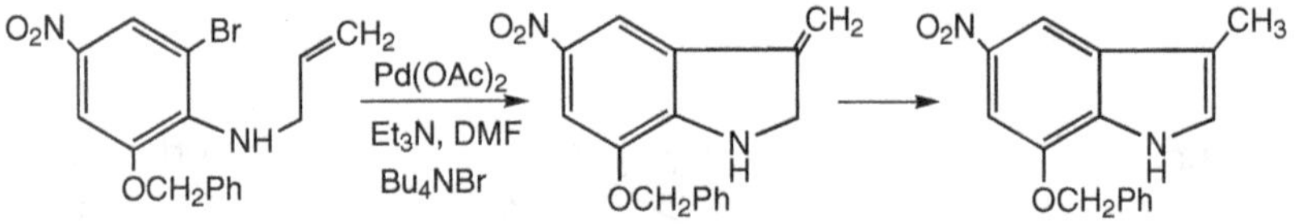

Scheme 4.36

To illustrate the usefulness of the Heck reaction, the synthesis of a tricyclic compound is shown in Scheme 4.37.[7]

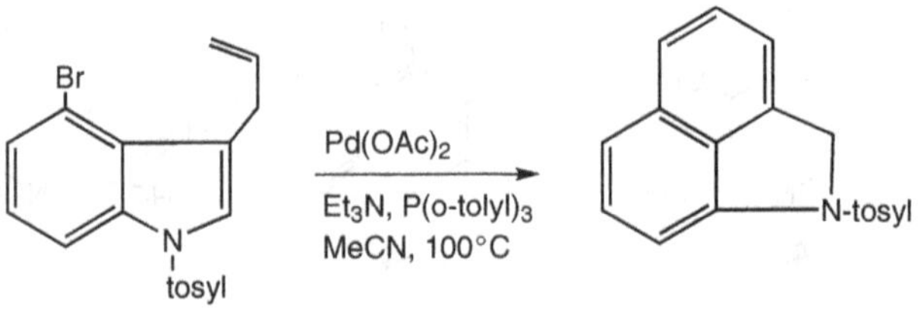

Scheme 4.37

The Heck intramolecular reaction can be applied to many other heterocycles; in Scheme 4.38, a benzofuran synthesis is shown.[8]

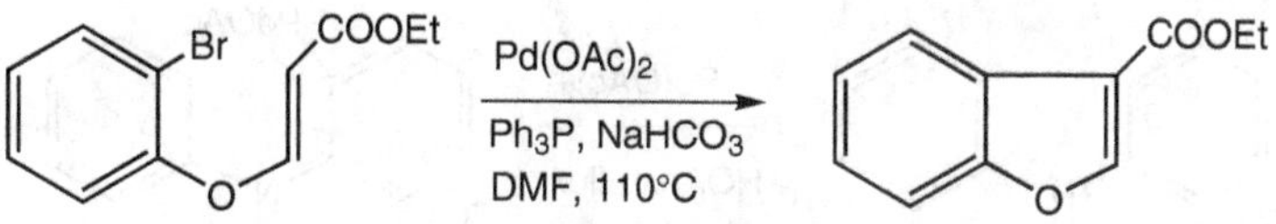

Br COOEt O → $Pd(OAc)_2$ / Ph_3P, $NaHCO_3$ / DMF, 110°C → COOEt O

Scheme 4.38

The mechanism of the Heck reaction with $Pd(OAc)_2$ probably involves the initial *in situ* reduction of the Pd^{++} ion to Pd(0), which then inserts in the C-X bond (recall the formation of the Grignard reagent from Mg and halides). The species R-PdX that is formed adds to the double bond in the sense of R and PdX, which in an intramolecular reaction will lead to a cyclic product. The adduct eliminates H-PdX to regenerate a double bond. This proposed mechanism is illustrated with the synthesis of the benzofuran **4.36** in Scheme 4.39.

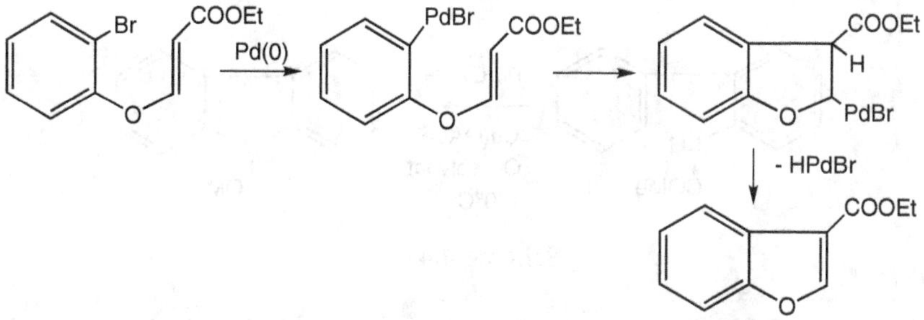

Br COOEt O → Pd(0) → PdBr COOEt O → COOEt H PdBr O → - HPdBr → COOEt O **4.36**

Scheme 4.39

Processes other than the Heck intramolecular coupling are also known that form heterocyclic rings. To illustrate, one of these is a useful carbazole synthesis that results from $Pd(OAc)_2$ oxidative coupling of the two benzene rings in a diphenylamine derivative. In its first form, this process was not a case of Pd catalysis but employed a full equivalent of $Pd(OAc)_2$ as a reactant. This performed electrophilic substitution on the ring, giving a Pd intermediate as shown in Scheme 4.40. After the coupling, the Pd was eliminated as the metallic element.[9] A modification has permitted a less expensive process by including an oxidizing agent ($Cu(OAc)_2$) to return the

Pd to the Pd^{++} state, and thus the process can be described as one employing $Pd(OAc)_2$ as a catalyst.[10]

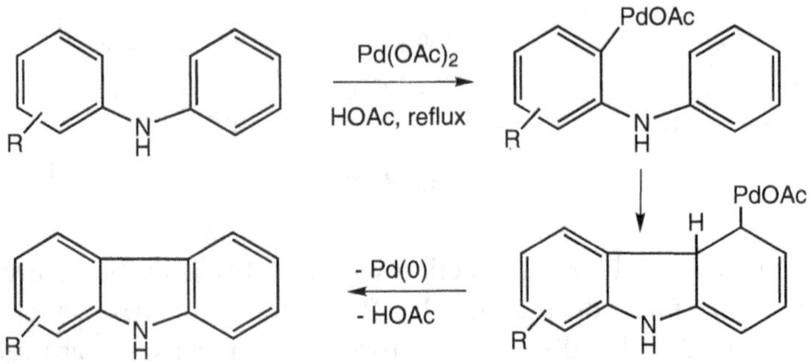

Scheme 4.40

The oxidative coupling can also be used to form the C-N bond. An example is the reaction of a diphenylamine derivative, where the coupling between an ortho-amino derivative and an ortho-carbon of the other phenyl ring would provide the carbazole system (Scheme 4.41).[11]

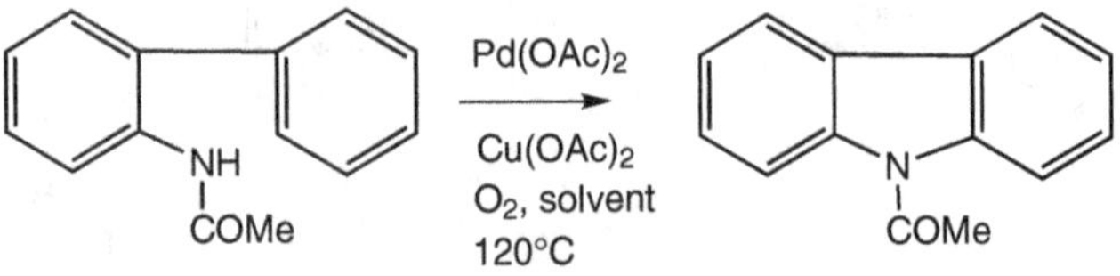

Scheme 4.41

The examples shown so far have been concerned with the synthesis of bicyclic or tricyclic heterocycles. A synthesis of a monocyclic heterocycle by the Heck reaction has been published in 2008.[12] The product is a tetrahydropyridine (Scheme 4.42). The nature of this reaction suggests the potential of making other monocyclics as well, which indicates the great power of palladium catalyzed reactions for heterocyclic synthesis.

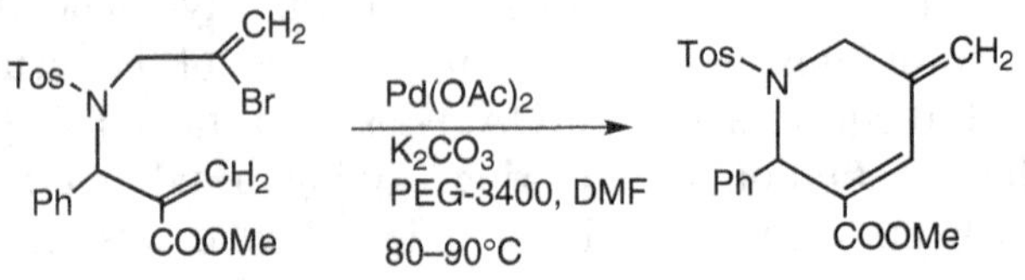

Scheme 4.42

The Pauson–Khand reaction provides another new approach to the metal-catalyzed synthesis of heterocycles. This reaction involves the interaction of the multiple bonds of an alkyne with an alkene and carbon monoxide in the presence of dicobalt octacarbonyl ($Co_2(CO)_8$), or with just this reagent as a source of CO. The overall process has been described as a [2 + 2 + 1] cycloaddition. Only a few applications to heterocyclic synthesis have been reported so far. A 2008 paper[13] that is illustrative of the process describes the use of this reaction for the construction of a heterocyclic ring that is part of an azabicyclo[3.3.1]nonane derivative. This ring system is present in the alkaloid (-)-alstonerine (**4.37**), which prompted this study.

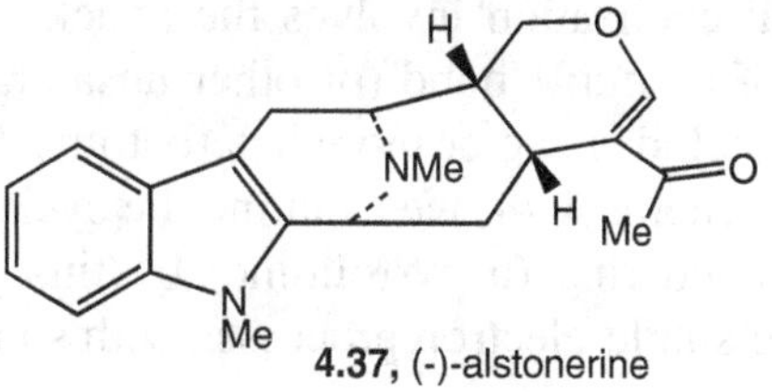

4.37, (-)-alstonerine

The Pauson-Khand reaction was first applied to the synthesis of several model compounds, e.g., the piperidine derivative **4.38**, as illustrated in Scheme 4.43. The dotted lines show the atoms that are to be connected in the reaction. The desired bicyclic structure was indeed obtained. Of significance was the stereochemistry of the reaction product, where the aza bridge and the proton on C-2 are trans-oriented as found in the alkaloid.

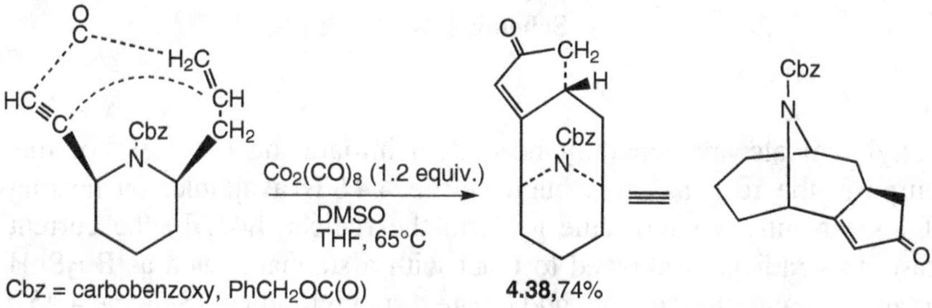

$Co_2(CO)_8$ (1.2 equiv.), DMSO, THF, 65°C

Cbz = carbobenzoxy, $PhCH_2OC(O)$

4.38,74%

Scheme 4.43

After these successful experiments, the authors went on to perform the total synthesis of alstonerine.

4.4. CYCLIZATIONS WITH RADICAL INTERMEDIATES

To this point, all reactions that have been used to effect cyclizations have involved steps that have been interpreted as depending on electron pair interactions. Much less common are syntheses in which free radicals are involved, but nevertheless such reactions are useful additions to the collection of methods available for ring formation. In most cases, radical reactions give rise to reduced or partially reduced heterocyclic systems. They proceed most readily when 5- or 6-membered rings are formed. We will consider only the most common types of radical cyclization, which are useful in illustrating some features of radical chemistry.

A typical radical cyclization involves the attack of a radical center on an sp^2 carbon of a double bond (or other unsaturated group) in the chain. If the chain includes one or more heteroatoms, then a heterocycle will be formed. In Scheme 4.44, we examine the cyclization of a radical to form a 5-membered ring (a pyrrolidine) by this process. Note the practice of showing single-electron processes with single-barbed arrows ("fish hooks").

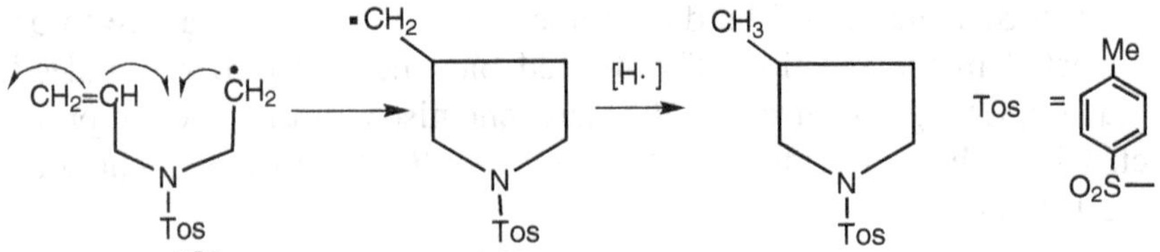

Scheme 4.44

Alkyl radicals are generally needed to initiate the process. For this purpose, the reagent azoisobutyronitrile **4.41** is available; on heating it loses a nitrogen molecule to form the radical **4.42**. In the current case, this radical is allowed to react with a stannane such as Bu_3SnH to abstract a hydrogen atom and create a stannyl radical (Scheme 4.45); this species is an effective agent for removing bromine from carbon to create a radical needed for cyclization.

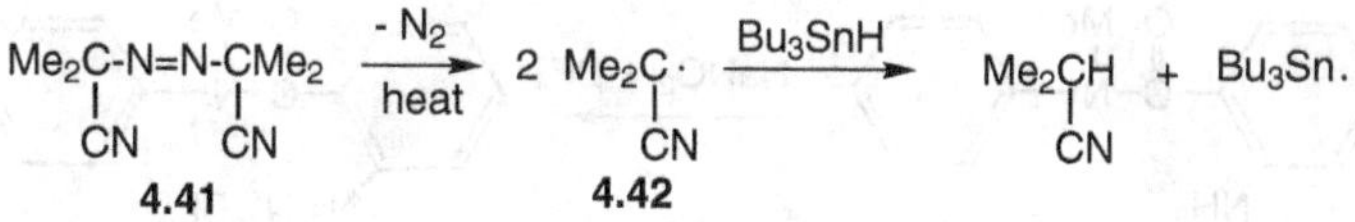

$$Me_2C(CN)\text{-}N{=}N\text{-}CMe_2(CN) \ (\mathbf{4.41}) \xrightarrow[\text{heat}]{-N_2} 2\ Me_2\dot{C}(CN) \ (\mathbf{4.42}) \xrightarrow{Bu_3SnH} Me_2CH(CN) + Bu_3Sn\cdot$$

Scheme 4.45

The reaction of the stannyl radical with the bromine atom is shown in Scheme 4.46, which also shows the regeneration of the stannyl radical and thus the creation of a chain reaction lasting until all of the stannane is consumed and converted to the bromostannane. Only a small fraction of an equivalent of initiator **4.41** is needed to get the process in operation.

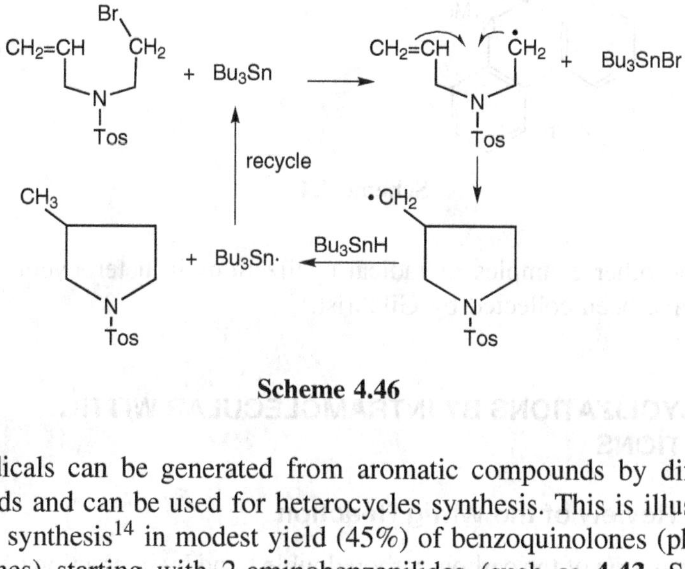

Br, $CH_2{=}CH$, CH_2, N, Tos + Bu_3Sn → $CH_2{=}CH$, $\dot{C}H_2$, N, Tos + Bu_3SnBr

recycle

CH_3, N, Tos + $Bu_3Sn\cdot$ ← Bu_3SnH — $\cdot CH_2$, N, Tos

Scheme 4.46

Radicals can be generated from aromatic compounds by different methods and can be used for heterocycles synthesis. This is illustrated by the synthesis[14] in modest yield (45%) of benzoquinolones (phenanthridones) starting with 2-aminobenzanilides (such as **4.43**, Scheme 4.47). The amino group is converted to the stable (even when dry) diazonium fluoroborate (**4.44**) from which an aryl radical is generated by action of metallic copper. The radical then adds to a double bond of the second benzene ring (Scheme 4.47) to form radical **4.45**, which is resonance delocalized. An oxidative step (even just exposure to air) is then required to achieve the fully aromatic system of the phenanthridone (**4.46**).

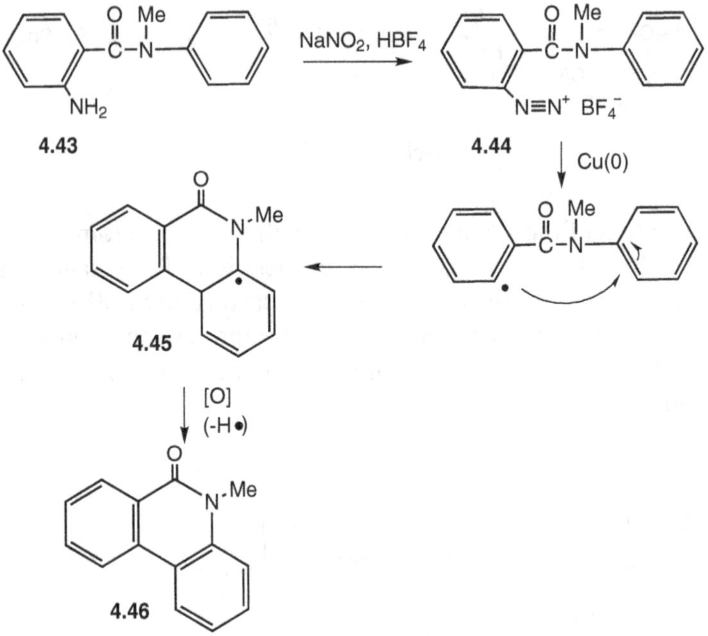

Scheme 4.47

Some other examples of radical cyclizations in heterocyclic chemistry have been collected by Gilchrist.[15]

4.5. CYCLIZATIONS BY INTRAMOLECULAR WITTIG REACTIONS

4.5.1. Review of the Wittig Reaction

The formation of the carbon-carbon double bond by interaction of ylides with carbonyl compounds was introduced into organic chemistry in the 1950s and rapidly became a major synthetic method. Many compounds have been made by this route. The discoverer, Georg Wittig of the University of Heidelberg, was awarded the Nobel Prize in 1979 for his extensive development of this method.

Ylides can be described as carbanions that are stabilized by attached positive groups, and hence they are dipolar. Most commonly, the phosphonium ion is used as the stabilizer and the usual expression of a Wittig reagent is shown as structure **4.47** in Scheme 4.48. The literature sometimes shows this structure as a member of a resonance hybrid with an ylene structure **4.48**, but the characteristics of Wittig reagents are clearly those of a carbanion. (In phosphorus chemistry, true double bonds to carbon are found only when phosphorus is in the tricovalent state, as in $RP{=}CR_2$.)

$$Ph_3\overset{+}{P}{-}\overset{-}{C}H{-}R \longleftrightarrow Ph_3P{=}CHR$$

4.47 **4.48**

Scheme 4.48

When treated with an aldehyde or ketone, the carbanion center adds to the carbonyl group in the same way observed in the aldol condensation. The adduct then decomposes with ejection of phosphorus in the form of a phosphine oxide ($R_3P(O)$), leaving behind the newly formed carbon-carbon double bond. Most commonly, the starting material for a Wittig synthesis is a quaternary phosphonium salt formed from triphenylphosphine and an alkyl halide. The salt is then reacted with a strong base, such as NaH, NaOEt, NaOH, etc., to remove a proton at the carbon attached to P, thus creating the dipolar ylide structure. The synthesis of the simple alkene 3-heptene by the Wittig process is shown in Scheme 4.49. (Phosphine oxides are commonly written $R_3P{=}O$, but as shown for Wittig reagents, they do not have a conventional double bond, and the noncommittal formula $R_3P(O)$ is in use to avoid the bonding issue.)

$$(1)\ Ph_3P{:} + CH_3CH_2CH_2Br \longrightarrow Ph_3\overset{+}{P}{-}CH_2CH_2CH_3 \quad Br^-$$

$$(2)\ Ph_3\overset{+}{P}{-}CH_2CH_2CH_3 + NaH \longrightarrow Ph_3\overset{+}{P}{-}\overset{-}{C}HCH_2CH_3 + H_2$$

$$(3)\ Ph_3\overset{+}{P}{-}\overset{-}{C}HCH_2CH_3 + O{=}CHCH_2CH_2CH_3 \longrightarrow CH_3CH_2CH{=}CHCH_2CH_2CH_3 + Ph_3P(O)$$

Scheme 4.49

In a more detailed view of the mechanism of the Wittig reaction, it has been shown that, after the carbanionic center adds to the carbonyl group, formation of a 4-membered ring with a P-C bond, an oxaphosphetane derivative (**4.49**), occurs. Such heterocycles are unstable when P is in the pentacoordinate state, but they can be detected by NMR spectroscopy at low temperatures. Their decomposition takes place by ejecting triphenylphosphine oxide (Scheme 4.50).

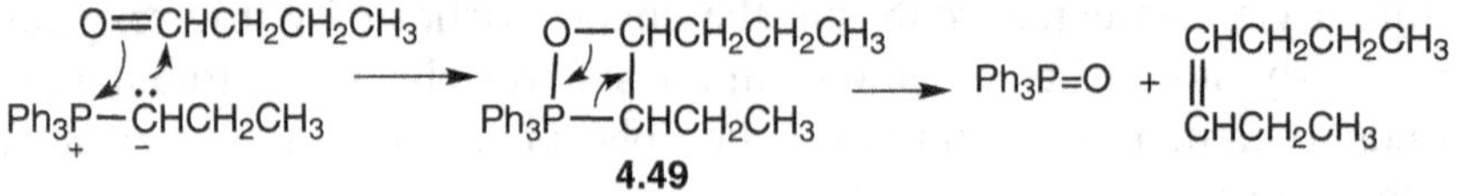

Scheme 4.50

4.5.2. The Aza–Wittig Reaction

In principle, the Wittig reaction could be used to form heterocycles if the ylide structure were created at the end of the chain of a heterosubstituted aldehyde or ketone, as exemplified by structure **4.50** in Scheme 4.51.

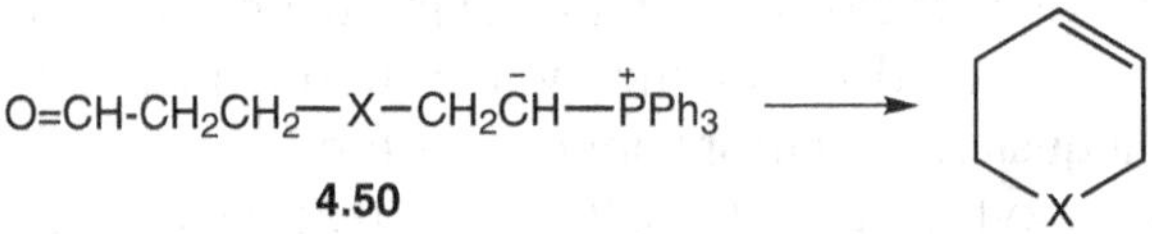

Scheme 4.51

However, of much greater significance to heterocyclic chemistry is the possibility of replacing the carbanionic center of the Wittig reagent with a nitrogen anion; when this species (e.g., **4.51**, an iminophosphorane or phosphinimine) is reacted with a carbonyl compound, the adduct has the oxazaphosphetane structure **4.52**, which then collapses to eliminate $Ph_3P{=}O$ as usual, but with the formation of the C=N bond rather than the C=C bond as in the normal Wittig reaction (Scheme 4.52). The iminophosphorane structure is formed readily in the reaction of a tertiary phosphine with an azide.

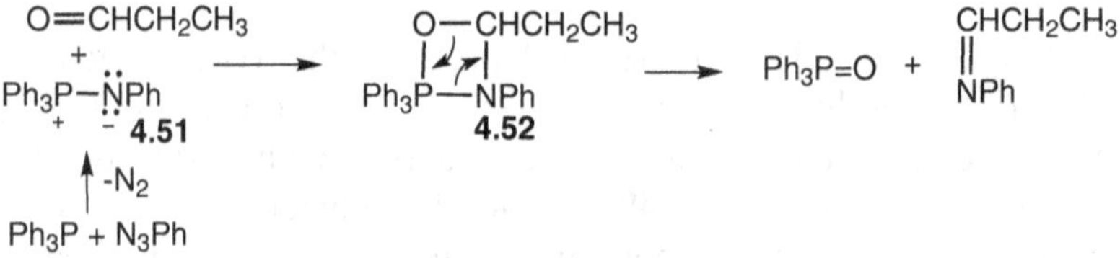

Scheme 4.52

Heterocyclic chemistry, of course, abounds with compounds containing C-N bonds, and the intramolecular "aza–Wittig" process has been adapted to form a great variety of such compounds, as indicated in the review of Molina and Vilaplana.[16]

4.5.3. Some Heterocycles Formed with Aza–Wittig Reagents

Here, we will review a few examples taken from the literature to demonstrate the versatility of the intramolecular aza–Wittig process in forming some common ring systems. Many other syntheses are known. Note the following characteristics: (1) ester and amide carbonyls can be used in the reaction with the iminophosphorane, (2) many reactions are general with various groups attached to phosphorus, (3) iminophosphoranes can be formed from trialkyl phosphites ($(RO)_3P$) as well as tertiary phosphines (R_3P), and (4) many reactions are so rapid that they occur at room temperature on mixing the reactants.

4.5.3.1. Pyrrolidines. Scheme 4.53.[17]

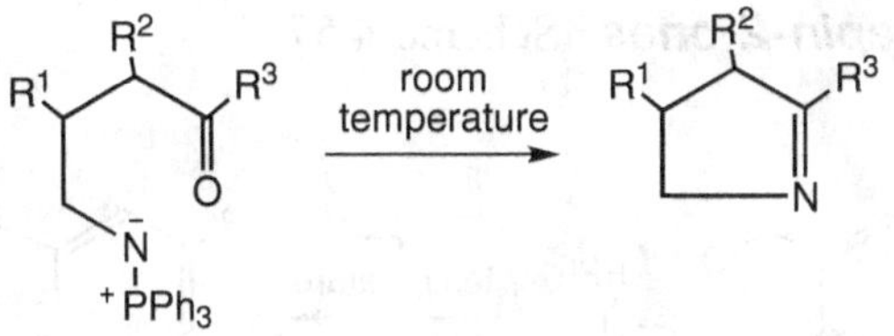

R2 R1 R3 O N $^+PPh_3$ room temperature R2 R1 R3 N

Scheme 4.53

4.5.3.2. Pyrroles. Scheme 4.54.[18]

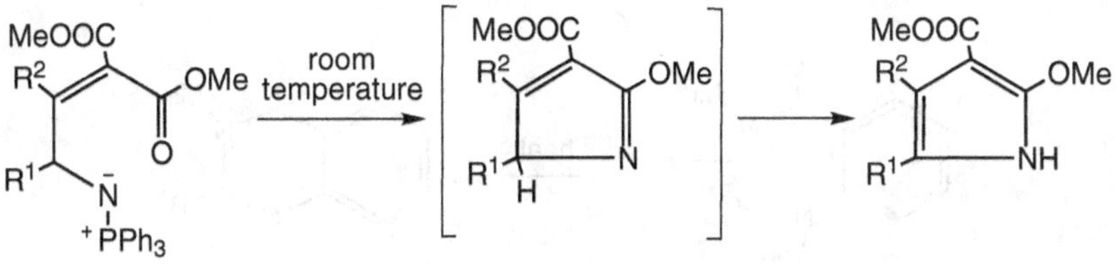

MeOOC R2 OMe O R1 N $^+PPh_3$ room temperature [MeOOC R2 OMe R1 H N] MeOOC R2 OMe R1 NH

Scheme 4.54

4.5.3.3. Isoquinolines. Scheme 4.55.[19]

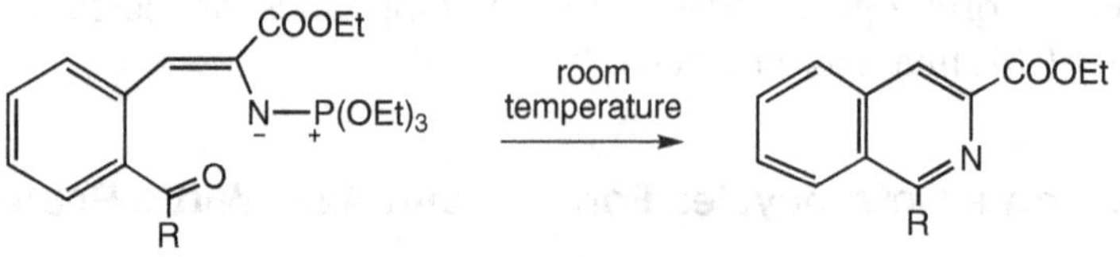

Scheme 4.55

4.5.3.4. 1,3-Oxazoles. Scheme 4.56.[20]

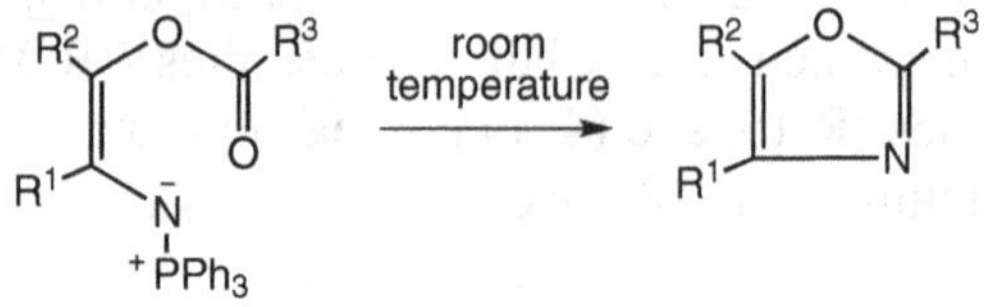

Scheme 4.56

4.5.3.5. Diazepin-2-ones. Scheme 4.57.[21]

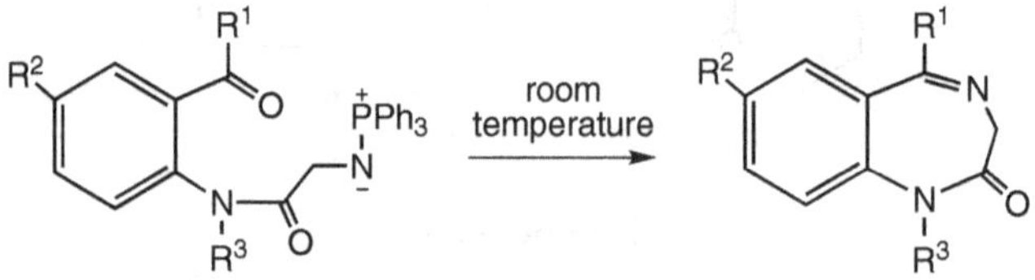

Scheme 4.57

4.5.3.6. Quinazolinones. Scheme 4.58.[22]

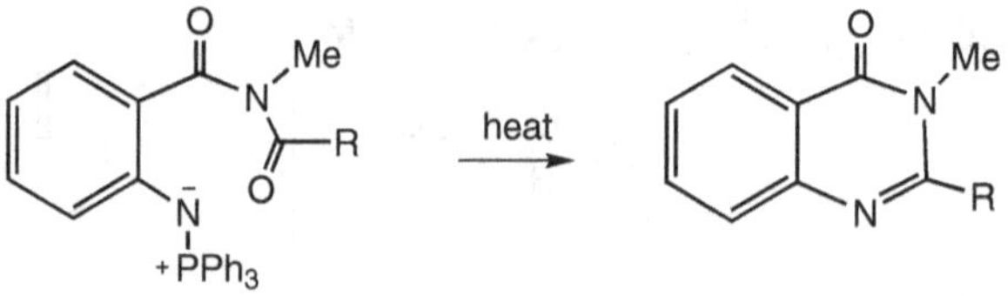

Scheme 4.58

4.6. SYNTHESIS OF HETEROCYCLES BY THE ALKENE METATHESIS REACTION

Alkene metathesis (from the Greek, meaning "change places") refers to the interaction of two olefinic groups whereby an interchange reaction occurs and two new alkenes are formed. Frequently, one of these alkenes is of low molecular weight and is removed in the gaseous state. The process is illustrated in its simplest form in Scheme 4.59 with metathesis of two terminal alkene groups. This results in the formation of ethylene, which escapes from the reaction mixture; this effect drives the reaction in the forward direction.

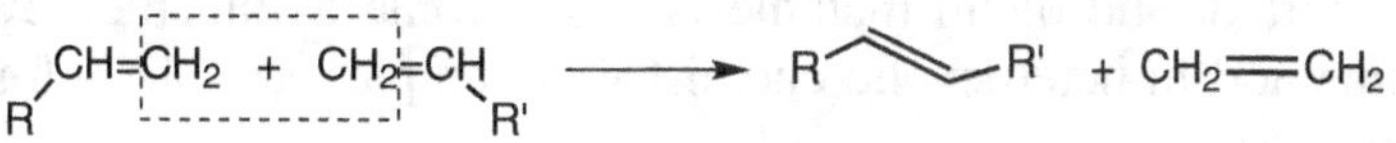

$$R{-}CH{=}CH_2 + CH_2{=}CH{-}R' \longrightarrow R{-}CH{=}CH{-}R' + CH_2{=}CH_2$$

Scheme 4.59

This chemistry had been studied for some years but was not a practical laboratory synthesis until it was discovered that certain metal coordination compounds were effective catalysts for promoting the interaction. Robert Grubbs at California Institute of Technology discovered that ruthenium complexes were peculiarly effective and developed several catalysts based on this element. Compound **4.53**, where Cy refers to cyclohexyl, was his first discovery; it is now referred to as Grubbs' catalyst because of its great utility for effecting practical olefin metathesis.[23] Among other contributions, Richard Schrock at Massachusetts Institute of Technology had found earlier that complexes based on molybdenum or tungsten also were useful catalysts.[24] These discoveries, along with earlier work by Yves Chauvin in France, made olefin metathesis into a great and widely used synthetic method, and of such value that Grubbs, Schrock, and Chauvin were awarded the Nobel Prize in Chemistry in 2005.

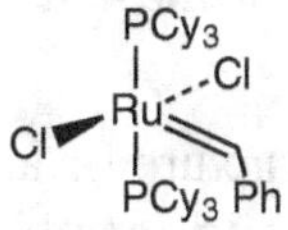

PCy_3, Cl, Ru, Cl, PCy_3, Ph

4.53

Alkene metathesis has been extended to include intramolecular interaction of two alkene groups, which results in a cyclic product

(Scheme 4.60). This process is known as ring-closing metathesis and is particularly important in the formation of 5- and 6-membered rings, as well as large-ring heterocycles.

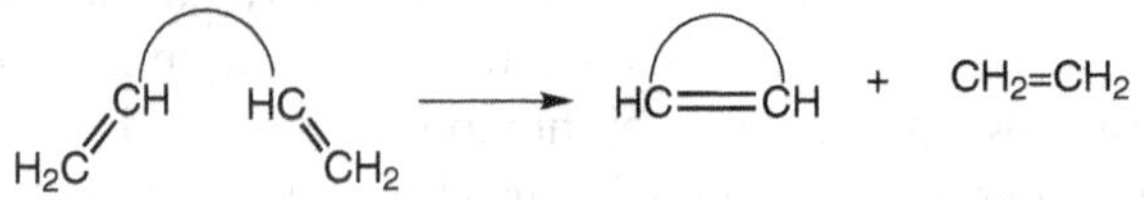

Scheme 4.60

If one or more heteroatom is present in the chain, then a heterocyclic product will be formed. Many examples of this process have been reported, and olefin metathesis has become a valuable new synthetic method in heterocyclic chemistry. A simple example is found in Scheme 4.61.

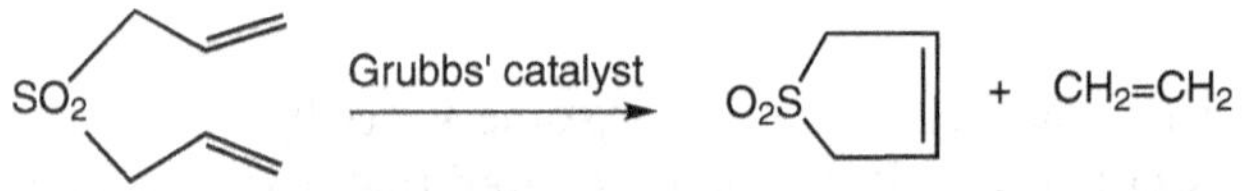

Scheme 4.61

The real value of the process, however, is its ability to produce the complex heterocyclic systems of natural products. To illustrate, the synthesis of a pyrrolizidine alkaloid, pyrrolam A, is shown in Scheme 4.62.[25]

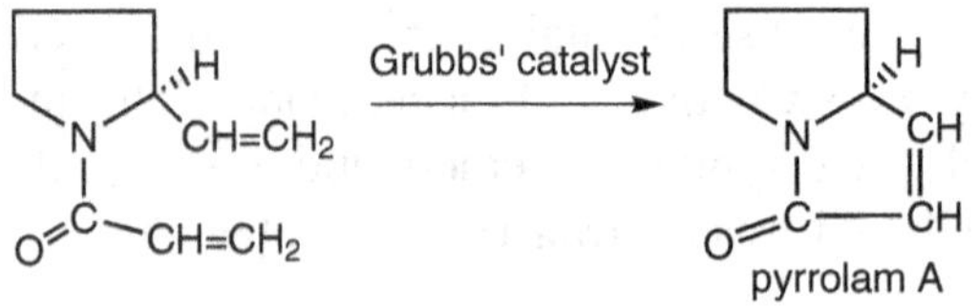

Scheme 4.62

Even more complex is the structure of a hepatitis C protease inhibitor called BILN 2061, which has a 15-membered ring containing two nitrogen atoms (**4.55**). Olefin metathesis was used in the final step of its synthesis to close the ring of generalized intermediate **4.54** where several different R groups were used (Scheme 4.63).[26]

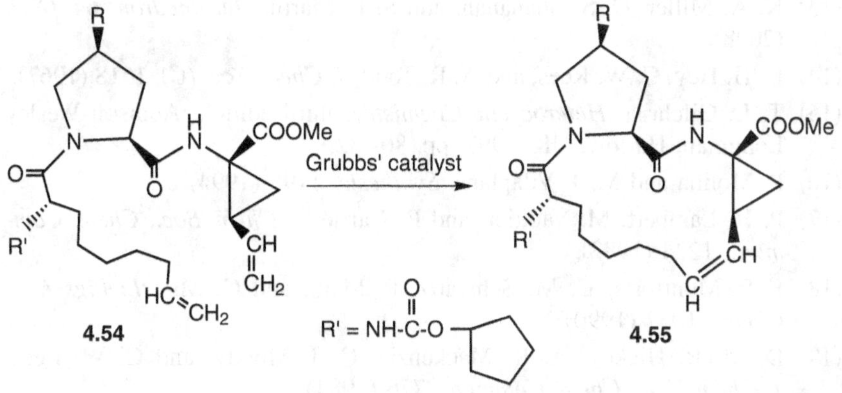

4.54 → (Grubbs' catalyst) → 4.55

R' = NH-C(O)-O-cyclopentyl

Scheme 4.63

REFERENCES

(1) A. R. Katritzky, D. L. Ostercamp, and T. I. Yousaf, *Tetrahedron*, **43**, 5171 (1987).

(2) S. Nagubandi and G. Fodor, *J. Heterocyclic Chem*., **17**, 1457 (1980).

(3) J. J. Li and G. W. Gribble, *Palladium in Heterocyclic Chemistry*, Elsevier, Amsterdam, The Netherlands, (2000).

(4) T. Mizoroki, K. Mori, and A. Ozaki, *Bull. Chem. Soc. Jpn*., **44**, 581 (1971).

(5) R. F. Heck and J. P. Nolley, Jr., *J. Org. Chem*., **37**, 2320 (1972).

(6) R. J. Sundberg and W. J. Pitts, *J. Org. Chem*., **56**, 3048 (1991).

(7) P. J. Harrington and L. S. Hegedus, *J. Org. Chem*., **49**, 2657 (1984).

(8) B. R. Henke, C. J. Aquino, L. S. Berkimo, D. K. Croom, R. W. Dougherty, G. N. Ervin, M. K. Grizzle, G. C. Hirst, M. K. James, M. F. Johnson, K. L. Queen, R. G. Sherrill, E. E. Sugg, E, M, Suh, J. W. Szewczyk, R. J. Unwalla, J. Yingling, and T. M. Willson, *J. Med. Chem*., **40**, 2706 (1997).

(9) B. Åkermark, L. Eberson, E. Jonsson, and E. Petterson, *J. Org. Chem*., **40**, 1365 (1975).

(10) H.-J. Knölker, K. R. Reddy, and A. Wagner, *Tetrahedron Lett*., **39**, 8267 (1998).

(11) W. C. Tsang, R. H. Munday, G. Brasche, N. Zheng, and S. L. Buchwald, *J. Org. Chem*., **73**, 7603 (2008).

(12) S. Gowrisankar, H. S. Lee, J. M. Kim, and J. N. Kim, *Tetrahedron Lett*., **49**, 1670 (2008).

(13) K. A. Miller, C. S. Shanahan, and S. F. Martin, *Tetrahedron*, **64**, 6884 (2008).
(14) D. H. Hey, C. W. Rees, and A. R. Todd, *J. Chem. Soc. (C)*, 1518 (1967).
(15) T. L. Gilchrist, *Heterocyclic Chemistry*, third edition, Addison Wesley Longman, Harlow, UK, 1997, pp. 80–82.
(16) P. Molina and M. J. Vilaplana, *Synthesis*, 1198 (1994).
(17) P. H. Lambert, M. Vaultier, and R. Carrié, *J. Chem. Soc., Chem. Commun.*, 1224 (1982).
(18) F.-P. Montforts, U. M. Schwartz, P. Maib, and G. Mai, *Liebigs Ann. Chem.*, 1037 (1990).
(19) D. M. B. Hickey, A. R. Mackenzie, C. J. Moody, and C. W. Rees, *J. Chem. Soc., Chem. Commun.*, 776 (1984).
(20) H. Takeuchi, S. Yanagida, T. Ozaki, S. Hagiwara, and S. Eguchi, *J. Org. Chem.*, **54**, 431 (1989).
(21) J. Ackrell, E. Galeazzi, J. M. Muchowski, and L. Tökés, *Can. J. Chem.*, **57**, 2696 (1979).
(22) H. Takeuchi and S. Eguchi, *Tetrahedron Lett.*, **30**, 3313 (1989).
(23) T. M. Trnka and R. H. Grubbs, *Accts. Chem. Res.*, **34**, 18 (2001).
(24) R. R. Schrock, *Angew. Chem., Int. Ed. Engl.*, **42**, 4592 (2003).
(25) M. Arisawa, E. Takezawa, and A. Nishida, *SynLett*, 1179 (1997).
(26) N. K. Yee, Y. Farina, J. N. Houpis, N. Haddad, R. P. Frutos, F. Gallou, X.-J. Wang, X. Wei, R. D. Simpson, X. Feng, V. Fuchs, Y. Xu, J. Tan, L. Zhang, J. Xu, L.-L. Smith-Keenan, J. Vitous, M. D. Ridges, E. M. Spinelli, M. J. Donsbach, T. Nicola, M. Brenner, E. Winter, P. Kreye, and W. Samstag, *J. Org. Chem.*, **71**, 7133 (2006).

REVIEW EXERCISES

4.1. Write equations showing the synthesis of each compound.

a.

O
NH
S
O N H

b.

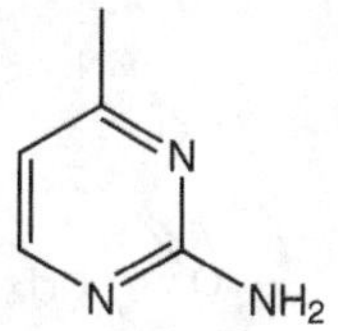

N NH2

4.2. Show the reaction and product for the following:

a. Urea + diethyl malonate

b. Synthesis of benzimidazole from 1,2-diaminobenzene

c. Synthesis of 2,3-dimethylpyrrole by the Paal–Knorr method

d. Michael reaction of aniline with the α–β unsaturated ketone followed by a Bischler–Napieralski reaction

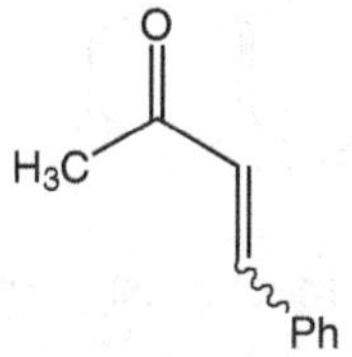

O

H_3C

Ph

e. thiourea and $NCCH_2CHO$

4.3. Predict the product:

a.

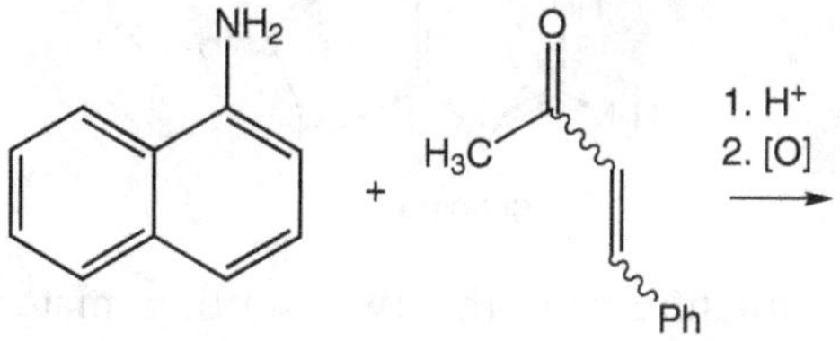

NH_2 + O H_3C Ph 1. H^+ 2. [O] →

b.

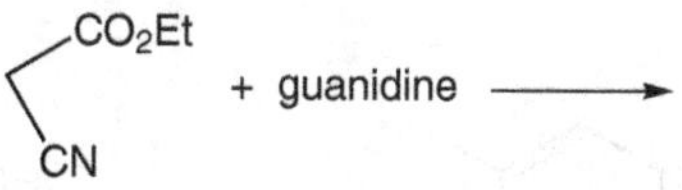

CO_2Et CN + guanidine →

c.

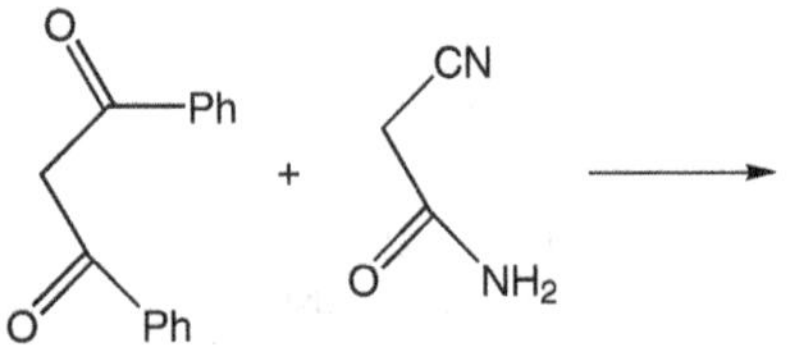

4.4. Write the synthesis (not the mechanism) for 2-methyl-3-ethylpyrrole from acyclic starting materials.

4.5. (J. C. Sloop, *J. Phys. Org. Chem.*, **22**, 110 (2009)). Sketch a synthesis for the following beginning with reactants containing 6 or fewer carbons.

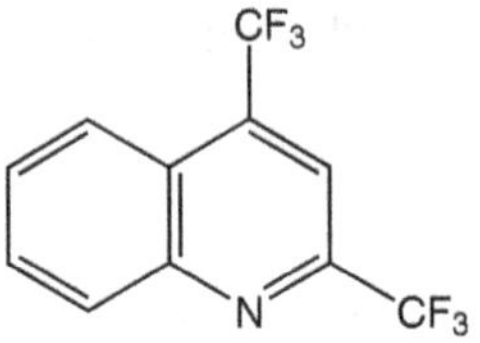

4.6. Give all steps and the mechanism for the formation of quinoline from aniline.

4.7. Write the starting materials for the synthesis of guanine by the Traube synthesis.

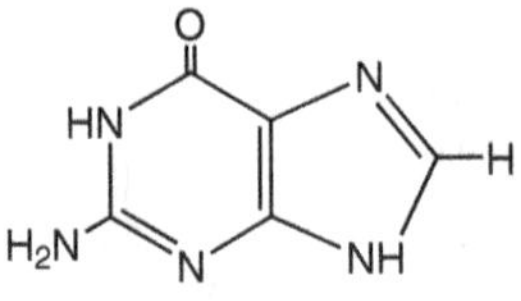

guanine

4.8. Write the structures for the two starting materials to prepare 2-methylfuran (from acyclic starting materials).

4.9. What is the likely product?

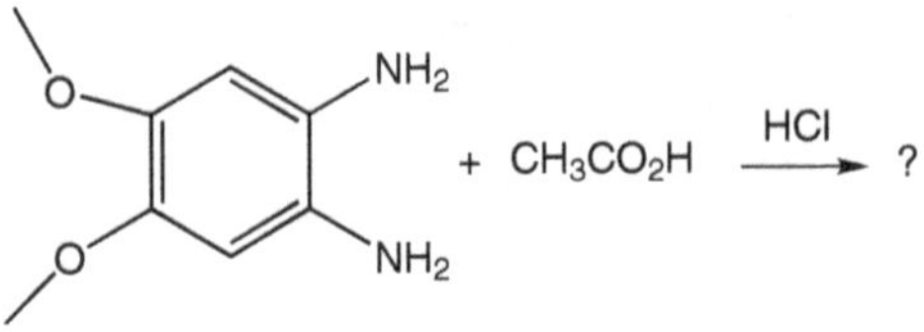

+ CH_3CO_2H $\xrightarrow{HCl}$?

4.10. What is the likely product of each step?

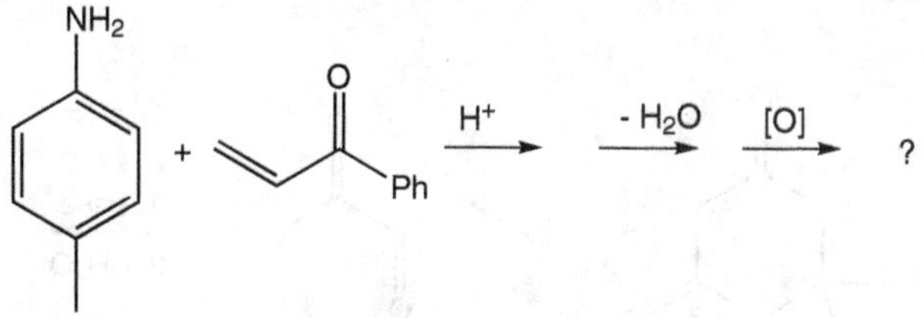

NH_2 + O Ph $\xrightarrow{H^+}$ $\xrightarrow{-H_2O}$ $\xrightarrow{[O]}$?

4.11. Give all steps and the mechanism for the synthesis of 3,4-dimethylpyrrole by the Paal–Knorr reaction.

4.12. Show the formation of a pyrimidine derivative from urea and a diketone.

4.13. Write the structures for the two starting materials to prepare phenobarbital (from nonheterocyclic starting materials).

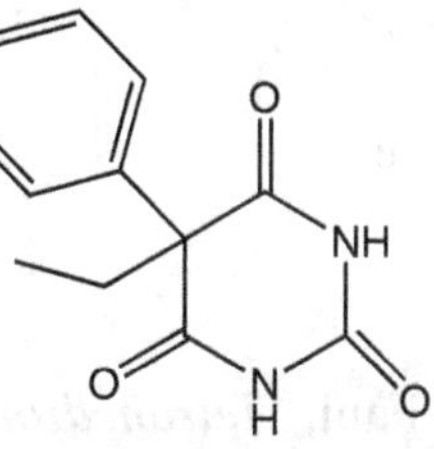

O NH O N H O

phenobarbital

4.14. Taddei et al., *Eur. J. Org. Chem.*, **939** (2005)). Identify compound a and compound b:

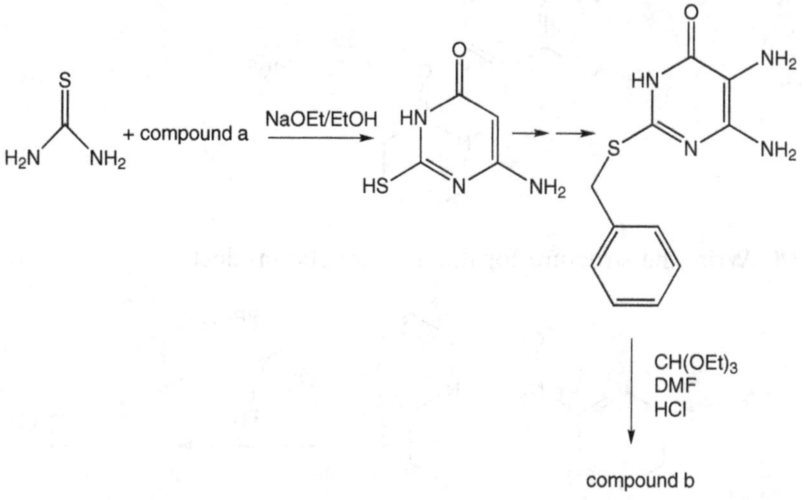

S H_2N NH_2 + compound a $\xrightarrow{NaOEt/EtOH}$ O HN HS N NH_2 → → O HN NH_2 S N NH_2

$CH(OEt)_3$
DMF
HCl

compound b

4.15. (Ranu et al., *Tetrahedron Lett.*, **49**, 4613 (2008)). Draw the product:

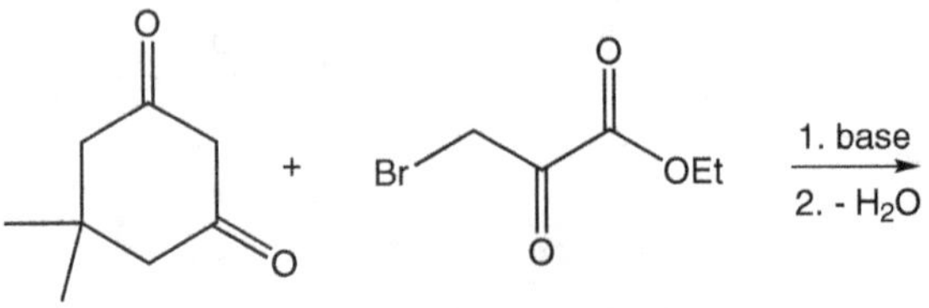

4.16. (El Kaim et al., *Org. Lett.*, **10** (2008)). This iodopyrimidine gives a product that then isomerizes. Show the structure prior to isomerization.

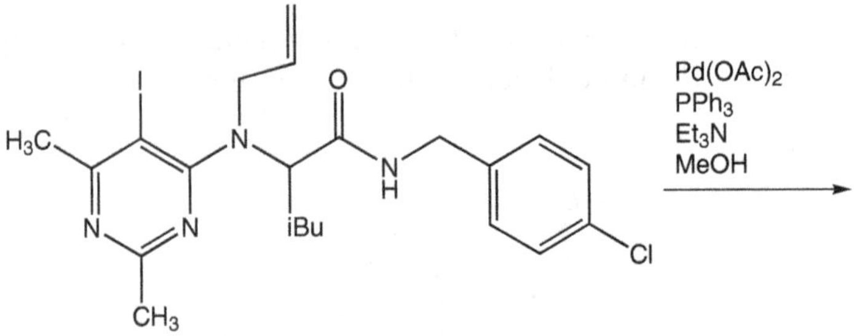

4.17. (J. Chan and M. Faul, *Tetrahedron Lett.*, **47** 3361 (2006); G. Hajós, and I. Nagy, *Curr. Org. Chem.*, **12**, (2008)). Draw the product:

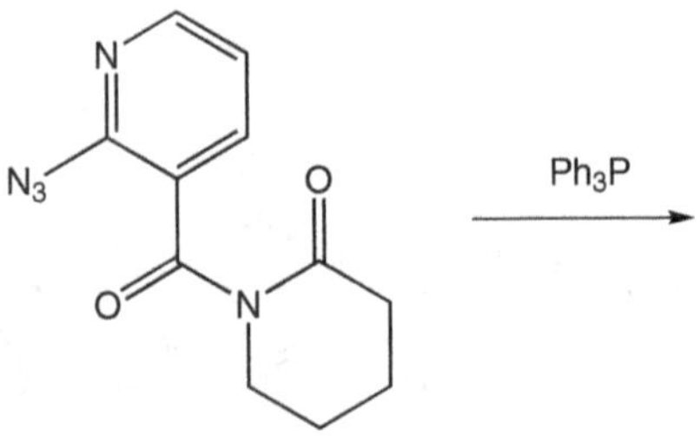

4.18. Write the structure for the heterocyclic product.

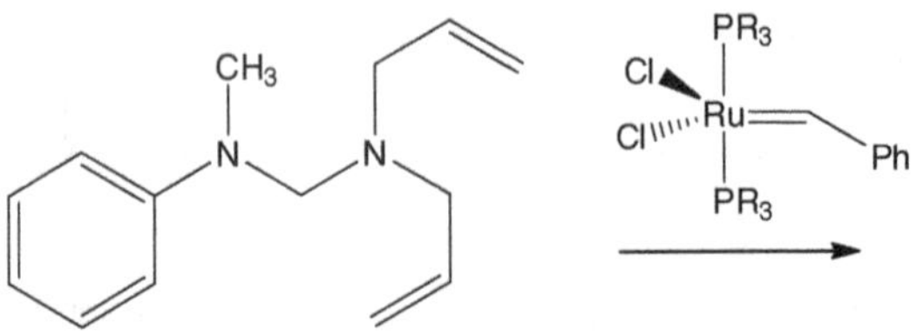

4.19. (V. M. de Almeida, et al., *Tetrahedron Lett.*, **50**, 684 (2009)).

a. Write the two major resonance forms for compound 1.

b. Write the structure for compound 2.

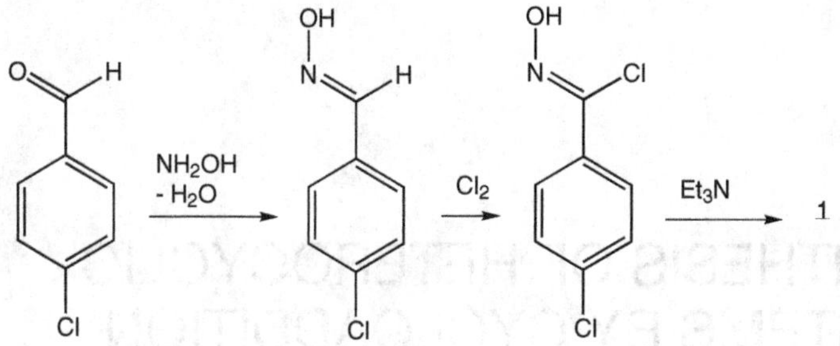

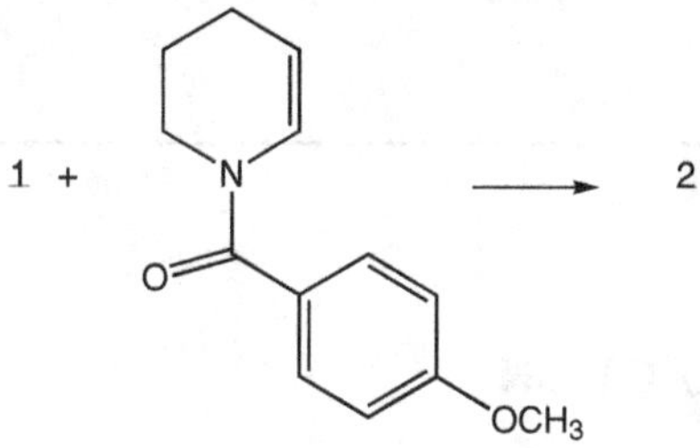

4.20. (E. Sobarzo-Sanchez, et al., *Synthetic Commun.*, **37**, 1331 (2007)). Provide the structures for 1, 2, and 3.

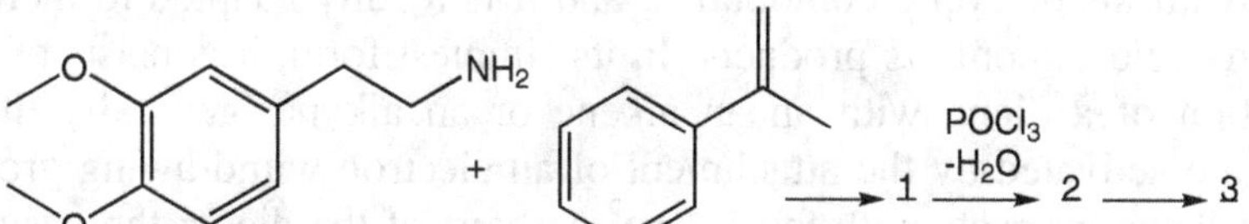

4.21. (B. C. Ranu, L. Adak, and S. Banerjee, *Tetrahedron Lett.*, **49**, 4613 (2008)). Draw a synthesis showing the formation of 2 from 1.

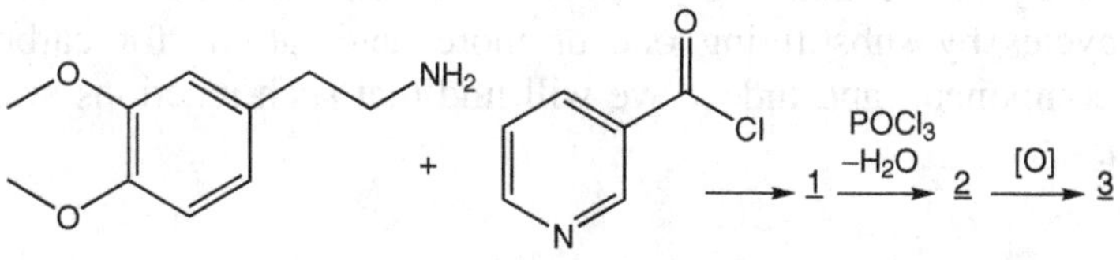

CHAPTER 5

SYNTHESIS OF HETEROCYCLIC SYSTEMS BY CYCLOADDITION REACTIONS

5.1. THE DIELS–ALDER REACTION

5.1.1. Review

The Diels–Alder reaction is one of the most important techniques for the synthesis of cyclic compounds, and it is readily adapted to include heterocycles among its products. In its simplest form, it consists of the reaction of a diene with an an alkene or an alkyne, generally those that are activated by the attachment of an electron withdrawing group. The alkene interacts with the 1,4-sp^2 carbons of the diene; the ensuing bond changes (not a mechanism) are shown in Scheme 5.1. Frequently, the activating group on the alkene is a carbonyl derivative, but cyano, nitro, sulfonyl, etc. may also be used. Alkenes and alkynes so activated are called dienophiles.

It is easy to see that many opportunities exist for the synthesis of heterocycles by substituting one or more heteroatoms for carbon in either component, and indeed we will find that such reactions are well known.

Fundamentals of Heterocyclic Chemistry: Importance in Nature and in the Synthesis of Pharmaceuticals,
By Louis D. Quin and John A. Tyrell Copyright © 2010 John Wiley & Sons, Inc.

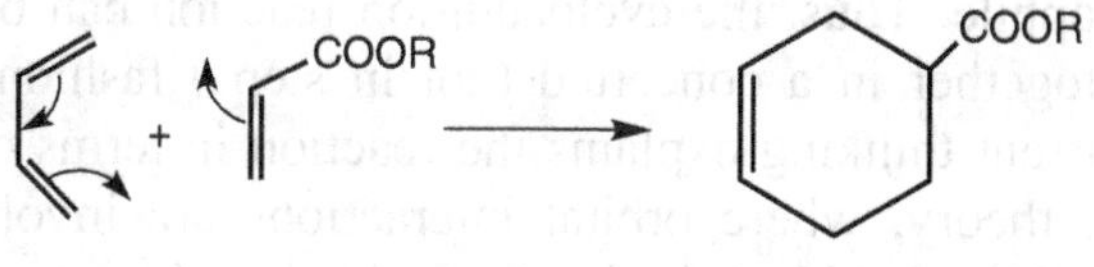

COOR + COOR → COOR

Scheme 5.1

When alkynes are used in the Diels–Alder reaction, the product is a cyclodiene, which generally is easily convertible to the fully unsaturated (aromatic) structure (Scheme 5.2).

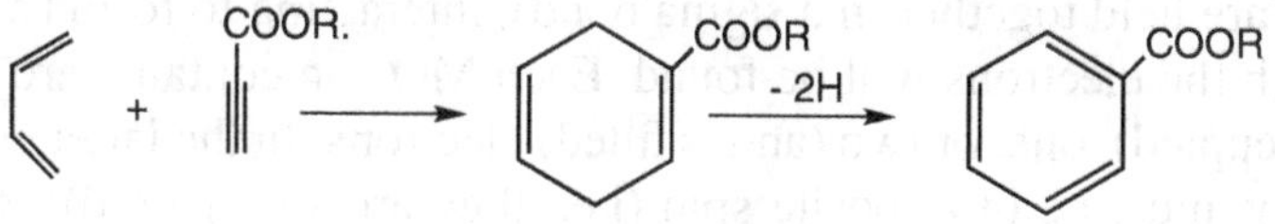

COOR. + → COOR - 2H → COOR

Scheme 5.2

Cyclic dienes are especially reactive to dienophiles and lead to valuable bicyclic products (Scheme 5.3). Here, isomers can be produced, wherein the substituent on the dienophile can be placed in the *exo* (**5.1**) or *endo* (**5.2**) position. Generally the *exo* isomer is formed faster, but the *endo* isomer is more stable and either predominates or can be made to predominate through reversibility of the addition reaction. Thus, the cycloaddition can be described as under thermodynamic (rather than kinetic) control.

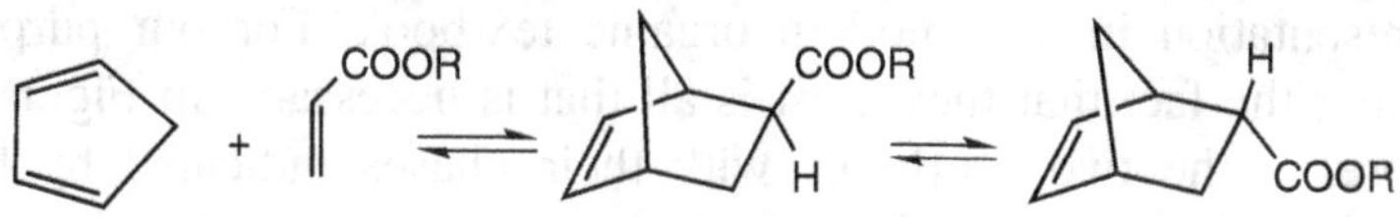

+ COOR ⇌ COOR H ⇌ H COOR

5.1, exo **5.2**, endo

Scheme 5.3

5.1.2. Mechanism of the Diels–Alder Reaction

The Diels–Alder reaction cannot be explained adequately by the familiar polar reaction mechanism involving electron pair interactions or by a free radical mechanism. In fact in older literature, it was sometimes described as a "no-mechanism" reaction. There are no intermediates, and kinetic studies show that the reaction is first order in both the diene and the dienophile. Thus, the cycloaddition reaction can be viewed as the coming together in a concerted (not in steps) fashion of the two reactants. Current thinking explains the reaction in terms of molecular orbital (MO) theory, where orbital interactions are involved. Before describing the Diels–Alder reaction from this standpoint, we must first review some of the basic ideas and terminology of MO theory in the nonmathematical way as used by organic chemists. Understanding MO theory will be of value also in the study of dipolar cycloadditions (section 5.2) and aromaticity of heterocycles (Chapters 6 and 7).

The molecular orbital view of a carbon–carbon double bond is that it is formed from the two atomic p-orbitals (AO) of the sp^2 carbons (which are held together in a sigma bond), interacting to form two MOs in which the electrons will be found. Each MO can contain zero (hence is unoccupied), one, or two (and is filled) electrons. In the latter case, the electrons must be of opposite spin (i.e., they are spin paired). Because there are only two AO's with one electron in each, there are not enough electrons to occupy both MO of the C-C bond. The electrons go first into the MO of lower energy, and we say that it becomes an occupied MO and that it is a bonding MO (also designated a Π MO). The higher energy MO is said to be unoccupied, and it is referred to as an antibonding orbital (also designated a Π* orbital). Thus, the MO energy picture for an alkene can be expressed as in Figure 5.1. The arrows in the bonding orbital are used to state the spin pairing of the two electrons.

In an atomic orbital view, the two p-orbitals are aligned and interact if they have the proper phase. There are two phases (lobes) to an orbital, which are usually designated by + and − (which are not to suggest polarities). To understand the phases, one would have to review the presentation in any modern organic textbook. For our purposes, accepting the fact that they exist is all that is necessary. In Figure 5.2, we observe the two p-orbitals with their phases indicated by black and white lobes, and the important notation that like phases interact to form the Π molecular orbital, where there is electron density above and below the plane of the double bond. When the AO have opposite phases, the orbitals repel each other, which gives the higher energy

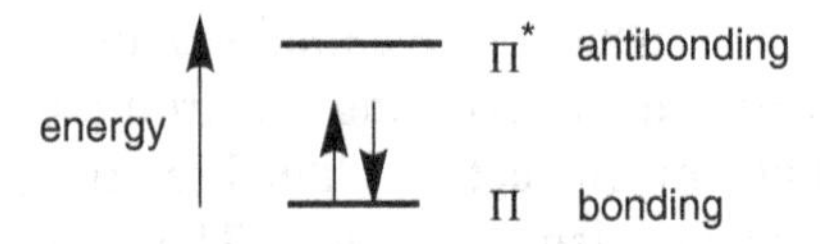

Figure 5.1. Alkene MOs.

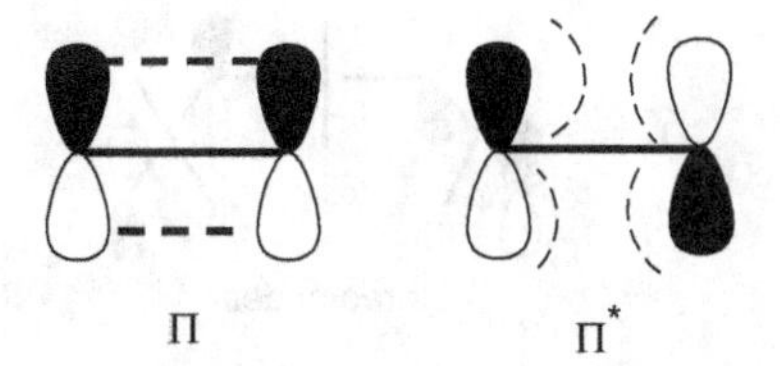

Figure 5.2. Alkene orbitals.

state of the antibonding Π* MO. At the midpoint of the two unlike phases, there is no electron density, and we say that a node is present. In a polyene, there can be several antibonding MOs, with increasing energy as the number of nodes increases. We will discuss this when we examine next the MO picture for a diene.

Conjugated dienes have four AO and, hence, four MO, all of which have different energies. The four electrons from the p-orbitals will be placed in the MO of lowest energy, always filling that MO before going into a MO of higher energy. This is depicted in Figure 5.3. HOMO in this figure refers to the highest occupied MO, and LUMO refers to the lowest unoccupied MO. Chemical reactions frequently occur with interaction of the HOMO of one reactant with the LUMO of the other.

This leads to two filled bonding MO (Π_1 and Π_2) and leaves two unoccupied MO (Π_3^* and Π_4^*). Π_3^* is important as the antibonding MO of lowest energy and is referred to as the LUMO. Π_2 is designated as the HOMO. The different energy levels are associated with an increase in the number of nodes as we go from Π_1 to Π_4^*. This is shown in the orbital interaction diagrams of Figure 5.4.

In MO theory, reactions are attributed to the interaction between MOs of the reactants with the same phase, creating new MOs for the new molecule. Reactions occur between those orbitals that have the lowest energy difference between them, which is called an energy gap. In the case of the Diels–Alder reaction between a diene and an alkene, the gap is smallest between the alkene LUMO and the HOMO (Π_2) of

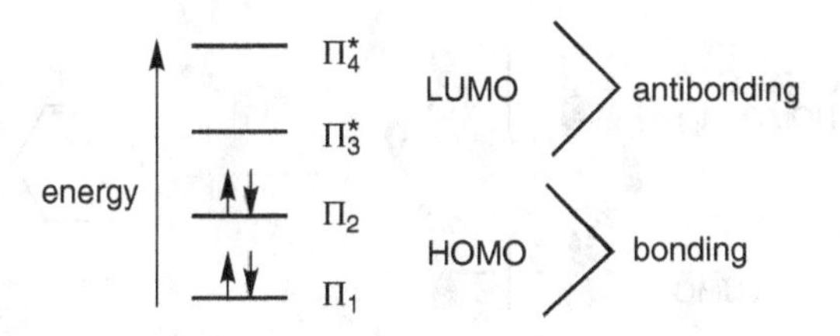

Figure 5.3. Diene MOs.

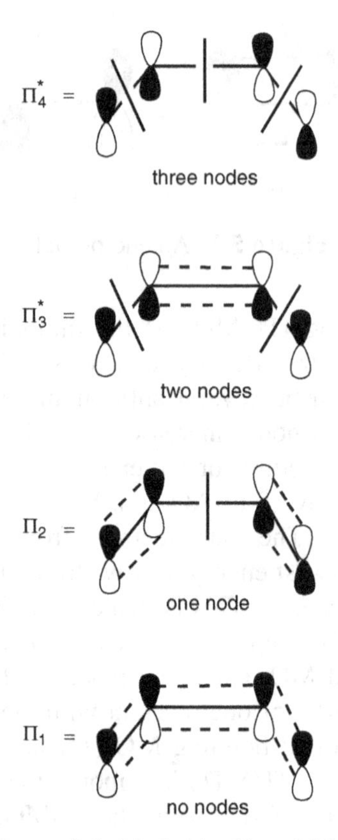

Figure 5.4. Nodes in diene MOs.

the diene. This is illustrated in Figure 5.5, where orbitals of like sign are observed to merge. The smaller is the energy gap, the faster is the cycloaddition reaction.

Two other points must be made about the Diels–Alder reaction. First, the reaction is successful only if the diene can adopt readily the cisoid conformation (**5.3**); if geometric or steric constraints are present to make the diene adopt the transoid conformation (**5.4**), the reaction will not

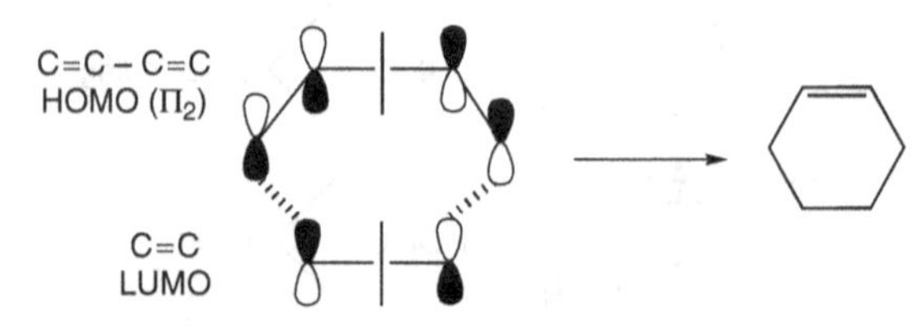

Figure 5.5. HOMO-LUMO interaction leading to a reaction.

occur, because this prevents the concerted interaction of the HOMO-LUMO orbitals (called the frontier orbitals). Second, the reaction with carbonyl-substituted alkenes can be catalyzed by Lewis acids such as BF_3, $AlCl_3$, and so on. These act by complexing with the carbonyl group as in structure **5.5** for BF_3, lowering its energy and reducing the energy gap between the frontier orbitals.

$F_3\bar{B}$—O+ ‖ C—OR

5.3 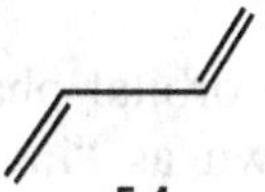 **5.4** 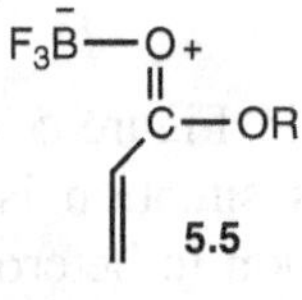 **5.5**

5.1.3. Applying the Diels–Alder Reaction to Heterocyclics Synthesis

5.1.3.1. Scope and Mechanism. In principle, replacing a carbon of a conjugated diene with a heteroatom does not prevent the Diels–Alder reaction from occurring. Some of the possible heterodienes based on N or O substitutions are shown in Figure 5.6; derivatives of most of these have been used in cycloadditions.

Similarly, heterosubstituted alkenes can be used as dienophiles. Some combinations include C=O, C=N, N=O, N=N, C=S, and CN. Hetero Diels–Alder reactions can occur by the usual mechanism of the LUMO of the dienophile interacting with the HOMO of the diene. However, one consequence of heteroatom substitution in the diene is some reduction in the reactivity because all hetero groups under consideration are electron withdrawing, which increases the energy gap of the HOMO-LUMO interaction. Thus, such reactions may require more forcing conditions than the all-carbon system. To compensate for this effect, it is the practice to increase the reactivity of the alkene by increasing its electron density with electron-releasing groups such as OR or NR_2. This has the effect of inverting the orbital combinations; the alkene is now of lower energy than the diene, so it will offer the HOMO for the reaction. The LUMO of the diene then must be used.

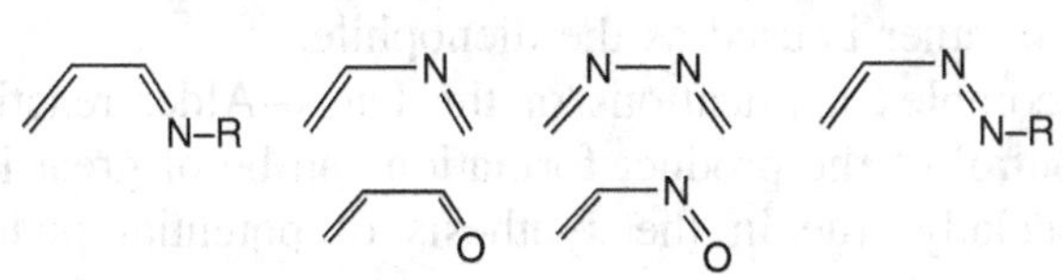

N–R N N—N N N–R O N O

Figure 5.6. Some heterodienes.

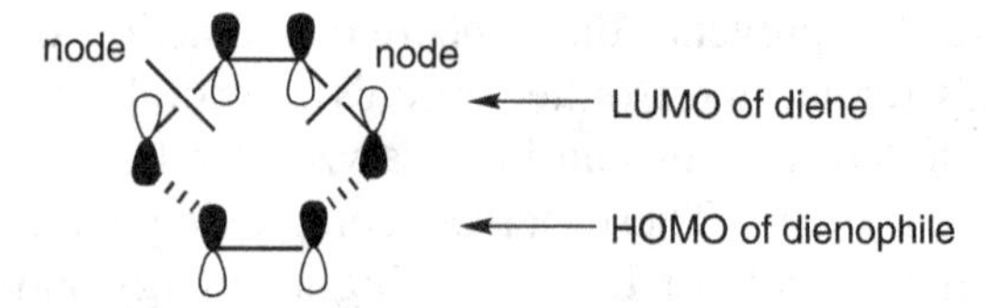

Figure 5.7. MOs in inverse electron demand.

As shown in Figure 5.7, the orbital phases match for the reaction to occur. This situation is known as "inverse electron demand" and is quite common in hetero Diels–Alder reactions.

In the sections to follow, various known hetero Diels–Alder reactions are shown. Many more are reported in the literature, and the reaction is of prime value in heterocyclic synthesis.

5.1.3.2. Synthesis of Pyridine Derivatives. Tetrahydropyridine derivatives are readily formed by Diels–Alder cycloadditions of a 1-azadiene with activated alkenes. A review[1] makes clear the many uses of this type of heterodiene in Diels–Alder and other heterocycle-forming processes. An example of tetrahydropyridine synthesis is shown in Scheme 5.4.

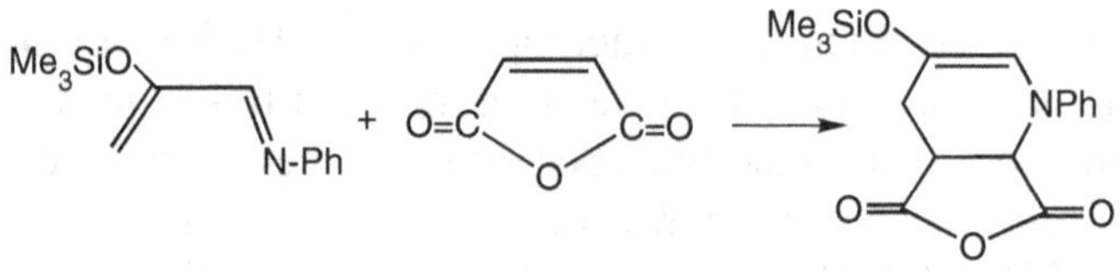

Scheme 5.4

Note here the use of an electron-releasing group ($OSiMe_3$) on the diene to compensate for the deactivating effect of C=N in the diene. Also, with its two carbonyl groups, maleic anhydride is a particularly reactive dienophile and is widely used.

Inverse electron demand in the dienophile is shown in Scheme 5.5, where an enol ether is used as the dienophile.

In more complex applications of the Diels–Alder reaction, stereochemical control of the product formation can be of great importance. This is especially true in the synthesis of potential pharmaceutical agents, where activity can be much greater (even exclusively) in one stereoisomeric form. This point will be considered in more depth in

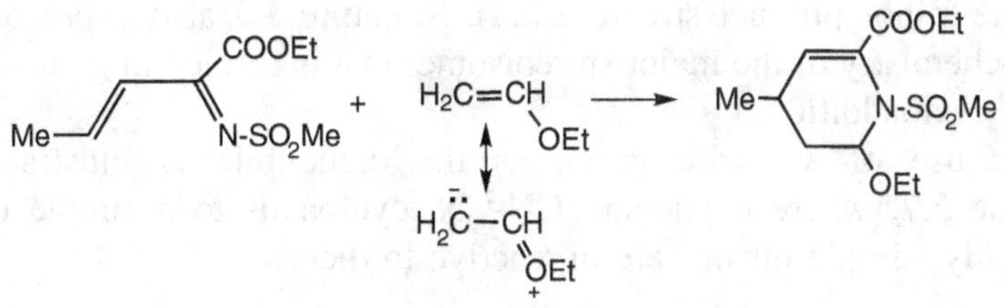

Scheme 5.5

Chapter 10. Without some form of control, equal amounts of the enantiomers (the R and S isomers) arising from creation of a chiral carbon will be formed, thus giving a racemic mixture. However, the presence of a chiral atom in either the R or S form in one reactant is well known to bias the course of a reaction so that the new chiral center will have an excess of one or the other enantiomer. This behavior is referred to as an asymmetric synthesis or as an enantioselective process, which is of great importance in the synthesis of complex natural products and chiral drugs (see Chapter 10). A relatively simple example of asymmetric synthesis in a Diels–Alder reaction is shown in Scheme 5.6, where an enol ether (**5.6**) being used in an inverse electron demand cycloaddition is in an optically active, single enantiomeric form. The cycloadduct will have three chiral centers, but at each center, one enantiomeric form will predominate and the product is optically active. The reaction occurs with high regioselectivity (the formation of one addition isomer over the other when two modes of addition are possible, as in the present case; here, the major regioisomer will have structure **5.7**

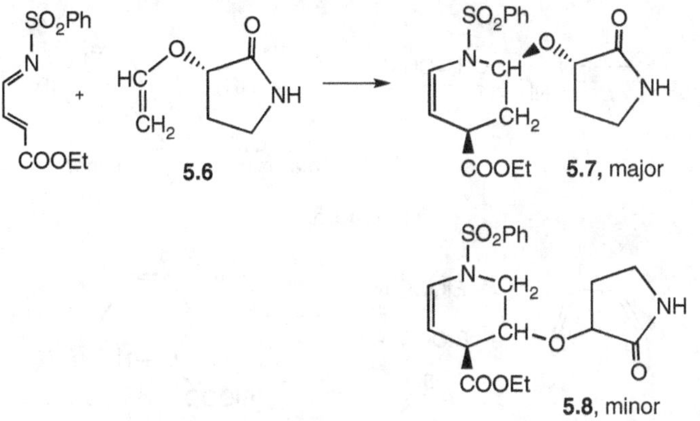

Scheme 5.6

and the minor product structure **5.8**). Structure **5.7** also expresses the stereochemistry of the major stereoisomer that is formed in great excess in the cycloaddition.[2]

The use of a nitrile group as the dienophile is illustrated in Scheme 5.7, where cyanogen, CN-CN, cycloadds to a simple diene. Normally, simple nitriles are unreactive to dienes.

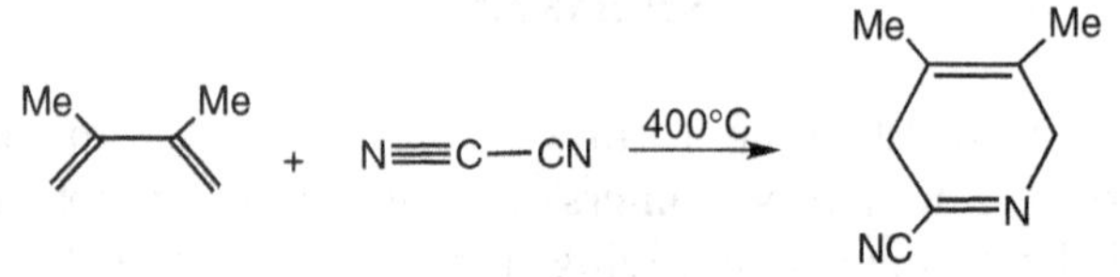

Me Me + N≡C—CN 400°C → Me Me NC N

Scheme 5.7

An imine is used as the dienophile in Scheme 5.8.

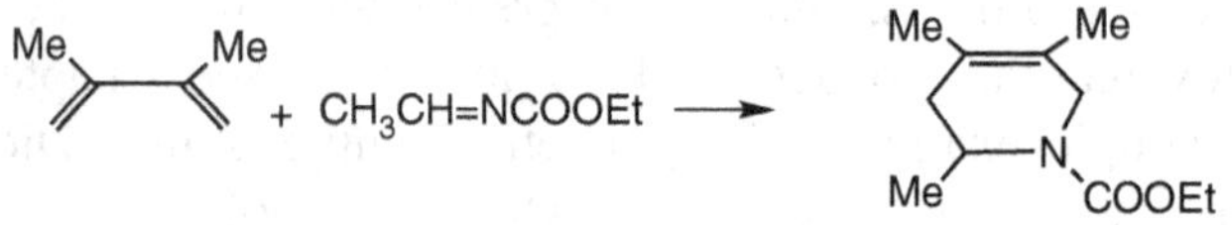

Me Me + $CH_3CH{=}NCOOEt$ → Me Me Me N COOEt

Scheme 5.8

5.1.3.3. Synthesis of Pyridazines. In some hetero Diels–Alder reactions, two isomers can be formed from the use of an unsymmetrical dienophile with an unsymmetrical diene. That was the case in Scheme 5.6, and it is shown again in a synthesis of a pyridazine (Scheme 5.9).

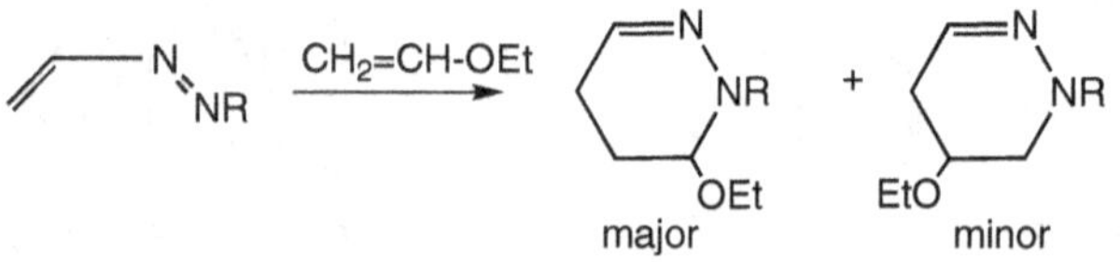

N NR $CH_2{=}CH\text{-}OEt$ → N NR OEt major + N NR EtO minor

Scheme 5.9

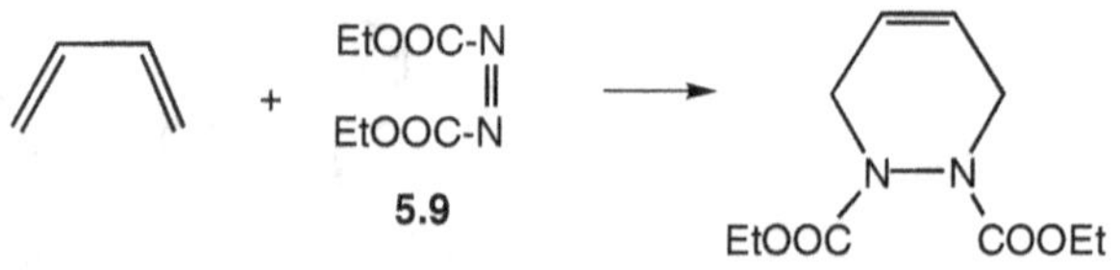

+ EtOOC-N ‖ EtOOC-N **5.9** → N—N EtOOC COOEt

Scheme 5.10

5.1.3.4. Synthesis of 1,2-Oxazine Derivatives. The N=O group in certain nitroso compounds (R–N=O) is well known to function as a dienophile, and it provides easy access to the 1,2-oxazine family. However, simple alkyl nitroso compounds are generally not stable and tend to dimerize. Aryl nitroso compounds are more stable and are useful in this process. The N=O group is also more stable with a carboalkoxy substituent on nitrogen, and it too is useful as a dienophile (Scheme 5.11).

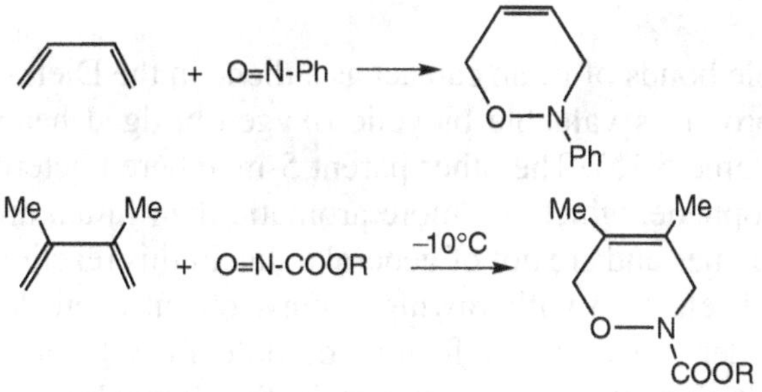

+ O=N-Ph → O—N Ph

Me Me + O=N-COOR –10°C → Me Me O—N COOR

Scheme 5.11

Another valuable approach to 1,2-oxazines makes use of the N=O group as part of the diene. Thus, nitroso ethylene (**5.10**) is reactive to an enol ether in another case of cycloaddition with inverse electron demand (Scheme 5.12).

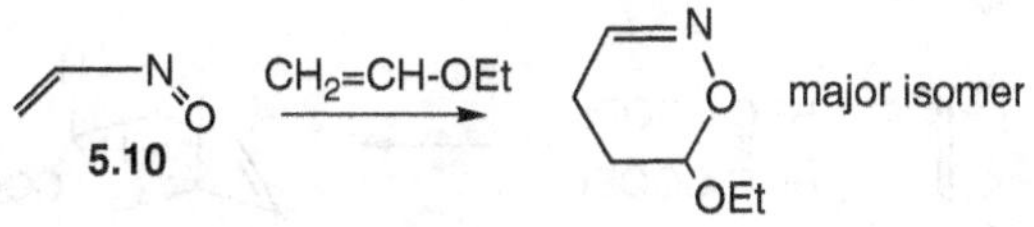

N O **5.10** CH_2=CH-OEt → N O OEt major isomer

Scheme 5.12

5.1.3.5. Synthesis of Oxygen Heterocycles. Alpha, beta-unsaturated ketones can cycloadd dienophiles of sufficient reactivity (thus with inverse electron demand). This is the case in the reaction in Scheme 5.13 where a pyran derivative is formed.

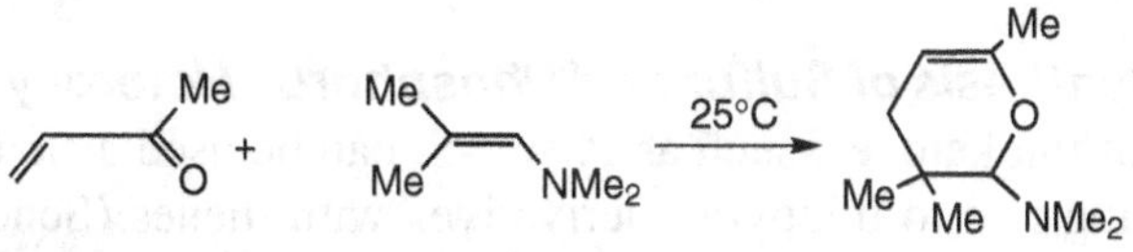

Me O + Me Me NMe_2 25°C → Me O Me Me NMe_2

Scheme 5.13

Pyran derivatives can also be prepared by condensing an activated carbonyl group, as in **5.11**, with dienes (Scheme 5.14).

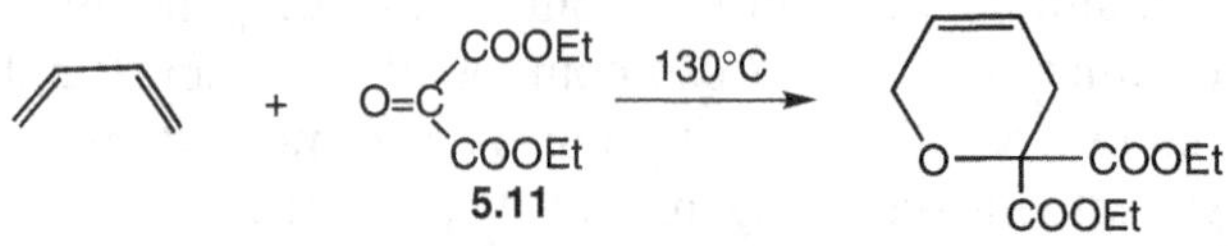

Scheme 5.14

The double bonds of furan can act as a diene in the Diels–Alder reaction. This produces valuable bicyclic oxygen-bridged heterocycles, as in **5.13** (Scheme 5.15). The other parent 5-membered heterocycles pyrrole and thiophene, which are more aromatic than furan, are much less reactive as dienes and are not of general value in this reaction. However, pyrroles with electron-withdrawing groups on nitrogen do participate in Diels–Alder reactions. A feature of note in Scheme 5.15 is that the cis-orientation of the ester groups in the dienophile (**5.12**, diethyl maleate) is preserved in the cycloadduct, where both groups are found in the *endo* positions. Preservation of stereochemistry is the rule in Diels–Alder reactions, which of course is consistent with the concerted mechanism of orbital interactions. The isomeric diethyl fumarate (**5.14**) would give the structure **5.15** with trans orientation of the ester groups.

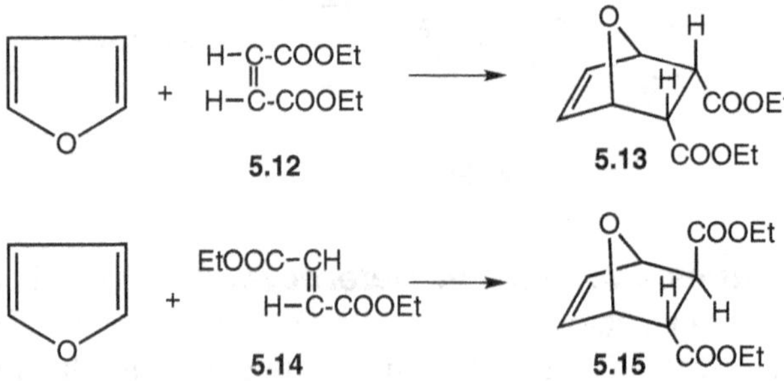

Scheme 5.15

5.1.3.6. Synthesis of Sulfur and Phosphorus Heterocycles. The C=S unit of thioketones, such as $Ph_2C{=}S$, can be used as a dienophilic center, giving rise to thiopyran derivatives with dienes (Scheme 5.16). Another useful dienophile is thiophosgene ($S{=}CCl_2$). Thioaldehydes as such are too unstable for use in this process because they dimerize readily, but they can be prepared in the reaction mixture containing a diene and then trapped as the Diels–Alder product before they dimerize.

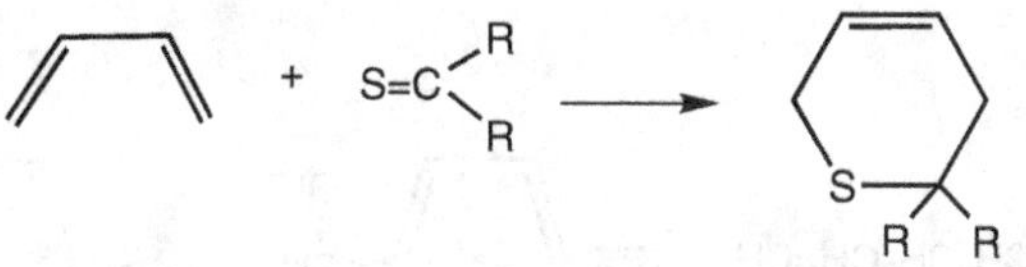

Scheme 5.16

Carbon-phosphorus double bonds also participate in the Diels–Alder reaction and are a source of a variety of tetrahydrophosphinine derivatives. The C=P unit has a relatively short history; the first compounds were not isolated until 1976.[3] Special characteristics must be present to give stable compounds that resist dimerization or polymerization. Strongly electron-attracting groups or large bulky groups that provide steric retardation of the dimerization or polymerization are required. The former type are the more useful in the Diels–Alder reaction, because steric effects from the bulky substituents can also reduce the reactivity of the C=P unit to additions. An example of a successful cycloaddition[4] is given in Scheme 5.17 (where TMS is Me_3Si). The literature on Diels–Alder reactions of C=P compounds is the subject of a 2008 review.[5]

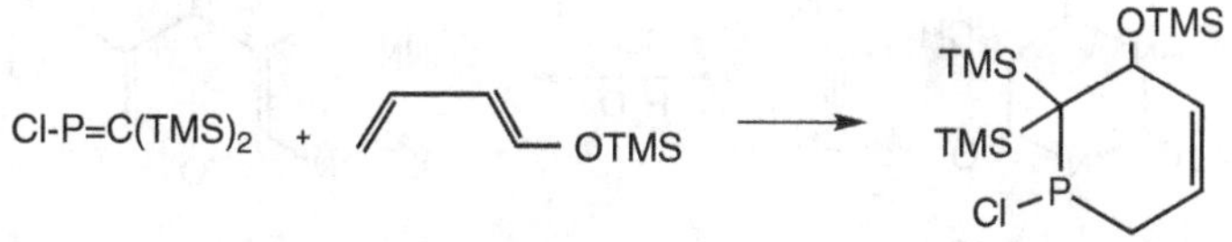

Scheme 5.17

5.1.3.7. Intramolecular Diels–Alder Reactions. The Diels–Alder reaction is not limited to the construction of monocyclic heterocycles as might be suggested by the examples above. Many complex structures have been created by causing *intramolecular* cycloaddition between the diene group and the dienophile that are contained in the same molecule. The situation is reminiscent of the olefin metathesis that occurs when the two olefinic groups are in the same complex molecule (Chapter 4, section 4.6). A limiting feature is that the interacting groups must have the geometric freedom to achieve the structure of the transition state of the cycloaddition. An example is provided in Scheme 5.18, where the furan ring acts as the diene in cycloaddition with a double bond.[6] Note that in this example, the double bond is isolated and lacks activation, but the reaction proceeds with ease, as is typical of intramolecular interactions.

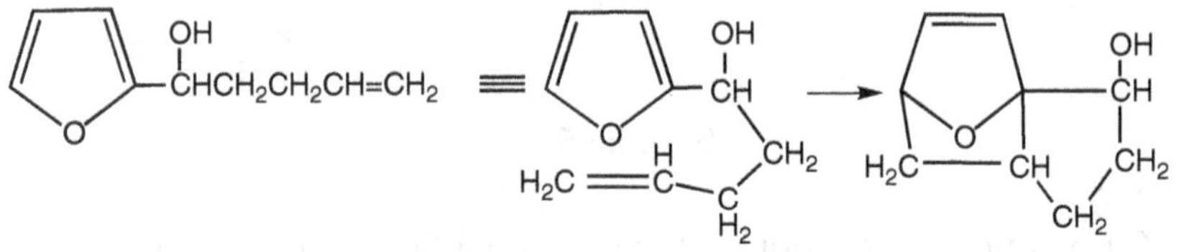

Scheme 5.18

In a more complex example (Scheme 5.19), the 1-oxa-1,3-butadiene unit consists of a carbonyl group on the pyrimidine ring with an exocyclic double bond as outlined in structure **5.16**. This molecule also contains a triple bond, and on refluxing in water with CuI as catalyst, an intramolecular cycloaddition takes place to form the tetracyclic compound **5.17**.[7]

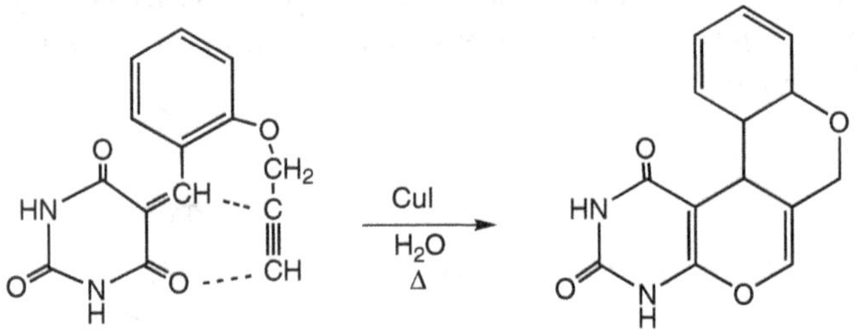

Scheme 5.19

An example employing an aza-alkene (an imine) in reaction with a diene is shown in Scheme 5.20. This is a reaction that has found use in the construction of alkaloids having nitrogen at a bridgehead position.

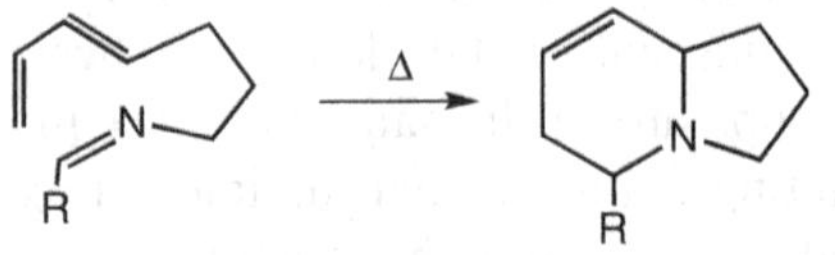

Scheme 5.20

5.1.3.8. Cheletropic Cycloadditions. If an atom can both donate an electron pair and accept an electron pair to move to a higher covalency, it can act as a dienophile to a diene. Such is the case with the sulfur atom in sulfur dioxide and with phosphorus in tricovalent halides. This type of cycloaddition is called cheletropic, and results in the formation of 5-membered rings. The reaction of butadiene with SO_2 is particularly important; it is conducted on a commercial scale to produce the cyclic sulfone **5.18** (called sulfolene). Its hydrogenation product **5.19** (sulfolane) is widely used as a nonaqueous highly polar solvent. Substituted dienes also participate in the cycloaddition. Here, the sulfur can donate its electron pair while accepting an electron pair to form a new carbon-sulfur bond.

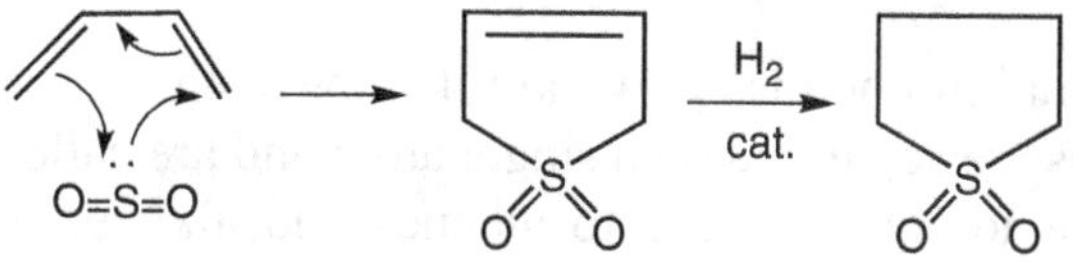

O=S=O
H_2
cat.
5.18
5.19

Scheme 5.21

Phosphorus forms many tricovalent compounds that will have an unshared electron pair. It also has available two higher covalency states (tetra and penta; hexacovalent compounds are known but not of importance in this context), and thus by valence expansion it can participate in a cheletropic cycloaddition. The most reactive compounds are of structure RPX_2 (called phosphonous dihalides or dihalophosphines) and PX_3 (phosphorus trihalides). Both bromine and chlorine derivatives can be used. The cycloaddition reaction, which is known in phosphorus chemistry as the McCormack reaction, is of great importance and many compounds have been prepared this way.[8] It is expressed in Scheme 5.22. The initial cycloadduct has pentacovalent phosphorus (**5.20**), but it dissociates rapidly to the ionic form (**5.21**). As is typical of most phosphorus-halogen compounds, these are highly reactive and not generally isolated. Instead, they are hydrolyzed to the corresponding phosphine oxides (**5.22**), which are stable and well known. Compounds of structure **5.22** are valuable catalysts for the commercial-scale production of carbodiimides (RN=C=NR) from isocyanates (RN=C=O).

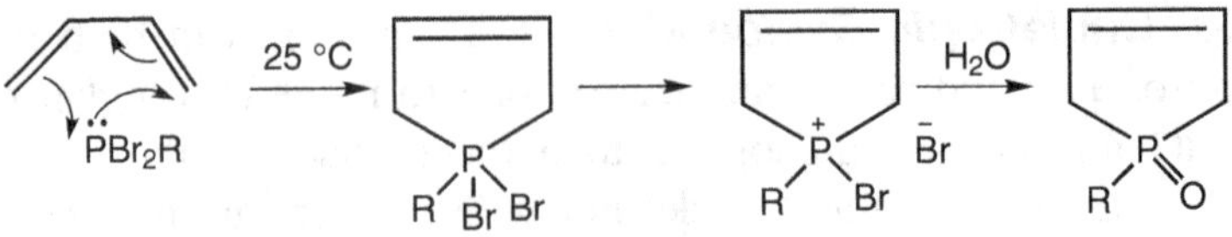

25 °C, H_2O, PBr_2R, 5.20, 5.21, 5.22

Scheme 5.22

5.2. DIPOLAR CYCLOADDITIONS

5.2.1. Definitions and Examples of 1,3-Dipoles

Many compounds, while neutral overall, have a positive and a negative atom and for which no resonance structure can be written that has no charges.*

In cycloaddition chemistry, we are dealing with the case where the charged atoms are separated by a single atom and are called 1,3-dipoles. These compounds are reactive to alkenes and alkynes, as well as to heteroatom derivatives of these, in a cycloaddition process that forms 5-membered heterocycles. These unsaturated participants are called dipolarophiles. Dipolar cycloadditions are perhaps the most versatile of all syntheses of 5-membered heterocycles.

1,3-Dipoles have another characteristic feature: The central atom must have an electron pair to stabilize the species by dispersal of the positive charge. The central atom therefore cannot be carbon; it is in fact usually nitrogen, although oxygen and sulfur are other possibilities. We can write a generalized structure for the dipole as **5.23**, and the corresponding resonance form as **5.24**. We show in **5.23** that atom *a* has a sextet of electrons and is a cation; atom *c* has an octet and is an anion; *b* has an octet with a free electron pair.

5.23 ⟷ 5.24

*To aid in recognizing the formal charges on atoms, use may be made of the following expression:

Formal charge = (Number of valence electrons on the atom)−1/2(Number of electrons used in bonding)−(Number of nonbonding electrons). Thus, if *c* in **5.23** were an oxygen atom with six valence electrons, we would have $6 - 1/2(2) - 6 = -1$ for the charge.

Another configuration has a multiple *a* to *b* bond, as in hybrid **5.25**.

$$:\overset{\oplus}{a}{=}\ddot{b}{-}\ddot{\underset{..}{c}}:^{\ominus} \longleftrightarrow :a{\equiv}\overset{\oplus}{b}{-}\ddot{\underset{..}{c}}:^{\ominus}$$

5.25

Note that the *c* atom always bears a negative charge and is not involved in the resonance delocalization. Another general feature where nitrogen is present is that N with four bonds must always bear a positive charge; with three bonds it is neutral and with two bonds it must be negative.

We can visualize (but not to be taken as a mechanism) the cycloaddition to a double bond occurring with the form bearing charges in the 1,3 positions (Scheme 5.23), although most dipoles are better represented by the resonance form with charges on the 1,2-positions.

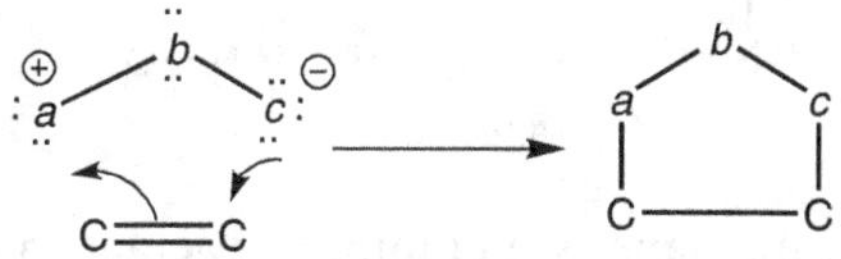

Scheme 5.23

Many structures meet the electronic requirements to function as a 1,3-dipole. R. Huisgen at the University of Munich first interpreted the older structures in the literature in terms of 1,3-dipoles.[9] He (and others) went on to synthesize many new dipoles and demonstrated their cycloaddition with various dipolarophiles. New structures are still being created, and the field is so large that a two-volume treatise of the subject has been published.[10]

The simplest example of a dipole is probably ozone, which is well known for its reaction with alkenes. Its resonance hybrid is shown as **5.26**.

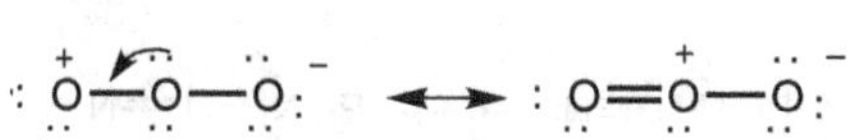

$$:\overset{+}{\ddot{O}}{-}\ddot{O}{-}\ddot{O}:^{-} \longleftrightarrow :O{=}\overset{+}{\ddot{O}}{-}\ddot{O}:^{-}$$

5.26

Other examples are summarized in the next section.

Nitrones. These can be understood as N-oxides of imines. However, they are not synthesized by oxidation of imines but by less direct methods, as is shown in Scheme 5.24. The former shows the

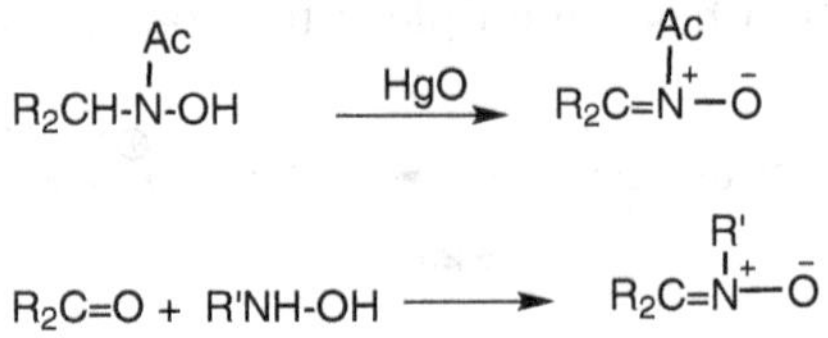

$$R_2CH\text{-}N(Ac)\text{-}OH \xrightarrow{HgO} R_2C{=}\overset{+}{N}(Ac){-}\overset{-}{O}$$

$$R_2C{=}O + R'NH\text{-}OH \longrightarrow R_2C{=}\overset{+}{N}(R'){-}\overset{-}{O}$$

Scheme 5.24

oxidation of a hydroxylamine derivative, and the latter shows the reaction of a hydroxylamine with an aldehyde or ketone. Nitrones are generally prepared at the time they are needed, as they have low stability.

The resonance hybrid for a typical nitrone is shown as **5.27**.

$$Ph\text{-}CH{=}\overset{+}{N}(Ph){-}\overset{-}{O} \longleftrightarrow Ph\text{-}\overset{+}{C}H\text{-}\ddot{N}(Ph){-}\overset{-}{O}$$

5.27

Diazoalkanes. Diazomethane is a common reactant in cycloaddition syntheses. It is a gas, which is generated as needed and introduced directly into a reaction mixture. Its resonance hybrid is shown as **5.28**. This is our first example of a 1,3-dipole originating from a triply bonded grouping in the 1,2-dipolar form. Alkyl and phenyl diazo compounds are also useful reactants.

$$:\overset{+}{N}{=}\ddot{N}{-}\overset{-}{\ddot{C}}H_2 \longleftrightarrow :N{\equiv}\overset{+}{N}\text{-}\overset{-}{\ddot{C}}H_2$$

5.28

Azides. Frequently used as reactants are compounds with the triply bonded azide structure, which is shown in its resonance formulation **5.29**. Some azides, especially aryl derivatives, are sufficiently stable that they can be bought commercially.

$$Ph\text{-}\overset{-}{\ddot{N}}\text{-}\overset{+}{N}{\equiv}N: \longleftrightarrow Ph\text{-}\overset{-}{\ddot{N}}{-}\ddot{N}{=}\overset{+}{N}$$

5.29

Nitrile Oxides. Yet another common reactant type is the family of nitrile oxides. Like nitrones, they are not made by direct oxidation of nitriles but by an indirect method shown in Scheme 5.25, starting with the oxime of an aldehyde. The aldehydic hydrogen is replaced by chlorine, and then HCl is eliminated by reaction

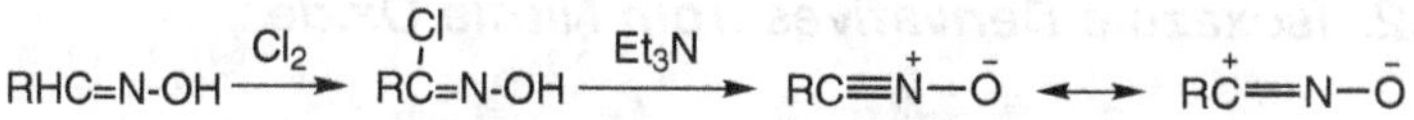

$$\mathrm{RHC{=}N{-}OH \xrightarrow{Cl_2} RC(Cl){=}N{-}OH \xrightarrow{Et_3N} RC{\equiv}\overset{+}{N}{-}\overset{-}{O} \longleftrightarrow R\overset{+}{C}{=}N{-}\overset{-}{O}}$$

Scheme 5.25

with base. Nitrile oxides are reasonably stable compounds. Nitrile sulfides are also reactive as 1,3-dipoles.

Other 1,3-Dipoles. Several less commonly used dipoles are known. Some of these are shown here in their 1,3-dipolar resonance form.

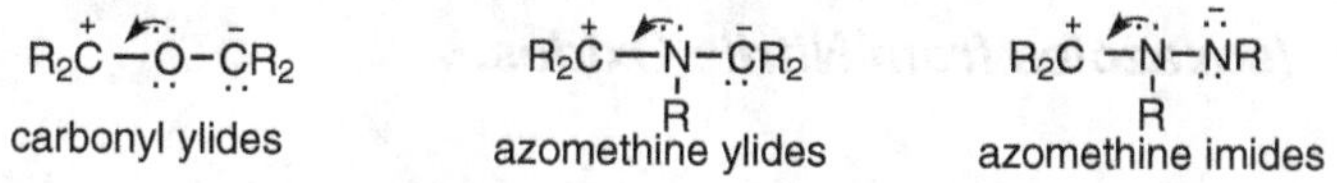

$\mathrm{R_2\overset{+}{C}{-}O{-}\overset{-}{C}R_2}$ carbonyl ylides

$\mathrm{R_2\overset{+}{C}{-}N(R){-}\overset{-}{C}R_2}$ azomethine ylides

$\mathrm{R_2\overset{+}{C}{-}N(R){-}\overset{-}{N}R}$ azomethine imides

Note that the fragment—CR_2^- is the same as found in Wittig reagents, and it is responsible for the "ylide" name.

5.2.2. Cycloadditions of Alkenes and Alkynes with 1,3-Dipoles

A great variety of heterocyclic compounds can be made by using the dipoles discussed above in cycloadditions with the carbon–carbon double bond and triple bond. Some of these are illustrated in the equations to follow. In most cases, regioisomers are possible, but frequently the major product is that with the least steric interaction between ring substituents. Other factors can play a role in product orientation, however (see section 5.2.5).

5.2.2.1. 1,2,3-Triazole Derivatives from Azides.

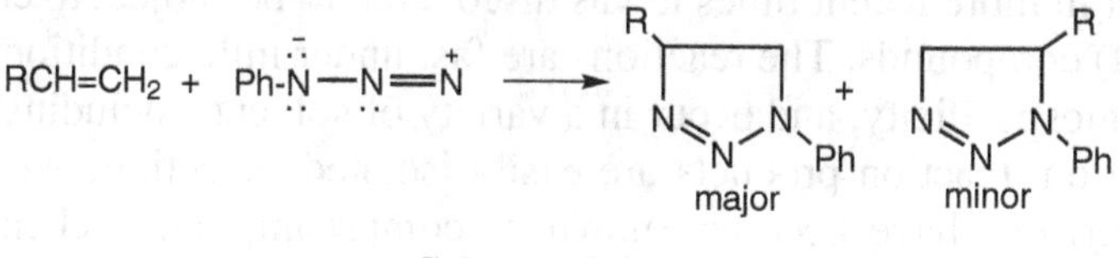

$$\mathrm{RCH{=}CH_2 + Ph{-}\overset{-}{N}{-}N{\equiv}\overset{+}{N} \longrightarrow}$$ major + minor

Scheme 5.26

Such triazolines are unstable and smoothly eliminate N_2, leaving an aziridine as a product. This is a useful synthetic method for making aziridines.

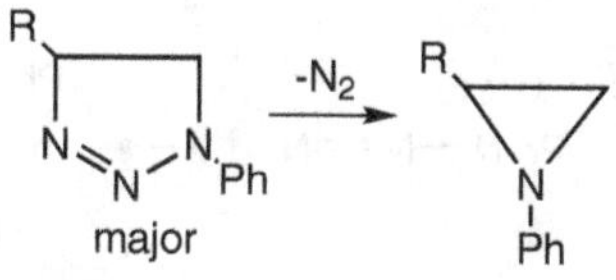

major $\xrightarrow{-N_2}$

Scheme 5.27

5.2.2.2. Isoxazole Derivatives from Nitrile Oxides.

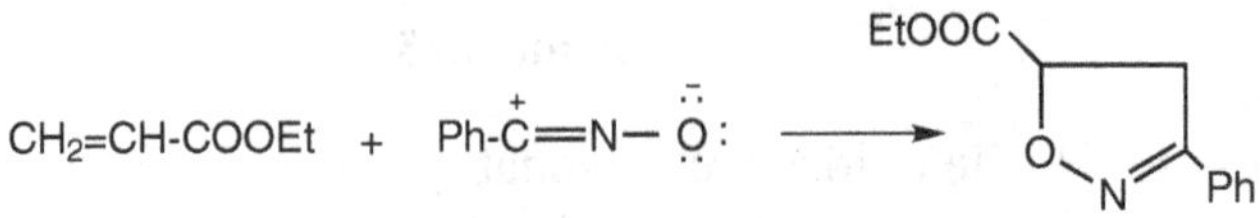

CH_2=CH-COOEt + Ph-C≡N−O: ⟶ (isoxazoline: EtOOC, O, N, Ph)

Scheme 5.28

5.2.2.3. Isoxazoles from Nitrile Oxides.

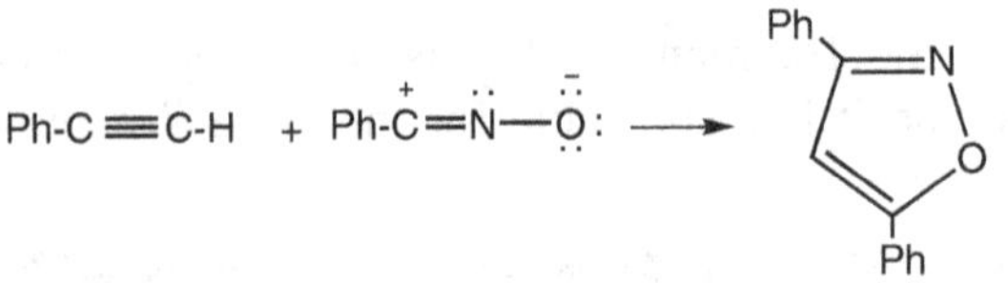

Ph-C≡C-H + Ph-C≡N—O: ⟶ (isoxazole: Ph, N, O, Ph)

Scheme 5.29

Note the useful feature that using an alkyne as the dipolarophile with an unsaturated dipole leads to the valuable fully unsaturated, aromatic ring system. See also Scheme 5.30 for another example of heteroaromatic synthesis.

5.2.2.4. Triazoles from Azides. Cycloadditions of this type constitute a valuable synthetic route to the triazole ring system. This is shown in Scheme 5.30. This combination dates back to the early work of Huisgen, but in more recent times it was discovered to be subject to catalysis by Cu(I) compounds. The reactions are fast under mild conditions, have high regiospecificity, and occur in a variety of solvents including water. In addition, reaction products are easily isolated. Reactions with these characteristics have become known as comprising “click chemistry”; this term was coined by K. B. Sharpless.[11] The first and most commonly used reaction referred to by this name is indeed the azide-alkyne cycloaddition, and new interest has developed in triazole hemistry since the discovery of the copper catalysis. In addition to its use in organic

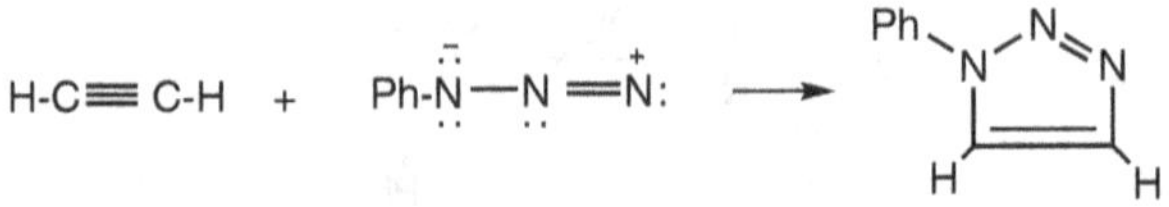

H-C≡C-H + Ph-N—N≡N: ⟶ (triazole: Ph, N, N, N, H, H)

Scheme 5.30

synthesis, a particularly important application of the process is in the fields of polymer and materials chemistry, and many papers have been published in these fields. Polymers have been made where triazole rings are repeating units in the chain or branches thereon or are used in postfunctionalization processes. This exciting and still developing new aspect of 1,3-dipolar cycloadditions is the subject of a 2007 review.[12]

5.2.2.5. Isoxazole Derivatives from Nitrones. With alkenes, nitrones cycloadd readily to give saturated isoxazolidines (Scheme 5.31).

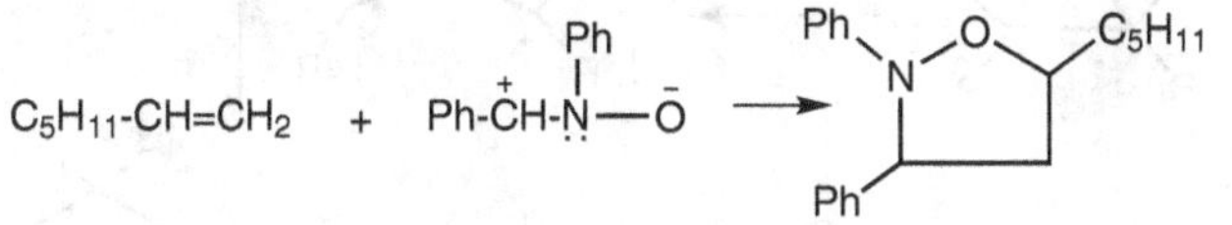

C_5H_{11}-CH=CH_2 + Ph-$\overset{+}{C}$H-N(Ph)—$\bar{O}$ ⟶ (isoxazolidine with Ph on N, C_5H_{11} at C-5, Ph at C-3)

Scheme 5.31

Cycloadditions of nitrones with the triple bond are also valuable and give isoxazoline derivatives. An advanced use of this cycloaddition is reported in a paper directed to the synthesis of a portion of the structure of the natural product cortistatin.[13] Here, the triple bond is generated in the form of a benzyne (**5.30**, Scheme 5.32 A) in the presence of a nitrone. Benzynes, of course, are not stable at room temperature and must be generated as needed in a reaction mixture. In this particular case, the nitrone reactant had this group conjugated to a double bond, as shown in structure **5.31**. The cycloaddition proceeded normally to give the adduct **5.32**.

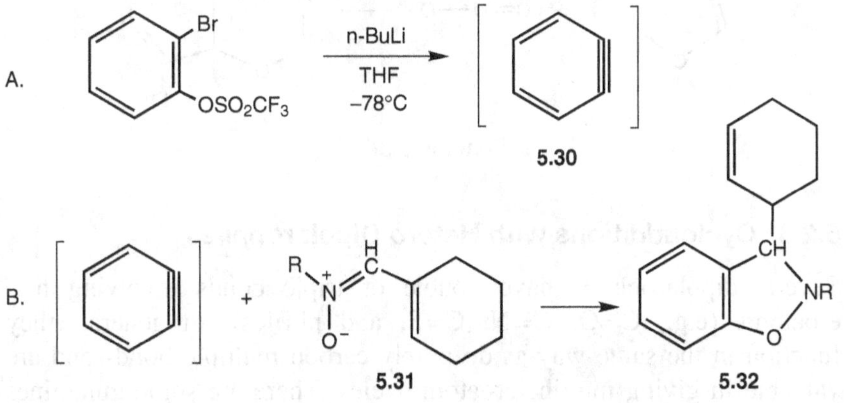

A. Br, OSO_2CF_3 (benzene) —n-BuLi, THF, −78°C⟶ [benzyne] **5.30**

B. [benzyne] + R-$\overset{+}{N}$(O⁻)=CH-(cyclohexenyl) **5.31** ⟶ CH, NR, O (benzisoxazoline with cyclohexenyl) **5.32**

Scheme 5.32

5.2.2.6. Tetrahydrofurans from Carbonyl Ylides. Carbonyl ylides are unusual species that must be generated as needed. One method is to break the C-C bond of certain epoxides thermally or photochemically, which yields an open chain. Thereby one carbon is positively charged and the other is negatively charged, which is the structural feature of carbonyl ylides. This is shown in Scheme 5.33 A. Cycloaddition with an alkene follows immediately to give tetrahydrofuran derivatives (Scheme 5.33).

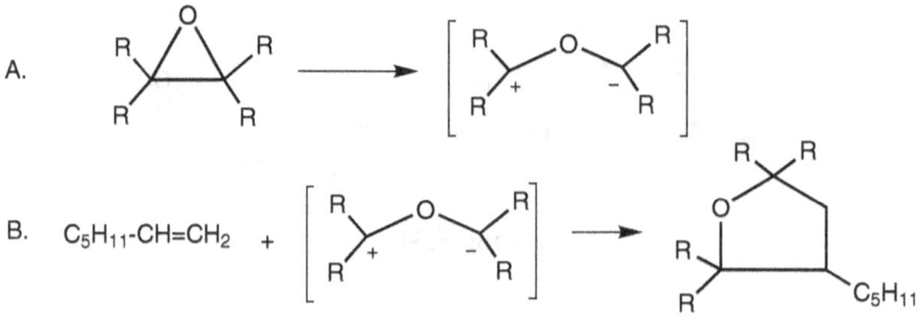

A.

B. C_5H_{11}-CH=CH_2 +

Scheme 5.33

5.2.2.7. Fused Heterocycles from Cyclic Dipolarophiles. Furan provides a useful example of a diene acting instead as a dipolarophile. As will be discussed in Chapter 6, this molecule has the lowest aromaticity of the 5-membered heterocycles. It is also useful as a diene in Diels–Alder cycloadditions, but here we will observe that one double bond can react with a 1,3-dipole (Scheme 5.34).

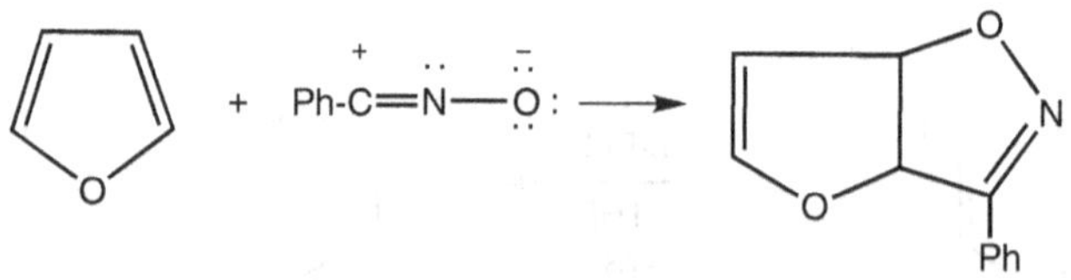

Scheme 5.34

5.2.3. Cycloadditions with Hetero Dipolarophiles

Several dipolarophiles have double or triple bonds involving heteroatoms (e.g., C=O, C=N, C=S, and nitriles). In general, they function in the same way as do purely carbon multiple bonds and are valuable in giving multiheteroatom cycles. There are some guidelines in predicting the preferred regioisomer from some combinations, because there will always be two possibilities. Some hetero-hetero structures have weak sigma bond stability and their formations are disfavored. Thus, a structure where three heteroatoms are bonded will be less stable than where two bonds are present. Also, the formation of an oxygen–oxygen bond is to be avoided because of inherent instability of peroxide links. Where such features are absent, steric interactions may be a disfavoring factor. However, as will be discussed in section 5.4, MO theory can be useful in making predictions. One structural feature that is not reliable in product prediction is the charge of the atoms to be involved in the new bonding. One would think that, for example, the positive end of the dipole might bond to the more negative atom of the dipolarophile, and vice versa; on occasion this is the case, but it is not always true. Many examples of cycloadditions with heterodipolarophiles are known; a few illustrations are shown in the next section.

1,2,4-Oxadiazoles from Nitriles and Nitrile Oxides.

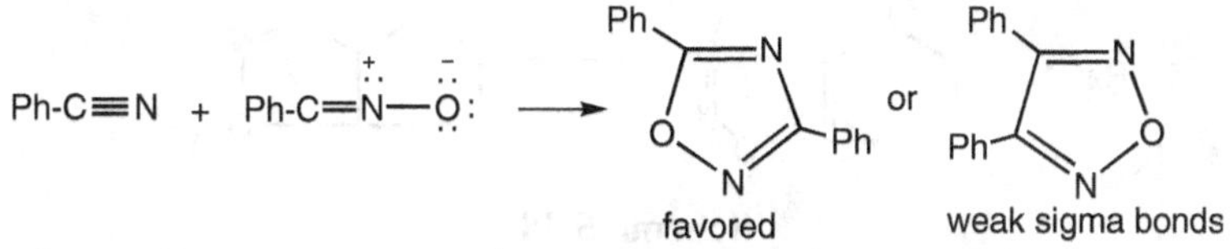

Ph-C≡N + Ph-C=N⁺—O⁻ ⟶ favored or weak sigma bonds

Scheme 5.35

1,2,3-Dioxazoles from Aldehydes and Nitrile Oxides.

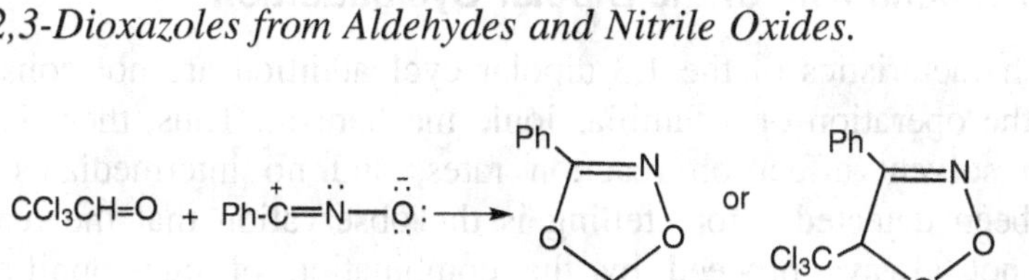

$CCl_3CH=O$ + Ph-C=N⁺—O⁻ ⟶ favored or weak sigma bonds

Scheme 5.36

1,2,3,4-Tetrazoles from Cyanamide and Hydrazoic Acid.

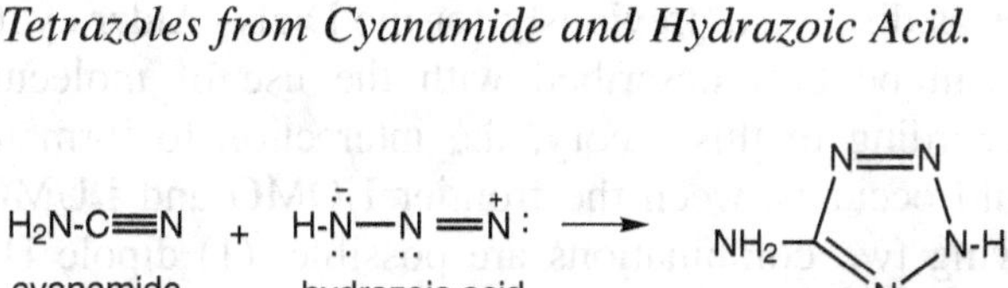

H_2N-C≡N (cyanamide) + H-N⁻—N⁺≡N (hydrazoic acid) ⟶ 5-aminotetrazole

Scheme 5.37

5.2.4. Intramolecular Cycloadditions

Just as was shown for the Diels–Alder cycloadditions, it is possible for the two participating groups in a 1,3-dipolar cycloaddition to be in the same molecule. No steric constraints must be present, and the two groups must be capable of being brought into the needed orientation. There are many cases in the literature of complex heterocyclic systems being synthesized by this method. Two examples are given in Scheme 5.38.

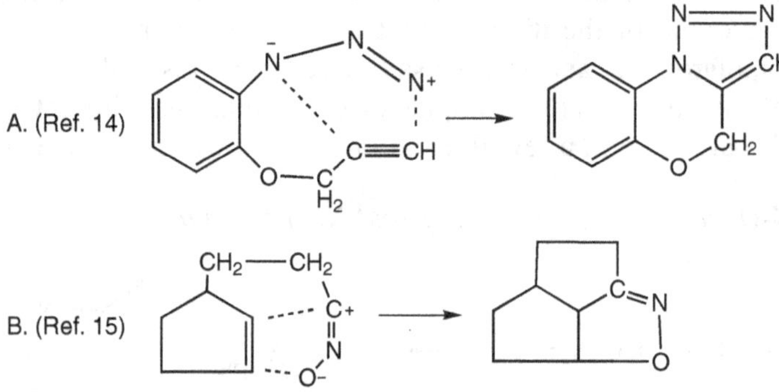

A. (Ref. 14)

B. (Ref. 15)

Scheme 5.38

5.2.5. Mechanism of the Dipolar Cycloaddition

The characteristics of the 1,3-dipolar cycloaddition are not consistent with the operation of a familiar ionic mechanism. Thus, there is little or no solvent effect on reaction rates, and no intermediates have ever been detected. More telling is the observation that the reaction does not always proceed by the combination of nucleophilic and electrophilic centers. The notion that a free radical mechanism is involved was given consideration but is no longer in favor. The best explanation of the mechanism is that the reaction is a concerted (i.e., a single step without intermediates) cycloaddition and depends on orbital interactions as discussed previously for the Diels–Alder cycloaddition. Hence, it can be best described with the use of molecular orbital theory. According to this theory, the interaction to form new sigma bonds would occur between the frontier HOMO and LUMO orbitals. The following two combinations are possible: (1) dipole HOMO and dipolarophile LUMO, when electron-withdrawing groups are present on the dipolarophile, and (2) dipole LUMO and dipolarophile HOMO, when electron-releasing groups are on the dipolarophile. That phases match properly in these combinations can be observed in Figure 5.8, which depicts only the frontier orbitals of the two cases. This theory does not preclude the possibility that the formation of one new sigma bond may be more advanced than the other and that the transition state may be unsymmetrical.

The choice of one set of frontier orbitals over the other can be established by calculations of the orbital energies of each combination. That combination with the smallest energy difference will be the preferred combination. As we will discuss, such calculations are of practical importance in giving predictions of the outcome of cycloadditions where two regioisomers can be formed. Orbital energies for many of the common reactants of dipolar cycloadditions are available elsewhere.[16] We will use the energies for one typical reaction to determine how they can be applied. The energies for the two possible frontier orbital interactions for the reaction of ethyl acrylate with diazomethane as the dipole are shown in Figure 5.9. The energy gap is the smaller (9 eV vs. 12.8 eV) for the interaction of the HOMO of the dipole and the LUMO of the dipolarophile, which is consistent with the generality stated above that the presence of an electron withdrawing group (here COOEt) on the dipolarophile will favor this combination.

More important from a practical standpoint is the prediction by MO theory of the regiochemistry of the cycloaddition (i.e., which of two possible isomers will be the major product). Steric hindrance effects can certainly influence the course of an addition, and we have

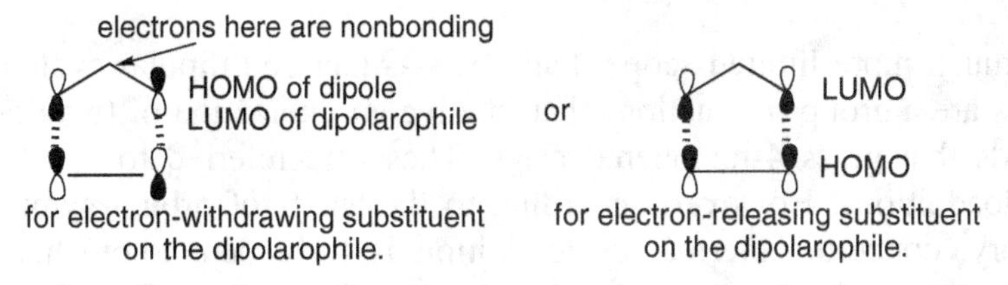

Figure 5.8. Dipole-dipolarophile MO interaction.

H_2C N N COOEt

HOMO −9 eV	LUMO +1.8 eV
LUMO 0 eV	HOMO −11 eV
difference 9 eV (favored)	difference 12.8 eV

Figure 5.9. HOMO-LUMO energy differences.

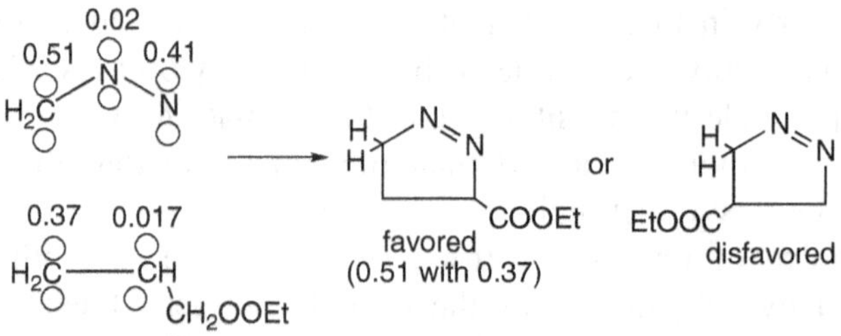

Figure 5.10. Prediction from atomic orbital coefficients.

already shown that weak sigma bond energies are to be avoided. In the absence of these influences, atomic orbital coefficients (AOCs) calculated from MO theory can be useful. Data on the coefficients are available, which makes possible the determination of the regiochemistry of many cycloadditions[17]; here again we will use only one set of data to illustrate their use in making a prediction. Again describing the ethyl acrylate-diazomethane cycloaddition, the pertinent coefficients for the interacting orbitals are shown in Figure 5.10. The interaction will occur between the largest AOC of each reactant. Here the largest AOC of the dipole is 0.51 and that of the dipolarophile is 0.37. This leads to the prediction shown in Figure 5.10 for the major regioisomer, which is in fact the experimental result.

5.3. [2 + 2] CYCLOADDITIONS

Of much more limited scope than Diels–Alder and dipolar cycloadditions are a group of reactions that involve an interaction of two double bonds that gives 4-membered rings. These are referred to as [2 + 2] cycloadditions. However, according to the tenets of orbital symmetry theory, concerted thermal cycloaddition is forbidden. Photochemical cycloaddition is allowed and some processes have been affected by this technique. Another mechanism involves a two-step process with a noncyclic intermediate being formed, which then cyclizes. Four-membered rings can be less stable than 5- and 6-membered rings, and they are encountered much less commonly in practical organic chemistry. The first [2 + 2] cycloadditions were reported in 1907 by H. Staudinger by reaction of diphenylketene with various unsaturated compounds. In fact, these reactions have been said to be the first of any cycloadditions.[18] Staudinger had synthesized beta-lactams, but little work in this area was reported until the war-time discovery of the beta-lactam (azetidinone) ring in the penicillins (Chapter 3, section 3.2.5). This set off major research efforts, continuing to this day because of the valuable biological activity of compounds containing this moiety.

The participants in [2+2] cycloadditions for heterocyclics synthesis are of a limited scope. One component is generally a ketene or a heterocumulene, all of which are characterized by having a central carbon atom bearing two double bonds. Examples include the following:

$CH_2=C=O$	R-N=C=O	R-N=C=S	R-N=C=N-R
ketene	isocyanate	isothiocyanate	carbodiimide

The other component can have C=C, C=N, or C=O bonds but of rather specific types. Some examples of successful [2+2] combinations yielding heterocycles are shown in the next section.

Ketenes with Imines.

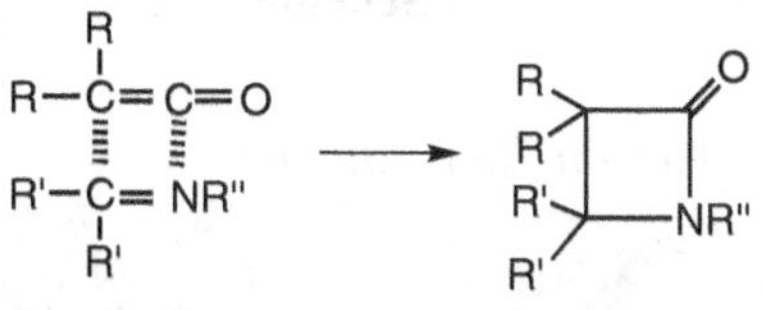

Scheme 5.39

This is an excellent way of synthesizing the beta-lactam structure, and it has been the subject of several reviews (e.g., Ref. 18). This is one of the reactions first reported by Staudinger, which is shown in Scheme 5.40. Many variants of this process have since been discovered.

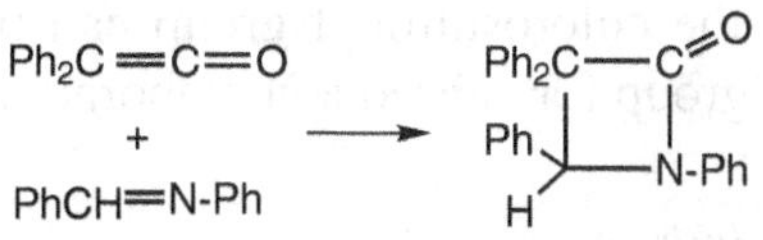

Scheme 5.40

An application arising from the extensive research on penicillin is outlined in Scheme 5.41.[19] Here, the synthesis of the bicyclic ring system found in penicillin was accomplished by generating ketene **5.33** in the presence of thiazoline **5.34**, which is participating as an imine. Ketenes are readily prepared by this process of dehydrohalogenation of an acyl chloride. The product **5.35** is a precursor of a penicillin analog called a penam.

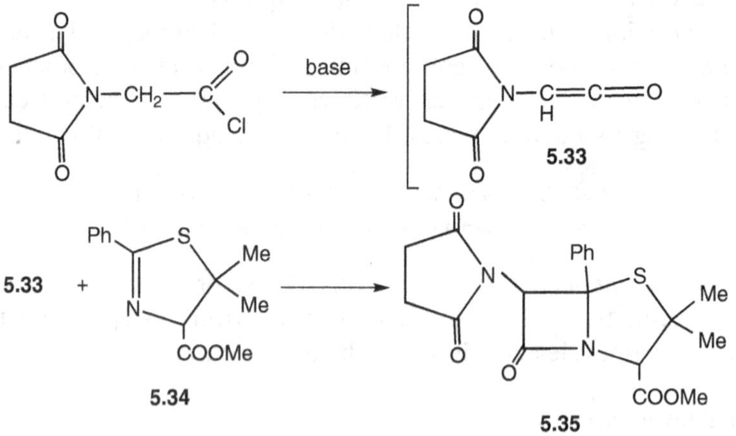

base

5.33

5.33 +

5.34

5.35

Scheme 5.41

Isocyanates with Electron-Rich Alkenes.

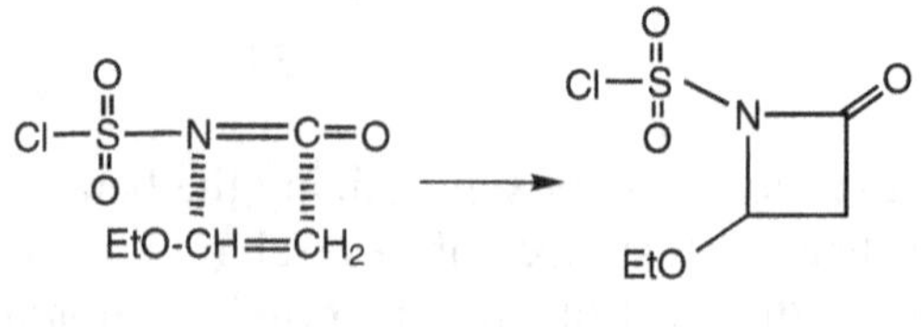

Scheme 5.42

This is a specialized way of constructing beta-lactams. It is of value because the chlorosulfonyl group can be readily removed, giving the NH group for subsequent elaboration of the beta-lactam moiety.

Isocyanates with Imines.

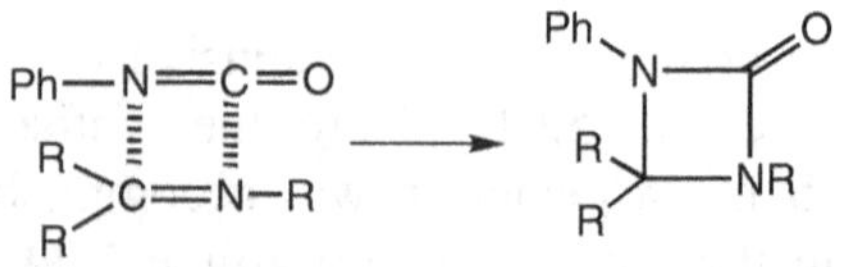

Scheme 5.43

Carbodiimides with Ketenes.

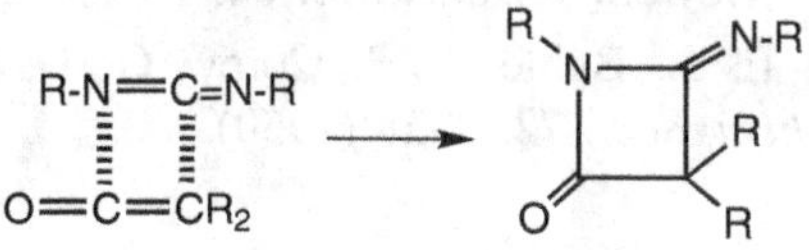

Scheme 5.44

Beta-lactams are again the product, but now bear an exocyclic imino group.

REFERENCES

(1) B. Groenendaal, E. Ruijter, and R. V. A. Orru, *Chem. Commun.*, 5474 (2008).

(2) R. C. Clark, S. S. Pfeiffer, and D. L. Boger, *J. Am. Chem. Soc.*, **128**, 2587 (2006).

(3) G. Becker, *Z. Anorg. Allg. Chem.*, **423**, 242 (1976).

(4) G. Märkl and W. Hölzl, *Tetrahedron Lett.*, **29**, 4535 (1988).

(5) R. K. Bansal and S. K. Kumawat, *Tetrahedron*, **64**, 10945 (2008).

(6) D. D. Sternbach, D. M. Rossana, and K. D. Onan, *Tetrahedron Lett.*, **26**, 591 (1985).

(7) M. J. Khoshkholg, S. Balaliae, R. Gleiter, and F. Rominger, *Tetrahedron*, **64**, 10924 (2008).

(8) L. D. Quin, *The Heterocyclic Chemistry of Phosphorus*, Wiley, New York, 1981.

(9) R. Huisgen, *Angew. Chem., Int. Ed. Engl.*, **2**, 565, 633 (1963).

(10) A. Padwa, *1,3-Dipolar Cycloaddition Chemistry*, Vols. 1 and 2, Wiley-Interscience, New York, 1984.

(11) H. C. Kolb, M. G. Finn, and K. B. Sharpless, *Angew. Chem., Int. Ed. Engl.*, **113**, 2004 (2001).

(12) J.-F. Lutz, *Angew. Chem., Int. Ed. Engl.* **46**, 1018 (2007).

(13) M. Dai, Z. Wang, and S. J. Danishefsky, *Tetrahedron Lett.*, **49**, 6613 (2008).

(14) R. Fusco, L. Garanti, and G. Zecchi, *J. Org. Chem.*, **40**, 1906 (1975).

(15) R. H. Wallenberg and J. E. Goldstein, *Synthesis*, 757 (1980).

(16) T. L. Gilchrist, *Heterocyclic Chemistry*, third edition, Addison Wesley Longman, Harlow, UK, 1997, p. 95.

(17) R. Huisgen, in *1,3-Dipolar Cycloaddition Chemistry*, Vol. **1**, A. Padwa, Editor, Wiley-Interscience, New York, 1984, pp. 130–141.

(18) N. Fu and T. T. Tidwell, *Tetrahedron*, **64**, 10465 (2008).

(19) J. C. Sheehan, E. L. Buhle, E. J. Corey, G. D. Laubach, and J. J. Ryan, J. *Am. Chem. Soc.*, **72**, 3828 (1950).

REVIEW EXERCISES

5.1. In a Diels–Alder reaction, a smaller difference in energy between the frontier orbitals (the reactant HOMO and LUMO) results in what?

a. No reaction

b. A faster reaction rate

c. A slower reaction rate

d. The requirement of antarafacial addition

5.2. Which reaction is likely to give product 1?

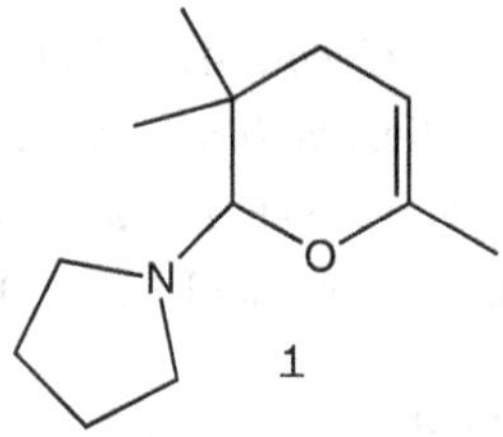

1

a.

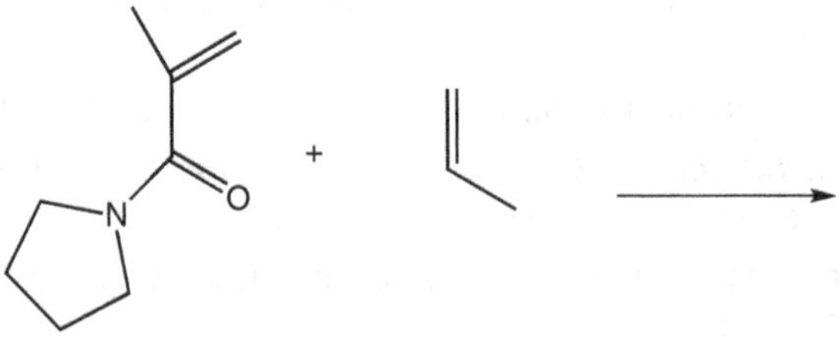

b.

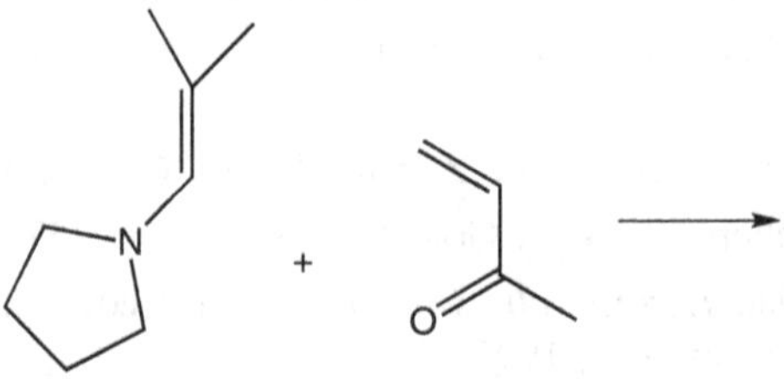

c.

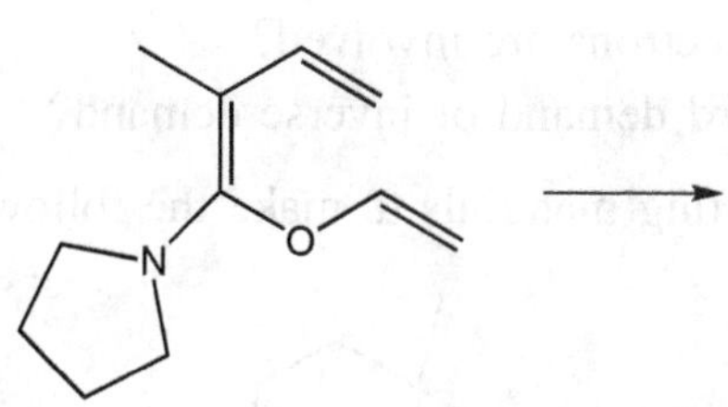

$\xrightarrow{\quad}$

d.

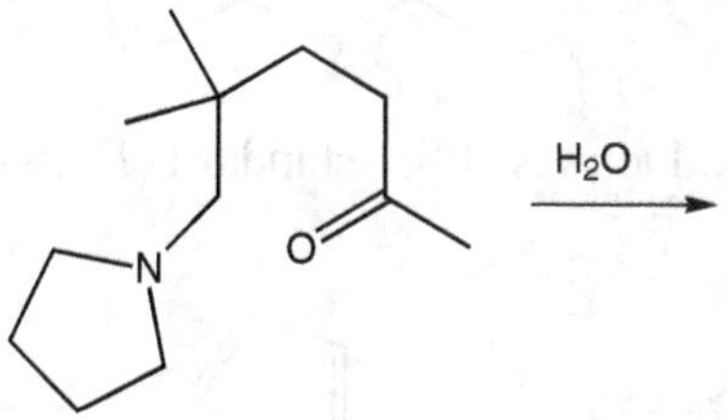

$\xrightarrow{H_2O}$

5.3. Sketch a synthesis of the bicyclo compound using starting materials containing six or fewer carbons.

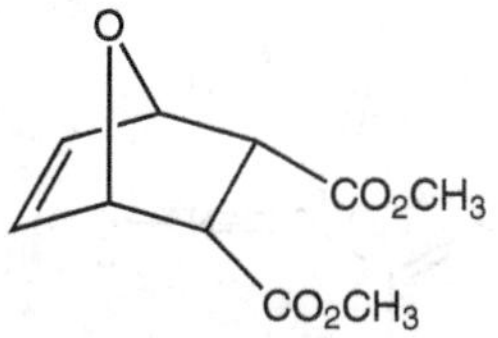

O

CO_2CH_3

CO_2CH_3

5.4. Sketch a synthesis starting with only nonheterocycles for the preparation of the following 1,2-oxazine.

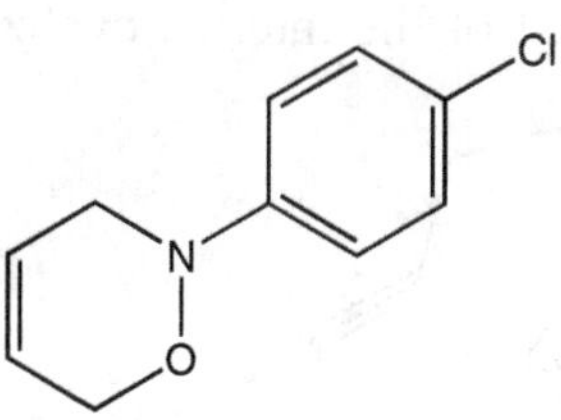

Cl

N

O

5.5. Consider the following Diels–Alder reaction:

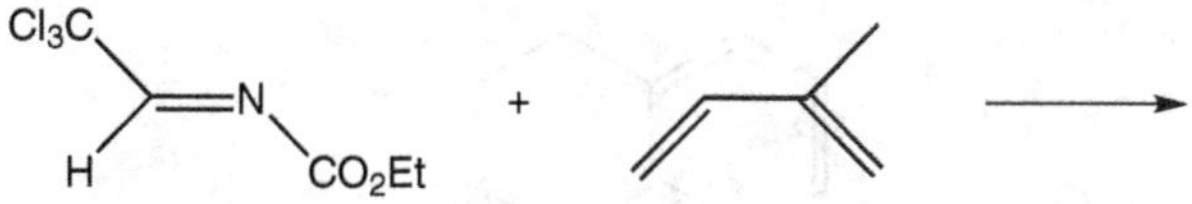

Cl_3C N H CO_2Et + $\longrightarrow$

Label the diene and the dienophile.

Write the product.

How many electrons are involved?
Is this standard demand or inverse demand?

5.6. Draw the starting materials to make the following pyran.

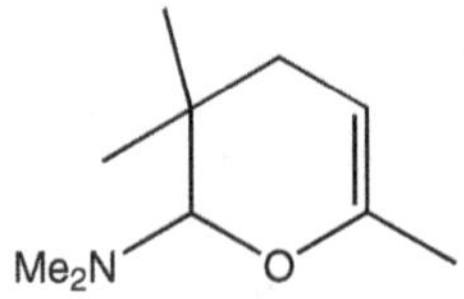

Me_2N O

5.7. Draw the product; is this standard Diels–Alder or inverse demand?

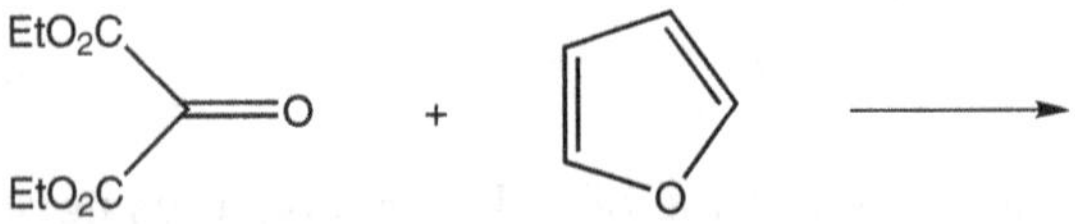

EtO_2C O + O →
EtO_2C

5.8. What is the formal charge on each of the indicated atoms.

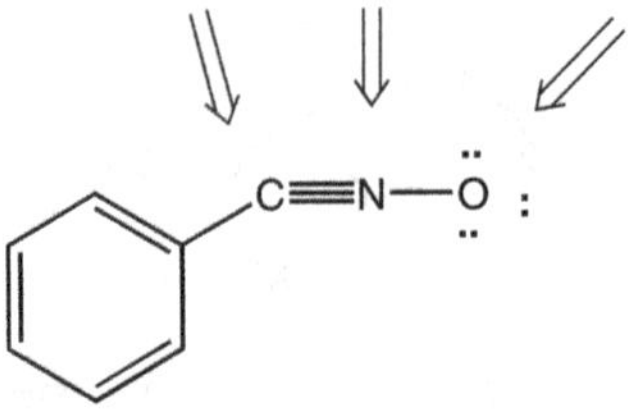

C≡N—O :

5.9. R. H. Wollenberg and J. E. Goldstein, *Synthesis*, **9**, 757 (1980). What is the product of the internal cyclization reaction?

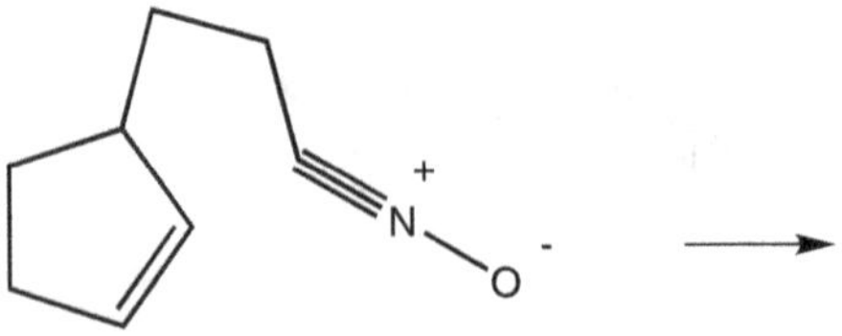

+ N O - →

5.10. Sketch a synthesis of the following compound.

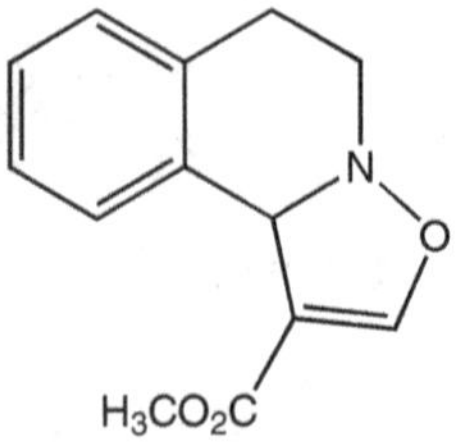

N O

H_3CO_2C

5.11. N. Chopin, et al., *J. Org. Chem.*, **74**, 1237-1246 (2009). Draw the expected product.

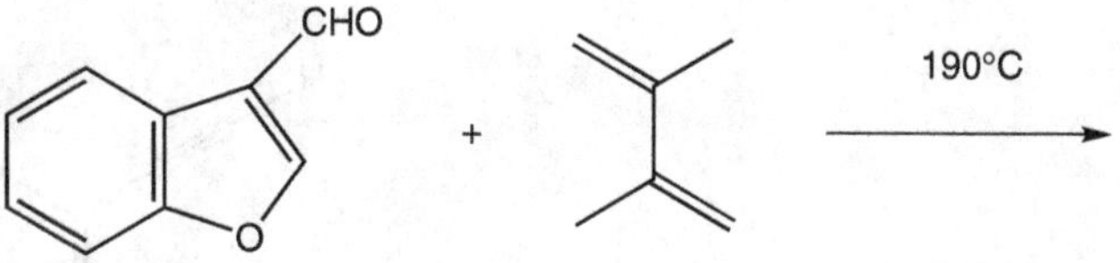

5.12. M. K. Manjula, et al., *Euro. J. Med. Chem.*, **44**, 280-288 (2009). The oxazine was evaluated as an antimicrobial. Sketch the synthesis from a nonheterocycle.

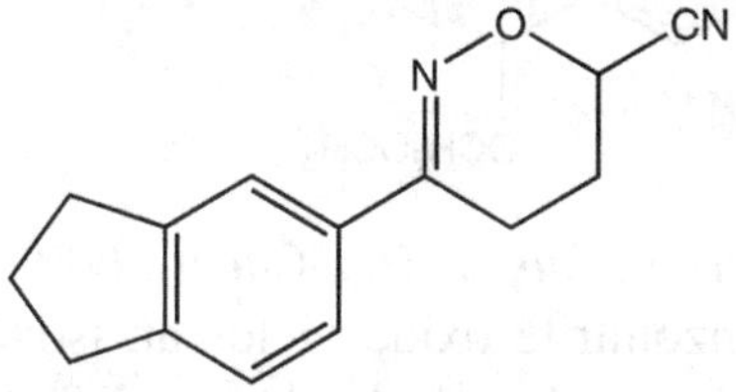

5.13. Y. Watanabe and T. Sakakibara, *Tetrahedron*, **65**, 599-606 (2009). Draw a reaction showing the conversion of ethyl thioxoacetate to the thiine.

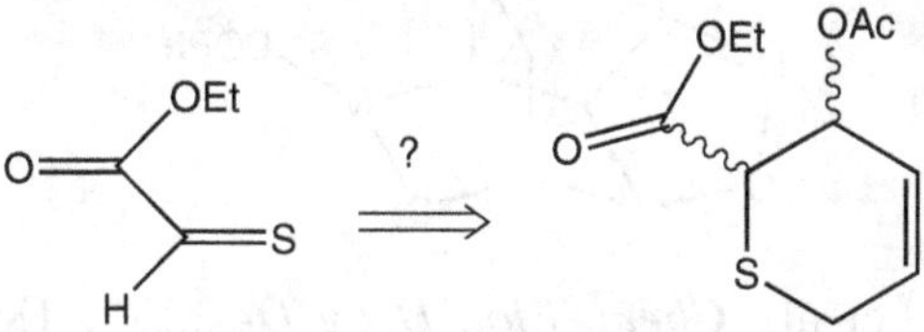

5.14. J. T. Fletcher, et al., *Organometallics*, **27** (21), 5430-5433 (2008). In this approach, the trimethylsilyl protecting group is removed in situ (replaced with a hydrogen) with H_2O/t-BuOH. Write the product.

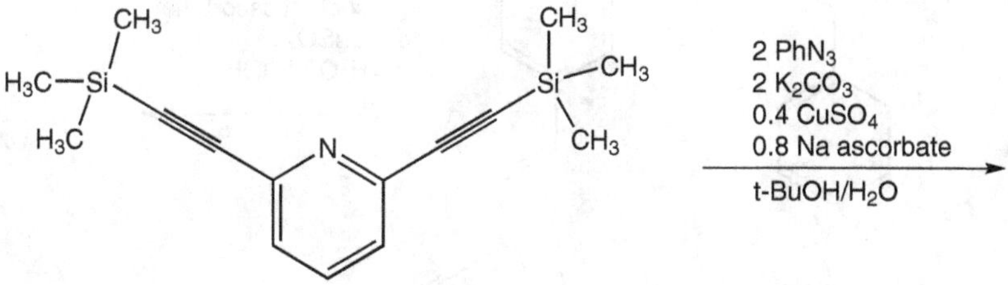

5.15. B. J. D. Wright, et al., *J. Am. Chem. Soc.*, **130** (49), 16786-16790 (2008). Draw the product, used as an intermediate in the synthesis of plurflavin A "Aglycone."

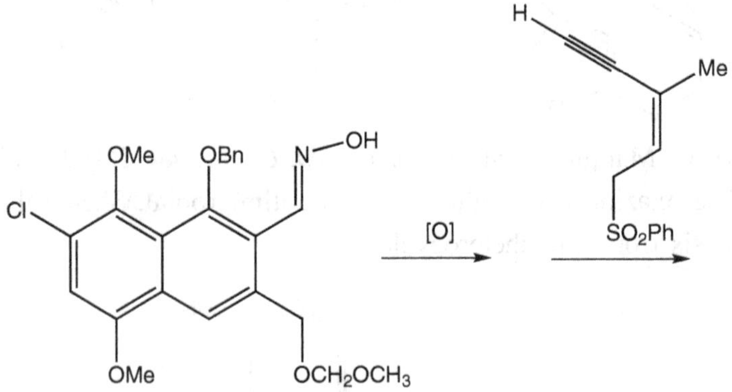

5.16. P. Quadrelli, et al., *Eur. J. Org Chem.*, 6003–6015 (2007). The reaction of benzonitrile oxide yields an isoxazoline shown in a mixture with three other cycloadducts. What is the structure of the starting 1,3-dipolarophile?

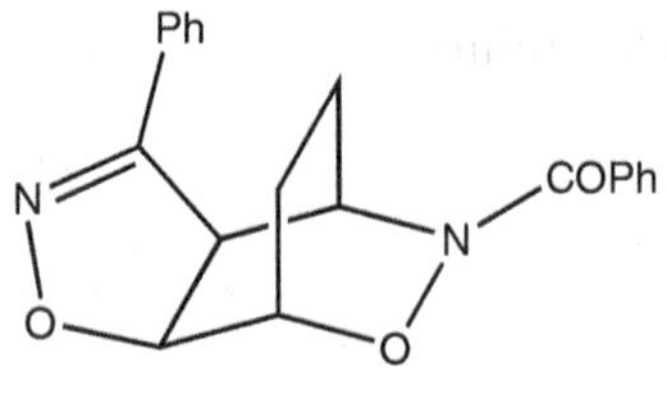

5.17. C. Pardin, et al., *Chem. Biol. Drug Des.*, **72**, 189–196 (2008). The product was evaluated as a tissue transglutaminase inhibitor. Draw the product.

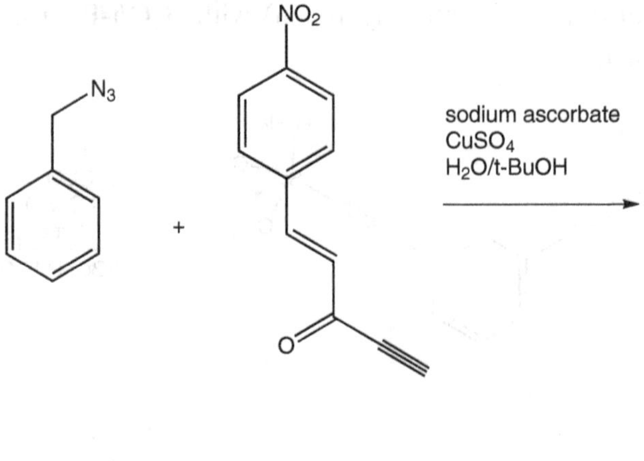

CHAPTER 6

AROMATICITY AND OTHER SPECIAL PROPERTIES OF HETEROCYCLES: PI-DEFICIENT RING SYSTEMS

6.1. GENERAL

When heterocyclic rings have the maximum degree of unsaturation, the properties of aromaticity can be present. This is the case for rings with the Hückel number ($4n + 2$, where n is a small whole number) of pi-electrons. All 6-membered rings where $n = 1$ containing one or more nitrogen or phosphorus atoms meet this specification, although P-containing rings are rare and exhibit special properties. Six-membered rings containing oxygen or sulfur can only have two double bonds and are not aromatic, although as will be discussed, the rings do have some characteristics of aromaticity if the heteroatom is positively charged (as in the oxonium ion). Five-membered rings with N, O, or S have four pi-electrons from the two double bonds and have two more non-bonded pi-electrons on the heteroatom. These rings are also considered to meet the Hückel specification. Larger rings where $n = 2, 3, 4$, etc., can also be considered to possess aromaticity.

Aromaticity can be defined in several ways. We will consider this point in connection with a review of benzene aromaticity in section 6.2. Suffice it to say that aromaticity has a profound effect on the properties of cyclic systems, controlling many aspects of their chemistry. In

Fundamentals of Heterocyclic Chemistry: Importance in Nature and in the Synthesis of Pharmaceuticals, By Louis D. Quin and John A. Tyrell Copyright © 2010 John Wiley & Sons, Inc.

this chapter, we will examine the properties of some fundamental aromatic heterocyclic systems. When rings are saturated or only partly unsaturated, their chemistry is much like that of nonaromatic carbocyclic counterparts. The features of stereochemistry become important in such heterocyclic systems; they will be considered mostly from this standpoint in Chapter 10.

There is a very convenient description of heterocycles as being pi-deficient or pi-excessive. The two types have different chemical reactivity. In a general way, one can predict the properties of a system if it can be recognized as fitting one of these classifications. We will show in a quantitative manner what it means to be pi-deficient or pi-excessive. Basically, the terms imply the level of pi-electron density relative to that of benzene, where every carbon has the density of one pi-electron. Electron-withdrawing heteroatoms decrease the pi-electron density at the carbon atoms and are thus pi-deficient relative to benzene. Heteroatoms using lone pairs as part of a Hückel system increase electron density in the ring relative to benzene, and they give pi-excessive systems. In this chapter, we will consider the pi-deficient systems; in Chapter 7, the pi-excessive systems will be studied. Also included in Chapter 7 will be yet another classification of 5-membered heterocycles called mesoionics, where positive and negative centers are present and no uncharged resonance structures can be written.

6.2. REVIEW OF THE AROMATICITY OF BENZENE

6.2.1. Theoretical Aspects

Developing the concept of aromaticity has had a long history in organic chemistry, and it cannot be considered here. Readers should review the discussions in any organic chemistry textbook. Here, we will pick up some of the main ideas and terminology from these discussions so that they can be applied to heterocycles.

We first need to recognize that aromaticity is not a well-defined physical property; various strategies have been used to measure it from its effects on other properties, and numerous attempts have been made to put numbers to it. The situation has been summarized in a review.[1] For our purposes, we can consider the depiction of aromaticity from three theoretical viewpoints, all of which address the main issue that the pi-electrons of benzene are delocalized and not at all like pi-electrons in isolated double bonds.

6.2.1.1. *The Resonance View.* The theory of resonance was developed in the period 1930–1940; although it is old, it is still a useful tool in organic chemistry, nowhere more so than in heterocyclic chemistry. We will use it extensively in the explanations of reaction mechanisms, spectral effects, and here in the study of aromaticity.

In resonance theory, the electronic structure of benzene can be expressed by two Kekulé formulas as in resonance hybrid **6.1** with the understanding that neither is real, but the electrons are delocalized and shared by the molecule as a whole. Thus there are no single or double bonds in benzene; all bonds have the same experimental length of 1.40 Å, which is between the values for an sp^2-sp^2 single bond (1.46 Å) and a double bond (1.34 Å). The ring is planar with all internal bond angles that of a hexagon, 120°.

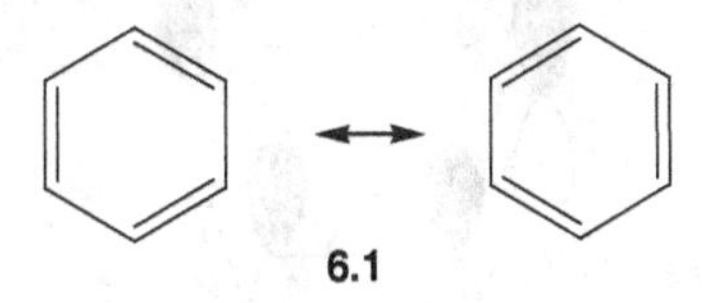

6.1

The presence of resonance (or delocalization) is universally an energy-reducing effect, and this is pronounced in the case of benzene. From experimental measurements of the heat of combustion or heat of hydrogenation, benzene has been found to have 36 kcal/mol less energy than expected for a molecule with three real double bonds. This energy value is commonly referred to as resonance energy, but it needs to be remembered that this is energy that a system *does not* have rather than what it *has*.

6.2.1.2. *The Orbital View.* The hybridization of the carbons in benzene is sp^2 (consistent with the bond angles in a hexagon of 120°). At each carbon, this would leave one electron (called a pi-electron) in a p-orbital above and below the plane established for the ring. This is depicted in Figure 6.1.

Overlap of the p-orbitals occurs to give a region of uniform pi-electron density above and below the ring. The overlapping orbitals are depicted in Figure 6.2, which results in a well-known view of benzene with a doughnut-like description of the overlapped orbitals above and below the ring plane (Figure 6.3).

The orbital view is helpful in understanding the functioning of benzene as a pi-donor in metal coordination and as a nucleophilic center in reactions with electrophiles.

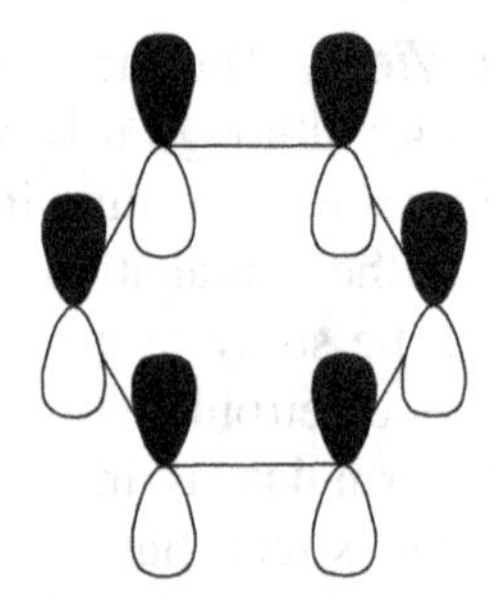

Figure 6.1. Benzene orbitals.

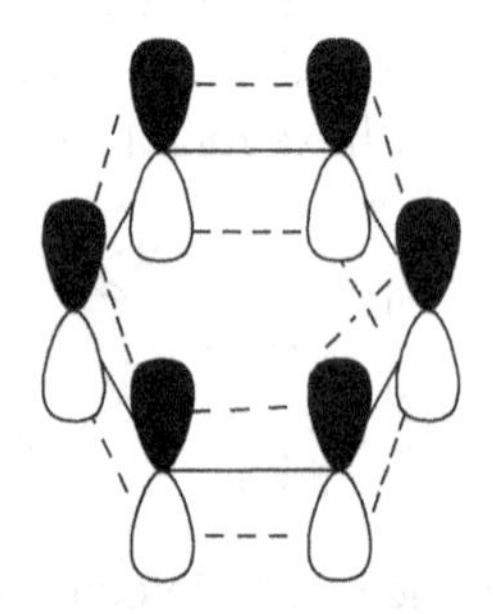

Figure 6.2. Benzene p-orbital interaction.

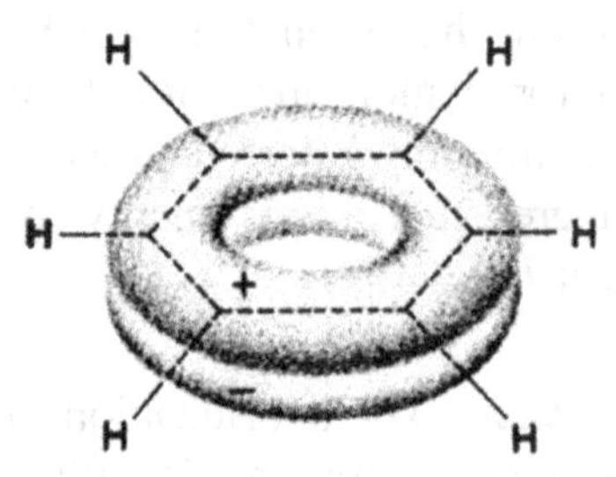

Figure 6.3. Benzene pi-clouds. Reprinted by permission of Wiley-Interscience from J. March, *Advanced Organic Chemistry*, Third Edition, 1985, p. 25, Figure 1a.

6.2.1.3. The Molecular Orbital View. According to molecular orbital (MO) theory as reviewed briefly in Chapter 5, the six atomic orbitals of benzene will form six molecular orbitals. There are three bonding MOs and three antibonding MOs. These are shown with their relative energies and electron distribution in Figure 6.4. Note that in the highest occupied MO (HOMO) (Π_2), there are two filled orbitals of equal energy. These orbitals are said to be "degenerate." Likewise,

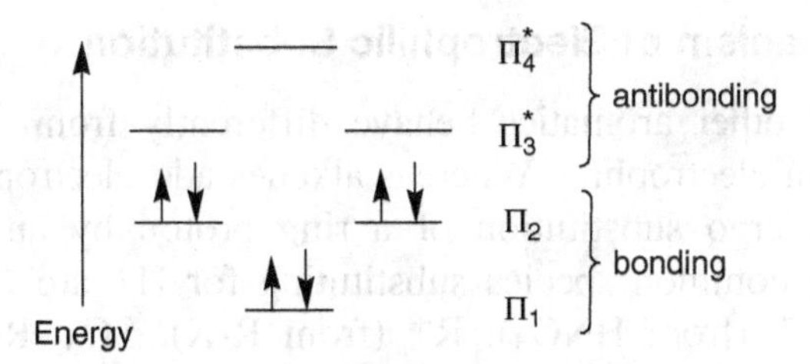

Figure 6.4. Benzene MO energies.

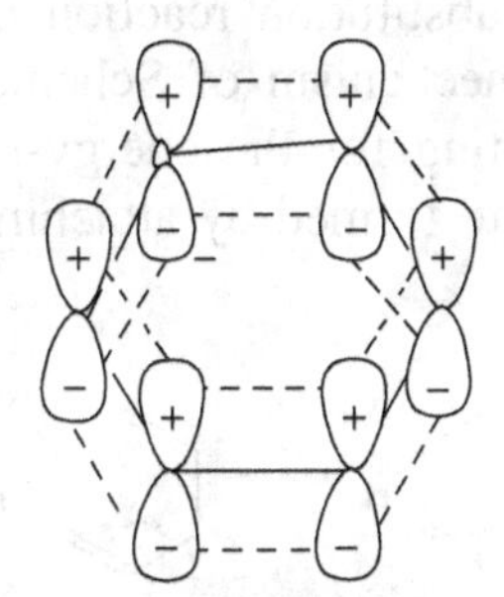

Figure 6.5. Benzene Π_1 MO.

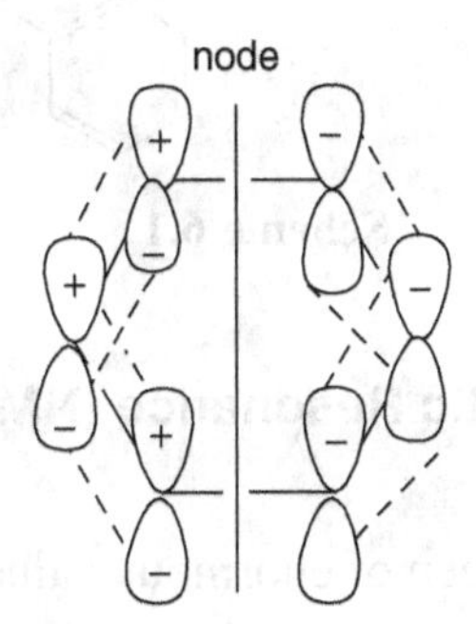

Figure 6.6. Benzene Π_2 MO.

the lowest unoccupied MO (LUMO) (Π_3^*) has two degenerate orbitals. Figure 6.5 shows the bonding orbital (Π_1) of lowest energy, whereas Figure 6.6 shows the HOMO (Π_2). The antibonding orbitals Π_3^* and Π_4^* have two and three nodes, respectively.

From MO theory can come a new definition of aromaticity: In a cyclic, fully unsaturated system, if there are no unfilled bonding molecular orbitals, the system is aromatic, but if these MOs are only partly filled, the system is antiaromatic. We will discuss in section 6.3.4 that this concept also applies to heterocycles.

6.2.2. Mechanism of Electrophilic Substitution

Benzene and other aromatics behave differently from alkenes when attacked by an electrophile. Whereas alkenes add electrophilic species, aromatics undergo substitution of a ring proton by an electrophilic center. Some common species substituting for H^+ are X^+ (from Cl_2 or Br_2), NO_2^+ (from HNO_3), R^+ (from R-X), SO_3, $R(O)C^+$ (from RCOX), etc. Reactions with these species are found also in aromatic heterocyclic chemistry, as will be discussed. Using X^+ as an example of an electrophile, the substitution reaction can be expressed in two steps according to the mechanism of Scheme 6.1. Resonance theory is useful here in accounting for the energy-reducing dispersal of the charge on the intermediate formed by attachment of the electrophile.

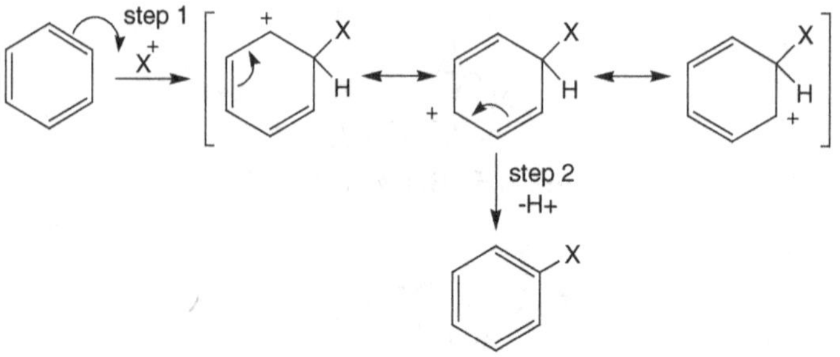

Scheme 6.1

6.2.3. Nuclear Magnetic Resonance (NMR) Spectral Properties

NMR spectroscopy has been of enormous value in the study of aromatic and heterocyclic compounds, and to understand the presentations in this book, it is recommended that the reader review this subject in any organic textbook. Here, we will use some of the ideas and terminology, but we cannot go far into the basics of NMR.

Protons in general have signals in the downfield (deshielded) direction from the reference tetramethylsilane (TMS, Me_4Si) taken as 0 ppm. Deshielding is a tiny effect, just several parts per million (ppm) of the applied magnetic field, but instruments can detect it. The normal degree of deshielding is 0–10 ppm but with exceptions. In saturated compounds, the deshielding is caused by the circulation of the sigma electrons in the applied magnetic field, which sets up a secondary magnetic field reinforcing the applied field. The magnitude of the secondary field is related to structural effects; these effects include, among others, electron density on the carbon bearing the proton as influenced by electron attracting substituents (most of the common heteroatoms) within 1–3 bonds, and distance- and angle-related anisotropic effects coming primarily from circulation of pi-electrons above and below the plane of double bonds and aromatic rings. The effect from pi-electrons is of greater strength than the effect from substituents; protons on double bonds are generally found at δ 5.0–5.5, whereas electron withdrawing substituents generally cause downfield shifts not exceeding about 3 ppm from the signal for alkane C-H at about δ 1. Coming now to protons attached to carbons of the benzene ring, we find deshielding taking place from the secondary magnetic field arising from circulation of the pi-electrons; this field reinforces the applied magnetic field in the region of the aromatic C-H bonds, so that the field that must be applied to achieve the resonance condition is diminished. The aromatic pi-electrons are considered to establish a current around the ring as they circulate in the applied field. This so-called "ring current" is more effective than the double bond effect in establishing a secondary magnetic field. The secondary magnetic field also has a region in the center of the ring, and above and below it, where the opposite is true, thus causing shielding of a proton and an upfield shift in these locations. All the protons on benzene are of course identical, and its proton NMR spectrum consists of a single peak at δ 7.37. It is easy, therefore, to distinguish aromatic proton signals from those of alkenes. Electron-withdrawing substituents on the ring reduce the electron density on ortho and para protons, which constitutes a deshielding effect. Thus, the spectra of most such compounds have a complex of overlapping signals from about δ 7.5 to 8.0. The ortho and para protons are the more downfield, with meta relatively upfield. This difference is explained readily by examining the resonance structures for benzaldehyde as a typical case; the partial positive charge can only appear in the ring at the ortho and para positions (Scheme 6.2).

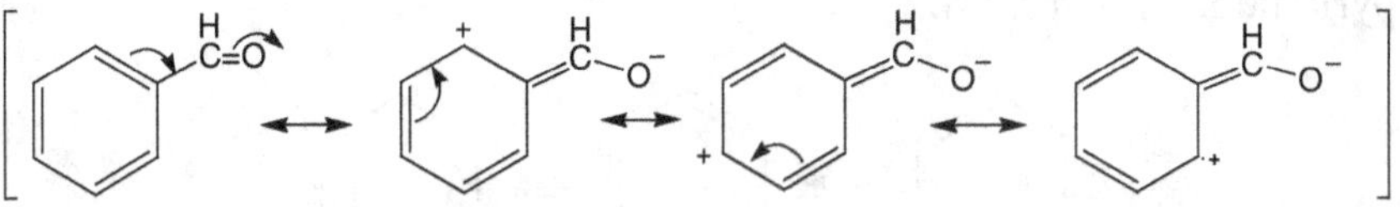

Scheme 6.2

Just like ^{1}H, the ^{13}C isotope has a quantum spin number (I) of 1/2 and thus has a low- and a high-energy spin state. When decoupled from protons, each C in a molecule will give a single, well-resolved peak in the NMR experiment. Because of the low natural abundance of the ^{13}C isotope (1.1%), the spectra are obtained by the Fourier transform technique employing many scans of the sample. Shifts are usually measured from TMS; the range of shifts is more than 200 ppm, with almost all signals appearing downfield of TMS. Saturated carbons are generally in the range δ 0–50, whereas sp^2 carbons are roughly at δ 100–200. The carbons of benzene are found at δ 128.5; these are close to the signals from alkenes, and thus here there are no special effects coming from the aromatic ring current. Substituents on the ring frequently cause large shift differences mostly from resonance effects, downfield if electron withdrawing, upfield if electron releasing. ^{13}C NMR is particularly useful with aromatic and heterocyclic compounds, as will be discussed. One important difference between ^{1}H and ^{13}C NMR is that increased positivity at carbon, whether from resonance or induction, acts to contract the p-orbitals at C, which is a deshielding effect.

6.3. PI-DEFICIENT AROMATIC HETEROCYCLES

6.3.1. A Study of Pyridine as an Illustrative Model

6.3.1.1. Some Physical Properties of Pyridine. Pyridine is a liquid with a rather low boiling point of 115°C. It is infinitely miscible with water, attesting to its polar character from inductive and resonance effects (section 6.3.1.2). Pyridine has a strong odor, which most find to be obnoxious.

6.3.1.2. Electronic Structure of Pyridine. The presence of the electron-withdrawing nitrogen atom in pyridine has a profound effect on its properties. The resonance picture of benzene can be applied to pyridine as in hybrid **6.2**.

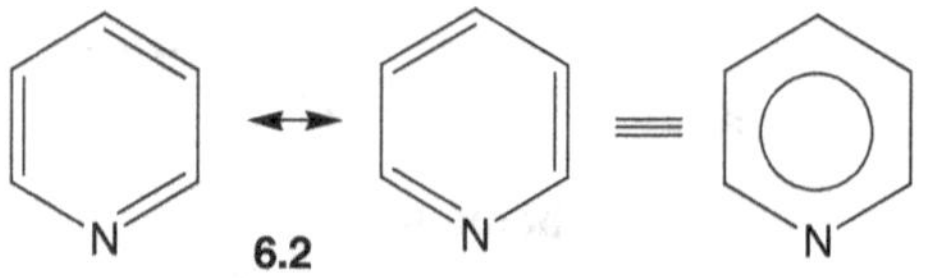

N ↔ N ≡ N

6.2

However, the electronegative nitrogen atom causes significant polarization of the molecule, as depicted in structure **6.3**.

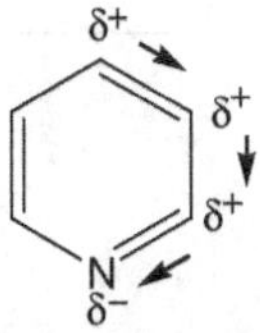

6.3

Also, additional resonance structures can be shown for pyridine, in which an electron pair from an attached double bond can be placed on nitrogen (Scheme 6.3).

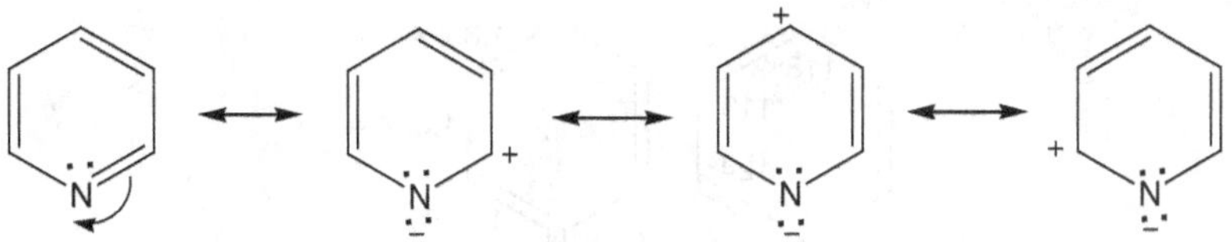

Scheme 6.3

This form has divalent but negatively charged N, which is a stable condition for N. The positive charge is dispersed to carbons around the ring, specifically to C-2 and C-4 (but not to C-3). The net effect is to reduce the pi-electron density in the ring relative to benzene, and this is the source for the description of pyridine as being pi-deficient. With benzene having the pi-electron density of 1.0 at all positions, calculations assign the densities as follows: N, 1.166; C-2, 0.866; C-3, 1.064; and C-4, 0.932. This effect is true of all 6-membered aromatic rings containing a C=N unit. As will be shown, the heteroatom in 5-membered rings does just the opposite, feeding pi-electron density into the ring, thus becoming pi-excessive. These effects are not trivial; there is a vast difference between the properties of the two types of rings. The resonance energy of pyridine has been found experimentally to fall in the range of 21–43 kcal/mol, and thus it is much like that of benzene. The range is wide because of experimental difficulties in performing the traditional experiments on heterocycles to obtain this value.

The molecular orbital picture of pyridine is shown in Figure 6.7. There is a difference between this picture and that for benzene. In benzene, the HOMO consists of two filled orbitals of equal energy (hence degenerate), but in pyridine there is only one filled HOMO. Importantly, none of the orbitals are degenerate.

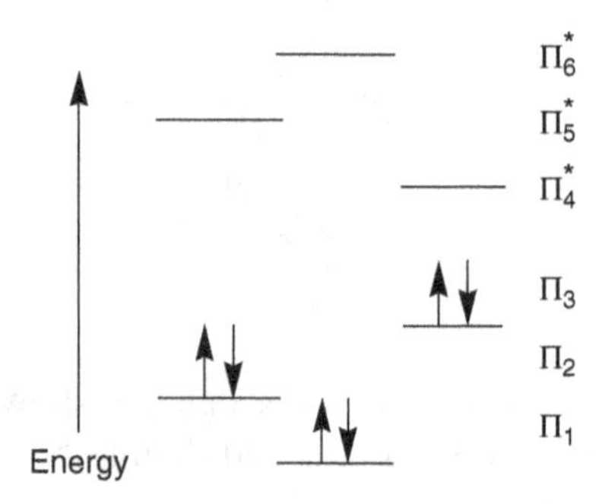

Figure 6.7. Pyridine MOs.

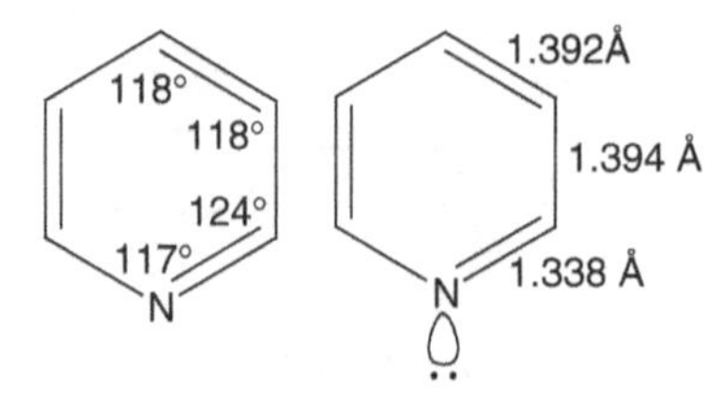

Figure 6.8. Pyridine bond angles and lengths.

The combined action of electron induction and resonance delocalization makes pyridine a molecule of considerable polarity. It has a dipole moment of 2.20 Debye (D) units. This value might be compared with that of piperidine (1.57 D), whose polarity is controlled only by induction. Pyridine finds many uses as a nonprotonic polar solvent.

6.3.1.3. Geometry of Pyridine. Because N is sp^2 hybridized and planar, the entire pyridine molecule is planar, and this is true for other aromatic heterocycles with more than one C=N unit. The C to N bonds are shorter than the C to C bonds of benzene; this causes the angles inside the ring (Figure 6.8) to be modified from the 120° observed for the perfect hexagon of benzene. The lone pair orbital on nitrogen falls in the plane of the ring, which makes it available for bonding to electrophiles or coordination to metal ions.

6.3.1.4. Special NMR Properties of Pyridine. Just as for benzene, the aromatic ring current causes deshielding at all protons attached to the ring carbons. However, the inductive effect of nitrogen (structure **6.3**) causes electron withdrawal from all carbons, but it diminishes as the number of bonds through which it must pass increases, resulting in additional deshielding especially at C-2,6 attached to N. Acting

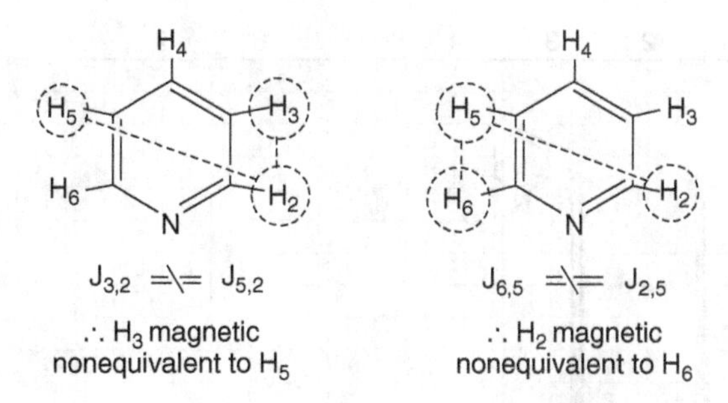

Figure 6.9. Test for magnetic nonequivalence.

in the same direction of electron withdrawal is the resonance effect (Scheme 6.3), which is selectively greater at C-2,6 and C-4. These effects combine to cause pronounced deshielding at C-2,6 (δ 8.52) relative to the shift for benzene (δ 7.37), and C-4 also shows some deshielding (δ 7.55). The consequence is that protons on the three ring positions are chemically nonequivalent, and this allows coupling between them. If that were all that was involved, the proton NMR spectrum would consist of three first-order multiplets. But the situation is more complicated than that. The protons at C-3 and C-5 are of course chemically equivalent, but they have a magnetic difference and are said to be magnetically nonequivalent: Each has a different coupling constant (J) to HC-2 (e.g., through three bonds for HC-3 and five bonds for HC-6), as well as to HC-5. The same analysis is true of HC-2 and HC-6. These relations are shown diagrammatically in Figure 6.9.

A spectrum where this effect is present is called a second-order spectrum. This gives rise to a complex spectrum, where much signal splitting is observed and a simple interpretation is not possible. The spectrum for pyridine is shown in Figure 6.10, and it is obvious that there is much more splitting than what would be expected for a first-order spectrum. The spectrum for pyridine can be described as of the AA′MM′X type. This terminology suggests that the shifts for nuclei A, M, and X are not close together and that there are two nuclei A that are magnetically nonequivalent, as are two nuclei M. Magnetic nonequivalence is not a rare occurrence; all proton NMR spectra of monosubstituted benzenes, 4- substituted pyridines, pyrroles, and even some noncyclic compounds can exhibit second-order spectra for the same reason. Protons on the ring carbons of other aromatic heterocycles with one or more C=N unit also resonate at a low field.

The carbon-13 NMR spectrum of pyridine of course also has three signals for the different carbons, but these show no coupling to each

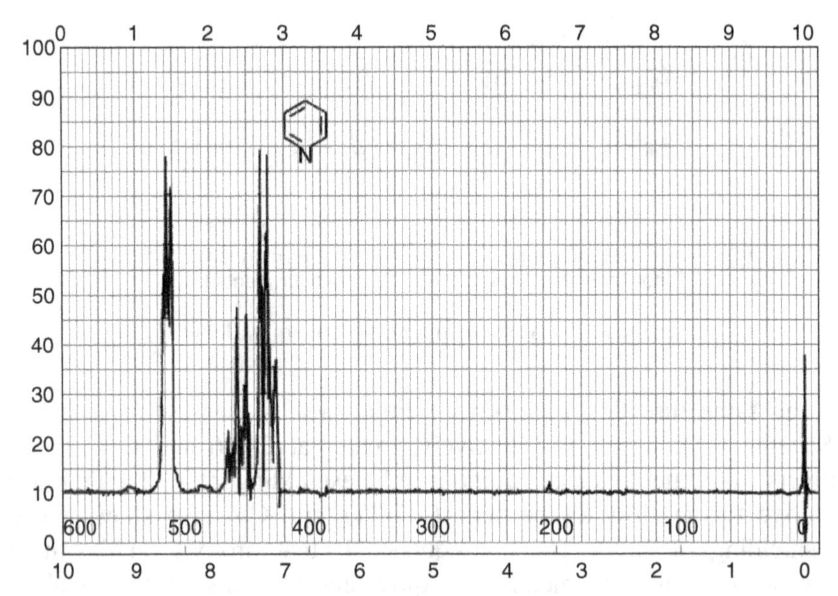

Figure 6.10. Proton MMR spectrum of pyridine. *Reprinted by permission of the Chemical Company from *Aldrich Catalog of NMR Spectra*, Vol. 2, p. 611A (1983).

other because the natural abundance of the ^{13}C isotope is so low (1.1%). However, the coupling is present in samples that have been enriched in this isotope. The ^{13}C signals at natural abundance are coupled to ring protons. But to simplify the spectra, this coupling is generally eliminated experimentally; such spectra are called proton decoupled, and each different ^{13}C then gives a separate sharp signal. The signals are well differentiated (Figure 6.11).

The combined inductive and resonance effects that cause increased positivity at the ring carbons act to contract the p-orbitals at C, which is a deshielding effect. These effects are striking in causing downfield shifts at both saturated and unsaturated carbon. Relative to benzene at δ 128, strong deshielding is observed at C-2,6 (δ 150.6) and at C-4

δ 136.4
C
C δ 124.5
C δ 150.6
N

Figure 6.11. ^{13}C NMR of pyridine.

(δ 136.4). In ^{1}H NMR, the explanation for deshielding at saturated C-H is different and involves the secondary magnetic fields set up when a magnetic field is applied to the system. This anisotropic effect is weak for ^{13}C and indeed for all magnetically active nuclei other than hydrogen, and the shifts are controlled instead by the p-orbital contraction effect. The aromatic ring current effect so important in ^{1}H NMR is also weak at carbons of heterocycles, and it is not considered to be a major factor. As noted, the carbons of benzene and the carbons of C-C double bonds give similar signals, all of which are derived from sp^2 carbons. In substituted pyridines, electron-withdrawing groups greatly increase the deshielding at C-2,6 and C-4, whereas groups that release electrons by the resonance effect cause substantial upfield shifts.

Nitrogen NMR can also be a valuable technique for characterizing heterocycles. This can be based on two different isotopes of nitrogen. The ^{14}N isotope is of course the dominant one in natural samples, but it has a spin quantum number of 1 and is quadrupolar (that is, it has four spin states). It gives rather broad NMR signals, but nevertheless these can be measured fairly easily. The range of nitrogen NMR shifts (usually expressed from ammonia as an upfield standard at 0 ppm) is more than 400 ppm, and the shifts are indicative of the structure at nitrogen. Saturated amines have signals in the region δ 0 to 60, whereas C=N units have far downfield shifts (e.g., $Ph_2C{=}NH$, δ 308). Note the similarity to ^{13}C NMR; the hybridization of the nucleus establishes a general position for the signal which then is subject to the shifts from groups attached to the atom. Thus, pyridine gives a signal at δ 315. The other NMR active isotope of nitrogen is ^{15}N. This isotope has a spin quantum number of 1/2 as do both the proton and the ^{13}C isotope (and also ^{31}P), and it gives a sharp signal at the same shift value as would be found for the ^{15}N isotope. However, ^{15}N has a natural abundance of only 0.365% and requires special techniques to obtain a good signal. The spectra must be run with proton decoupling [which as for ^{13}C NMR helps to sharpen the signal and enhances its size by the nuclear Overhauser effect (NOE)] using the Fourier transform technique of pulsed irradiation and the accumulation of the individual scans into a single spectrum. ^{15}N NMR spectroscopy has moved into an important position for the characterization of nitrogen heterocycles.

6.3.1.5. Other Spectral Properties of Pyridine. Also used in characterizing heterocycles are ultraviolet (UV) and infrared (IR) spectroscopy. The UV spectrum of pyridine is much like that of benzene in the normal range; it has a maximum at 251 nm (log ε 3.30) with a shoulder at 270 nm. Other aromatic heterocyclic rings can also be characterized by this technique, which historically was the first instrumental technique to attain prominence in structure characterization. Of course, it is far surpassed in usefulness by the NMR techniques, but it still finds use in special cases for structure elucidation. Its great sensitivity makes it also valuable for quantitative analysis. IR spectroscopy is used routinely to characterize heterocyclic compounds and to recognize or confirm the functional groups present. Mass spectroscopy is also of great importance in characterizing heterocycles. The parent aromatic rings generally give strong molecular ion signals (M^+); substituted heterocycles give characteristic fragmentation patterns, and much can be learned about the molecular structure from a study of the fragmentation of ions formed in the process.

6.3.1.6. Pyridine as a Base. The lone electron pair on nitrogen of C=N units is available for reaction with protic acids as well as Lewis acids, and this is certainly true for pyridine and other heterocycles. Pyridine is a weak base and is widely used as a proton acceptor; it has a K_b of 1.4×10^{-9} (pK_b 8.8; pK_a for the conjugate acid 5.2).This is much lower basicity than that of a saturated tertiary amine; triethylamine has K_b 5.6×10^{-4} (pK_b 3.2; pK_a for the conjugate acid 10.8). This is primarily because of a hybridization effect; with sp^2 hybridization (meaning 33% s-character in the orbitals) for C=N rather than sp^3 (25% s-character) for tertiary amines, the lone pair orbital is contracted toward the nucleus and is less available for bonding. Alkyl groups on the pyridine ring acting by induction and hyperconjugation increase electron density on N, and the three isomers of methyl pyridine (called alpha, beta, and gamma picolines) have greater basicity than does pyridine by 0.6–0.8 pK_b units. These too are useful as weak bases, as are some of the dimethylpyridines (lutidines) and 2,4,6-trimethylpyridine (collidine). Electron-withdrawing groups have a much more pronounced effect in the opposite direction; thus, a nitro group at the 2-position reduces the basicity of pyridine by 7.8 pk_b units. This is partly inductive but primarily a pronounced resonance effect.

Pyridine reacts with simple protic acids to form crystalline salts. Protonated pyridine is referred to as the pyridinium ion, which is shown in Scheme 6.4 from a reaction with hydrochloric acid. The pi-system of the ring is not affected by attachment of groups to nitrogen, and the aromaticity remains high.

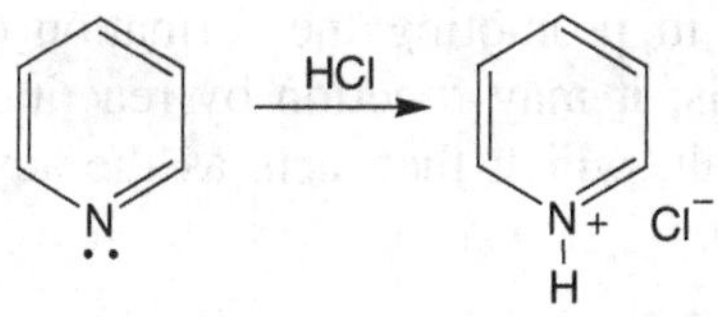

Scheme 6.4

Lewis acids such as BF_3, SO_3, $AlCl_3$, etc., readily react with pyridine to form Lewis salts (Scheme 6.5), and many coordination complexes have been prepared with metallic ions.

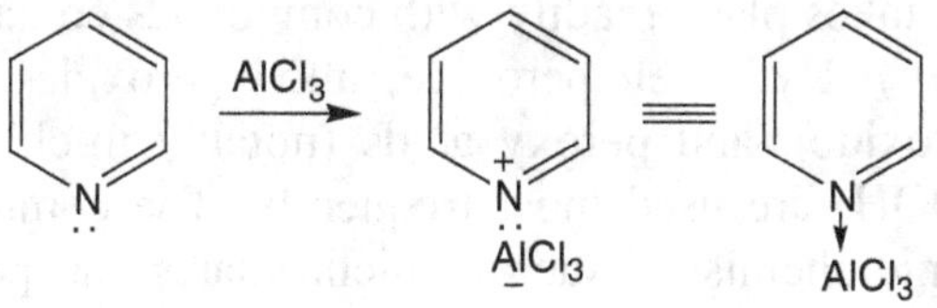

Scheme 6.5

6.3.1.7. Other Reactions Involving the Lone Electron Pair. Two reactions at nitrogen are of great importance in the chemistry of pi-deficient systems based on this element: quaternization and oxidation. Both, of course, are well known among nonaromatic amines. Pi-excessive nitrogen heterocycles generally do not participate in these reactions, and we have here one of the great dividing points between the two families of heterocycles.

Quaternization involves the reaction of a tertiary amine with an alkylating agent. Typically, primary and some secondary alkyl halides are used for this purpose, although alkyl sulfates have also been used. Tertiary halides are not useful because they undergo elimination rather than substitution. The reaction, which is an example of an S_N2 process, proceeds readily to give crystalline, stable, and still-aromatic quaternary salts (Scheme 6.6). These have been used for many years to characterize amines. Acyl halides also react in a similar manner with pyridine, but the resulting salts are unstable and generally not isolated.

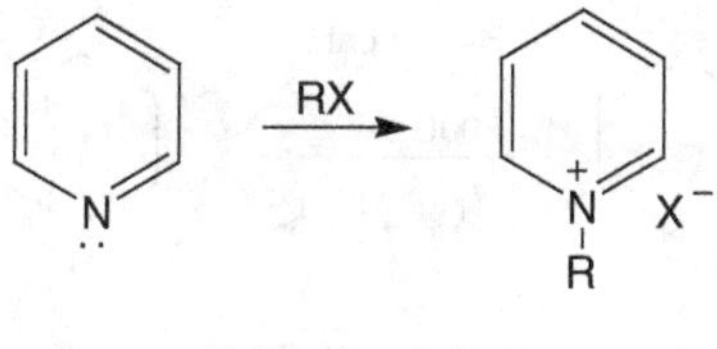

Scheme 6.6

Pyridine is useful in promoting the formation of esters from acyl chlorides and alcohols; it may function by reacting first with the acyl chloride to form a salt, which then acts as the acylating agent to the alcohol (Scheme 6.7).

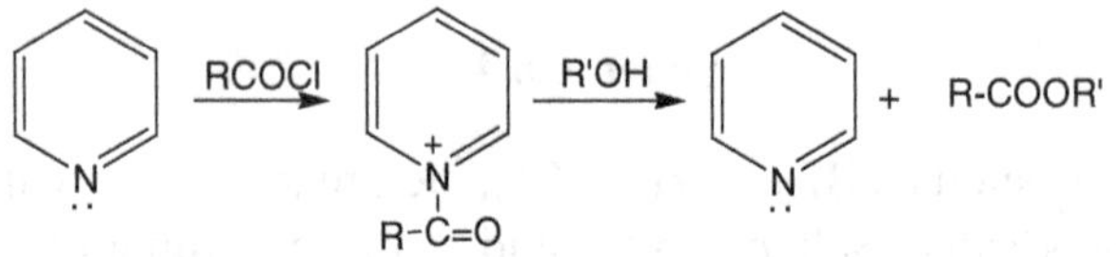

RCOCl R'OH + R-COOR' R-C=O

Scheme 6.7

N-Oxidation takes place readily with compounds containing the peroxy link (-O-O-). Hydrogen peroxide, alkyl peroxides (especially t-butyl hydroperoxide), and peroxy acids (notably m-chloroperbenzoic acid, ArC(O)OOH) are used most frequently. The common oxidizing agents of organic chemistry, such as dichromates and permanganates, have no effect on pyridine. In fact, a pyridine-chromic oxide mixture is used for the oxidation of alcohols to carbonyl compounds. Pyridine N-oxide and related compounds retain aromatic character but have some valuable properties, which will be discussed in section 6.3.1.12.

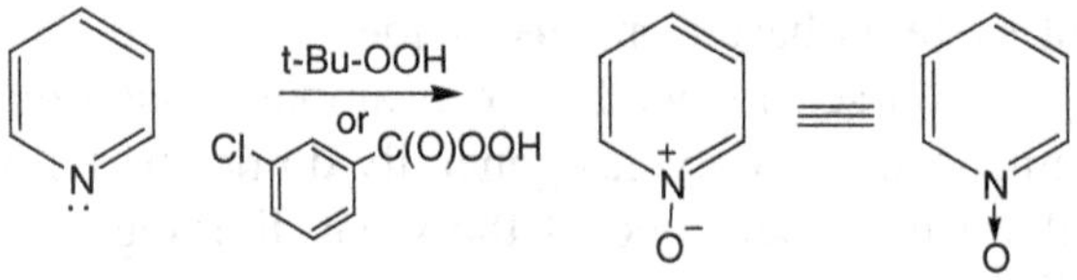

t-Bu-OOH or Cl C(O)OOH

Scheme 6.8

6.3.1.8. Reduction of Pyridine. Highly resonance-stabilized aromatic heterocycles are, like benzene, resistant to catalytic hydrogenation. The conditions for reducing pyridine to piperidine shown in Scheme 6.9 suggest this. However, chemical reduction, as with sodium and alcohol, is a much easier process and accomplishes the same goal under much milder conditions. This type of reduction is referred to as "dissolving metal reduction" and proceeds with free-radical intermediates.

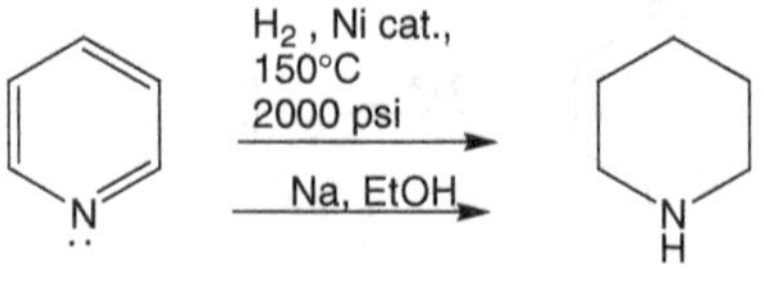

H_2, Ni cat., 150°C 2000 psi

Na, EtOH

Scheme 6.9

Hydrides such as $LiAlH_4$ and $NaBH_4$ can be used to synthesize partially reduced pyridines. The presence of substituents on the pyridine ring can greatly modify the conditions needed for hydrogenation. To illustrate, nicotinamide can be reduced readily, with the process having a clean stopping point at the tetrahydro stage (Scheme 6.10).

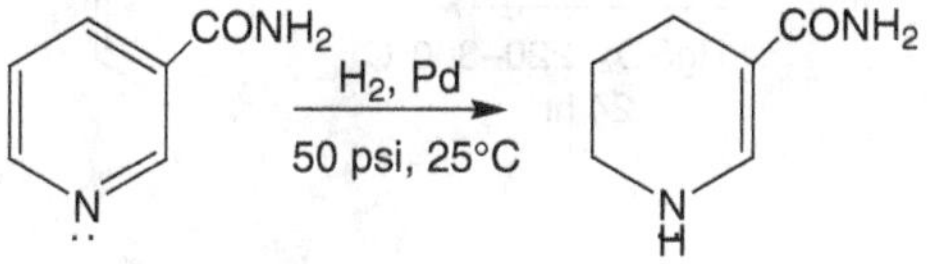

$CONH_2$ — H_2, Pd, 50 psi, 25°C → $CONH_2$, N–H

Scheme 6.10

6.3.1.9. Electrophilic Substitution of Pyridine. One hallmark of pi-deficient heterocyclic systems is their low reactivity with electrophilic agents. For example, pyridine is less reactive than benzene by a factor of 10^6 when subjected to conditions of nitration. The reactivity is on the order of that of nitrobenzene, which is well known to require much more drastic conditions than those for benzene itself. The general mechanism of electrophilic substitution reviewed in section 6.2.2 applies to the case of heterocyclic systems. A major part of the explanation for this low reactivity is that pyridine presents much lower pi-electron density to the attacking positive species. Consistent with this thought is that pi-excessive rings are much *more* reactive than benzene. Furthermore, protic reagents such as nitric and sulfuric acid probably perform protonation on nitrogen, which means that the actual species being attacked may be the pyridinium ion, with a still-aromatic ring system but subsequently reduced in pi-electron density by the induction arising from the positive nitrogen. However, this explanation does not address the fact that the three possible positions for attack on pyridine have greatly different reaction rates in electrophilic substitutions, most notably that the 3-position is much more favored than the 2,6- or 4-positions. This is exactly the situation found in nitrobenzene and other benzene derivatives with electron-withdrawing groups, which are all classed as meta-directing groups, and in a manner of speaking it can be said that the C=N unit of pyridine acts as a built-in meta director. The theory of resonance provides an explanation for the directive influences in the two families of compounds. But first, examine the experimental results of a few of the common electrophilic reactions to appreciate just how drastic is the reduction of reactivity in pyridine. Of course, this reactivity effect applies to all pi-deficient systems and is even more pronounced when additional C=N units are present in the ring.

Sulfonation.

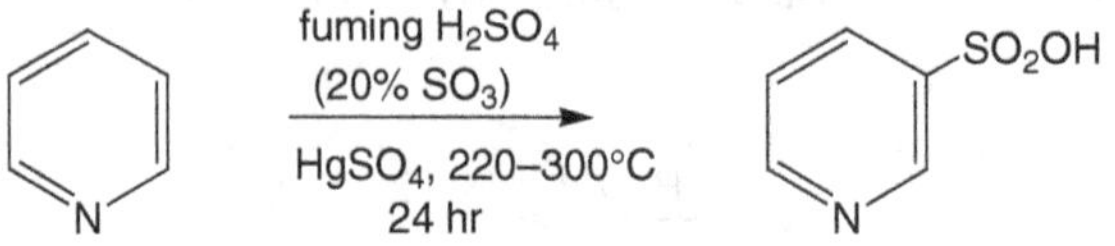

fuming H_2SO_4 (20% SO_3), $HgSO_4$, 220–300°C, 24 hr → SO_2OH

Nitration.

KNO_3, 100% H_2SO_4, 300°C → NO_2

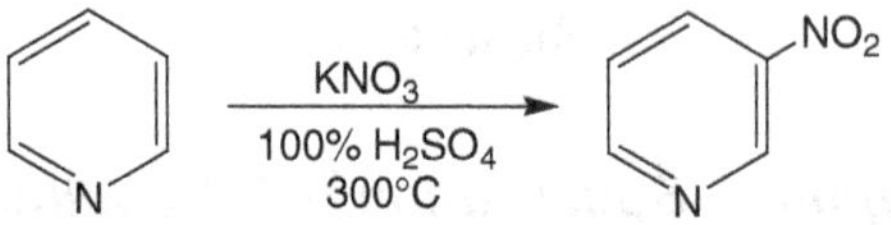

Friedel–Crafts Alkylation and Acylation. Just as for nitrobenzene, these reactions cannot be accomplished

Halogenation. With two moles of aluminum chloride (one that complexes on nitrogen and the other to activate the halogen) and a reaction temperature of 80–115°C, pyridine can be chlorinated at the 3-position. These conditions also give some 3,5-dichloropyridine. Bromination can be accomplished in the gas phase at high temperatures, but it gives the 2-bromo and 2,6-dibromo derivatives. These conditions suggest that a free radical mechanism is operative.

The role of resonance in the orientation of electrophilic substitution can be understood by examining the various hybrids that result from attack of a cationic species (E^+) at the three possible ring positions.

Attack at the 2-position

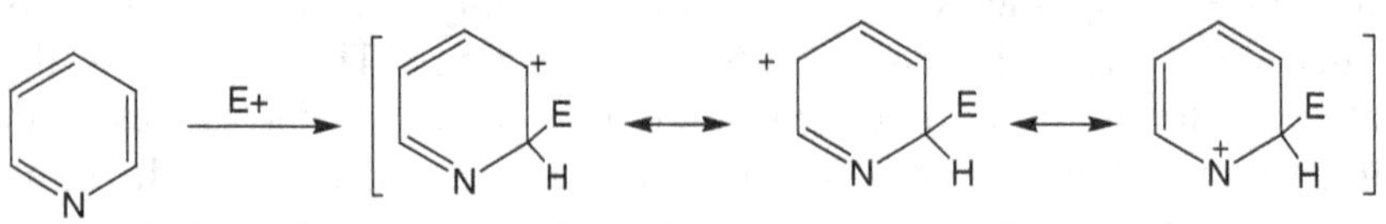

E+

Scheme 6.11

The third canonical structure has a positive charge on nitrogen that has only six electrons from the lone pair and two bonds. This is a high-energy form of nitrogen and is strongly disfavored.

Attack at the 4-position (Scheme 6.12)

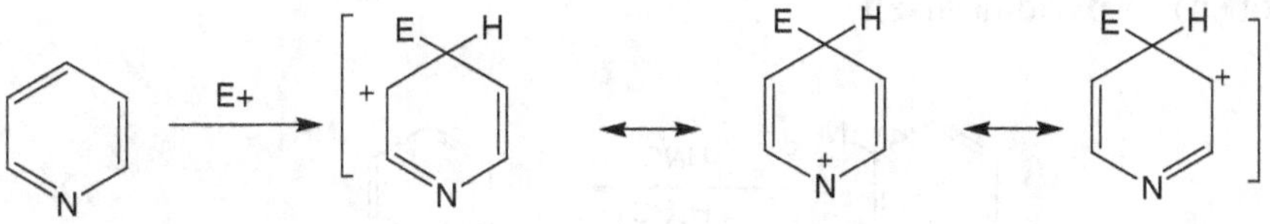

Scheme 6.12

The second canonical form has the same high-energy feature of positive N with 6 electrons as viewed in attack at the 2-position and is disfavored.

Attack at the 3-position (Scheme 6.13)

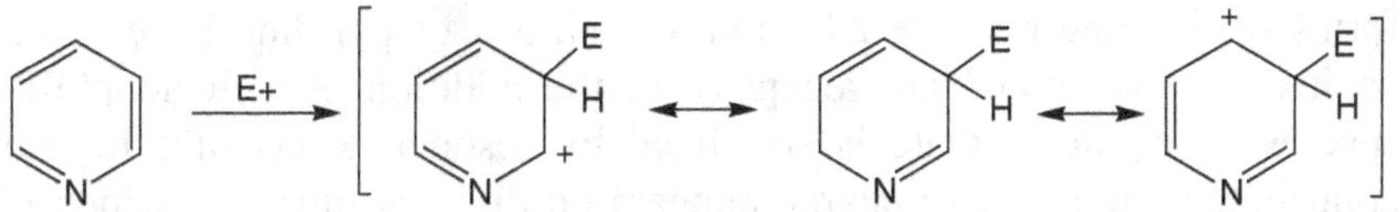

Scheme 6.13

There are no disfavored forms in this hybrid. It is of relatively lower energy than that of the other two hybrids, and this accounts for the fact that electrophilic attack occurs at the 3-position. As for nitrobenzene, the directive effect is not a matter of *activation* at the preferred position but of *less deactivation* relative to the other choices. Resonance also explains in part the reduction in reactivity of the pyridinium ion. Here, nitrogen will bear a full positive charge; in any of the three resonance structures from attack of an electrophile, there will be a positive charge either on the carbons adjacent to the protonated N or indeed even on this nitrogen; any of these, because of charge repulsion, will increase the energy level of the reaction intermediate.

Just as is true in benzene chemistry, the placement of electron-withdrawing groups on the pyridine ring will reduce the reactivity, whereas electron-releasing groups will greatly *increase* the reactivity, through stabilization of the positive reaction intermediate. Such groups, notably amino, alkoxy, and to a lesser extent alkyl, also control the orientation of the substitution by an electrophile. Both effects show up in the conditions and orientation in the nitration of 3-aminopyridine (Scheme 6.14); the reaction occurs under mild conditions, with the electrophile attacking the 2- position, ortho to amino, rather than the open 5-position, which would be favored (like the 3-position were it open) in pyridine itself.

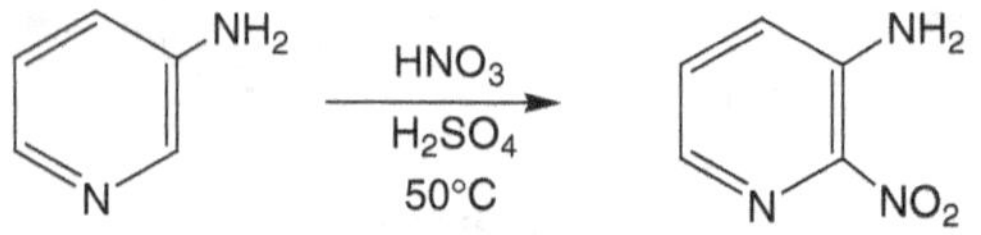

Scheme 6.14

6.3.1.10. Nucleophilic Substitution of Pyridine. The electron-withdrawing effect of the C=N unit has the opposite influence on the attack of a nucleophile; the reaction rates are greatly increased, and the nucleophilic attack occurs at the 2- or 4-positions. This orientation is easily understood with resonance theory. As shown in the resonance forms of Scheme 6.3, the 2,6- and 4-positions of pyridine have some positive character and are receptive to the addition of a nucleophilic species. The intermediate is stabilized by resonance, because in one contributor, the negative charge appears on divalent nitrogen, which is energetically favored. After attachment of the nucleophile, the hydrogen originally on the carbon of attachment must be eliminated as the hydride ion. The reaction sequence is shown as Scheme 6.15. Here, the reaction takes place in liquid ammonia as solvent; the hydride ion reacts with this compound to generate NH_2^- and H_2.

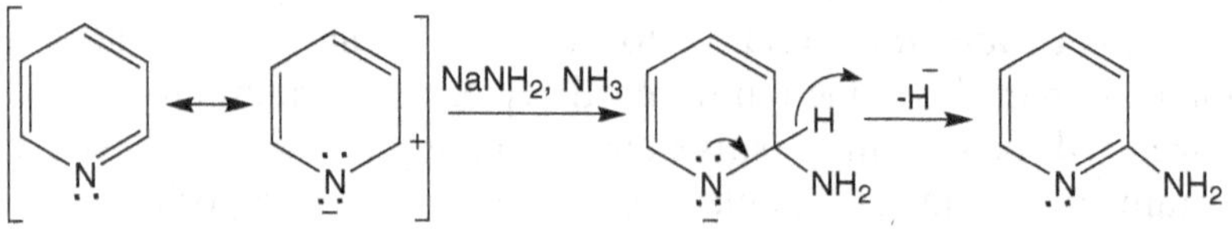

Scheme 6.15

The process of Scheme 6.15 is important in pyridine chemistry and indeed throughout the family of pi-deficient heterocycles, as amino derivatives are valuable. It is named the Chichibabin reaction after its discoverer. The reaction with pyridine is specific for the 2-position, and the formation of the isomeric 4-aminopyridine is insignificant.

The hydroxide ion also is an effective nucleophile toward pyridine (Scheme 6.16). However, the final product does not have the structure of a hydroxypyridine (**6.4**), but instead it has the structure of the tautomeric keto form (**6.5**). This compound is referred to as 2-pyridone. Nitrogen thus acquires the proton and now can be considered to be present as an amide link in the heterocycle.

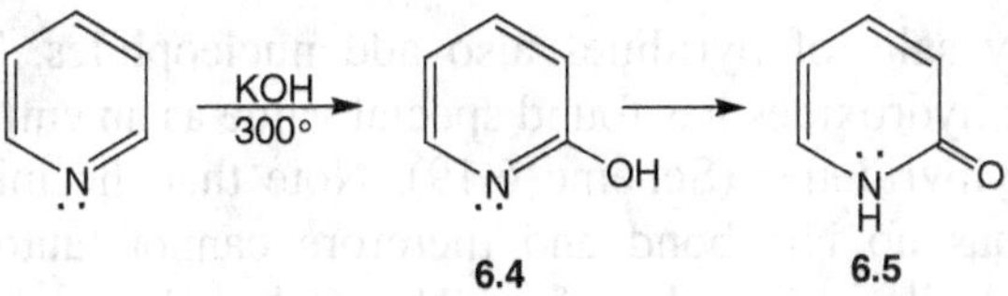

KOH
300°
OH
6.4
6.5

Scheme 6.16

This tautomerism of alpha-hydroxy groups (and also gamma-hydroxy but not beta-hydroxy) is found in all pi-deficient nitrogen heterocycles and plays an important role in their chemistry. It is discussed in more detail in section 6.3.5.

On the surface, it might seem that the resonance stabilization of pyridine is lost on conversion to a pyridone. However, as is true for all amides, pyridones can be expressed in a resonance form, and as shown in hybrid **6.6**, it retains the six pi-electron system of an aromatic species.

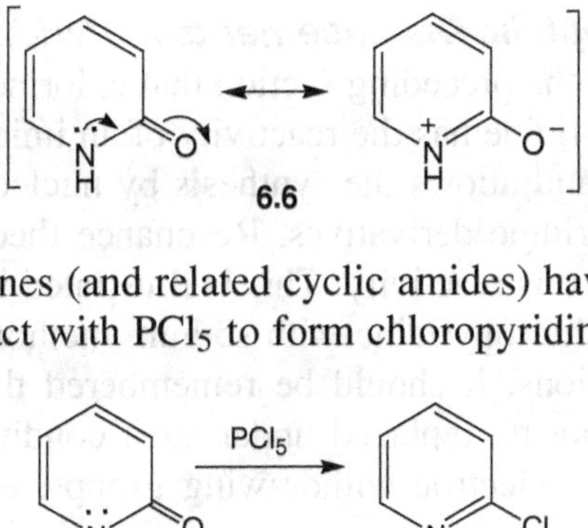

6.6

2- and 4-Pyridones (and related cyclic amides) have a most valuable property: They react with PCl_5 to form chloropyridines (Scheme 6.17).

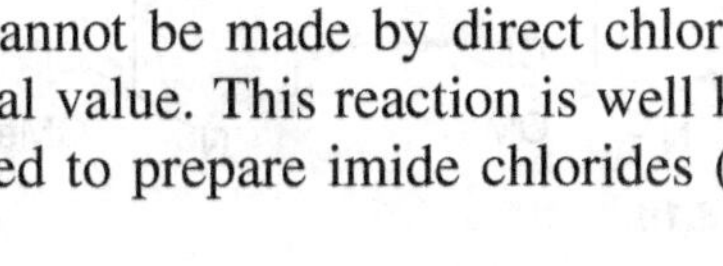

PCl_5

Scheme 6.17

These compounds cannot be made by direct chlorination and thus the process is of practical value. This reaction is well known for noncyclic amides, and it is used to prepare imide chlorides (**6.7**) and from them imino-esters (**6.8**).

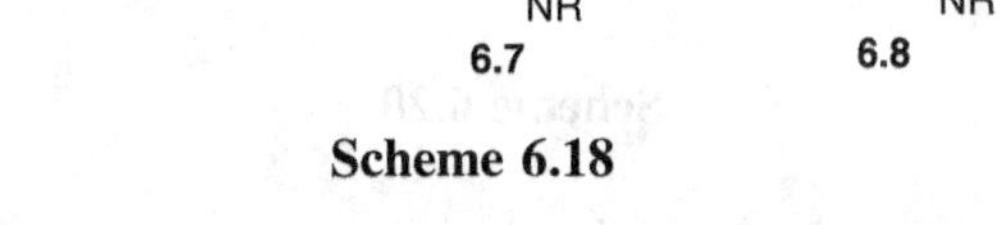

$$RC(=O)\text{-}NHR \xrightarrow{PCl_5} R\text{-}C(Cl)=NR \xrightarrow{ROH} R\text{-}C(OR)=NR$$

6.7 6.8

Scheme 6.18

2-Chloropyridine can be described as a cyclic imide chloride and will be found in the next section to have similar reactivity to this functionality.

Quaternary salts of pyridines also add nucleophiles. The reaction with metallic hydroxides has found special value as an entry to the family of N-alkylpyridones (Scheme 6.19). Note that the initial hydroxy adduct **6.9** has no NH bond and therefore cannot tautomerize. The adducts are easily oxidized to form the carbonyl group of the pyridone.

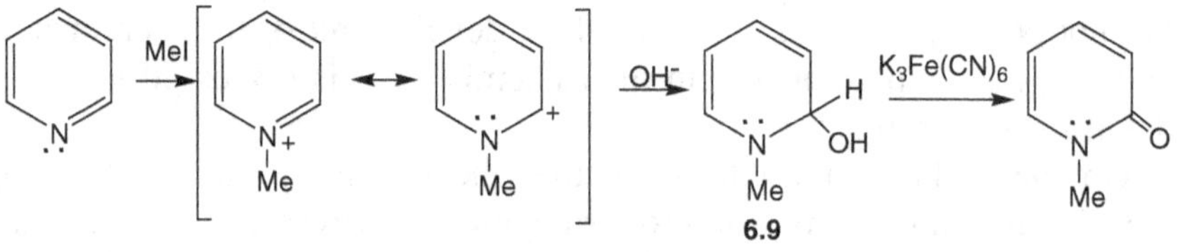

MeI
OH⁻
H
OH
$K_3Fe(CN)_6$
N+
Me
N
Me
N
Me
6.9
N
O
Me

Scheme 6.19

6.3.1.11. Nucleophilic Displacements in Pyridine Chemistry. It was pointed out in the preceding section that chlorine attached to the 2- or 4-position of pyridine has the reactivity of an imide chloride. This is a useful property and allows the synthesis by nucleophilic substitution of a variety of pyridine derivatives. Resonance theory again provides an explanation for this reactivity. This is illustrated in Scheme 6.20 for the reaction of 2-chloropyridine with sodium methoxide, which occurs under mild conditions. It should be remembered that halogen on the benzene ring cannot be replaced under such conditions, except when activated by strong electron-withdrawing groups, especially the nitro group.

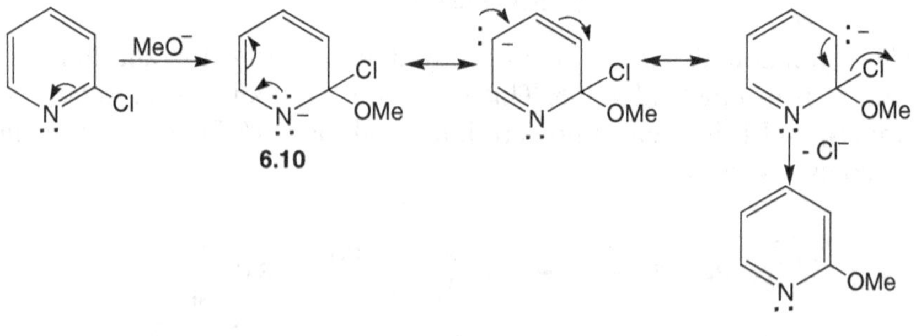

N
Cl
MeO⁻
N⁻
Cl
OMe
6.10
N
Cl
OMe
N
Cl
OMe
- Cl⁻
N
OMe

Scheme 6.20

Similar resonance forms can be drawn for attack at the 4-position. The difference with regard to benzene chemistry is that the nitrogen atom, in the divalent negative ion form (**6.10**), is effective in reducing the energy of the resonant system.

We observe a similar nitrogen-stabilizing effect when a negative charge is created on a carbon attached to the ring at the 2- or 4-positions. Thus, methyl and other alkyl groups can be reacted with strong bases, typically sodium amide, sodium hydride, or sodium hydroxide, to form carbanions. These can be alkylated with alkyl halides (one equivalent to avoid reaction on nitrogen) or condensed in the usual way with carbonyl groups. Such processes are outlined in Scheme 6.21. Similar chemistry is found with alkylbenzenes when the nitro group is present.

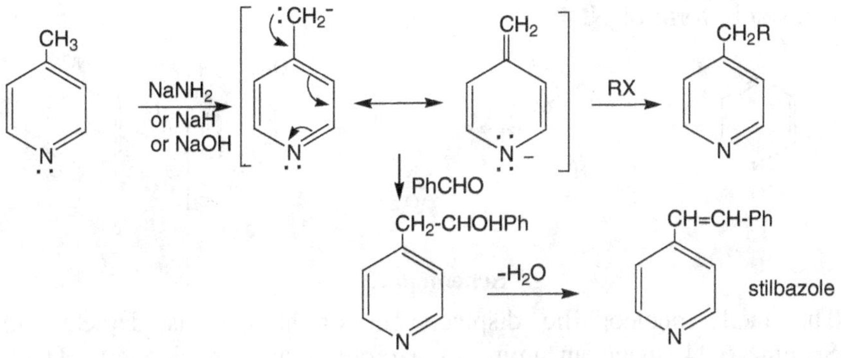

Scheme 6.21

6.3.1.12. Reactions of Pyridine N-Oxide. The attachment of oxygen to nitrogen in pi-deficient systems has a strong influence on the response of the ring to electrophilic substitution, to nucleophilic displacements of ring halogens, and so on. Electrophilic substitution occurs with ease, and conventional conditions from benzene chemistry are useful. Furthermore, the orientation of the substitution is changed; in pyridine, attack is favored at the 3-position, but in the N-oxides, attack at the 4-position is favored, with some attack also occurring at the 2-position (Scheme 6.22). As shown in Scheme 6.22, the N-oxide can be reduced, typically by tricovalent phosphorus compounds such as PCl_3 or Ph_3P, to give the substituted pyridines, which are unavailable by direct reactions. Catalytic hydrogenation is also used frequently to effect this reduction.

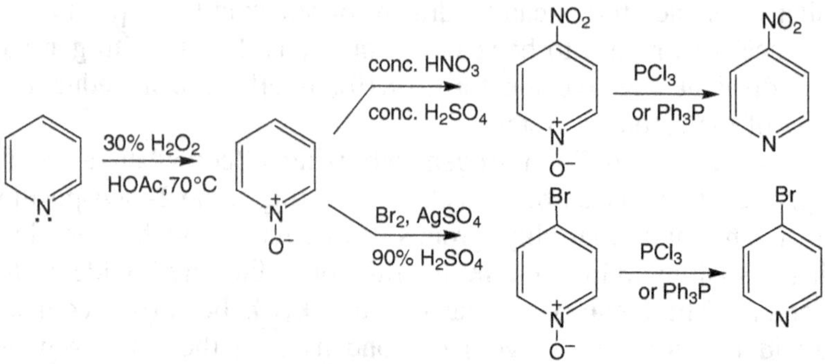

Scheme 6.22

The explanation once again can be found in resonance theory. This is illustrated in Scheme 6.23 with the nitration of pyridine N-oxide. Resonance form **6.11** has tetravalent N bearing a positive charge, which is a stable form of N.

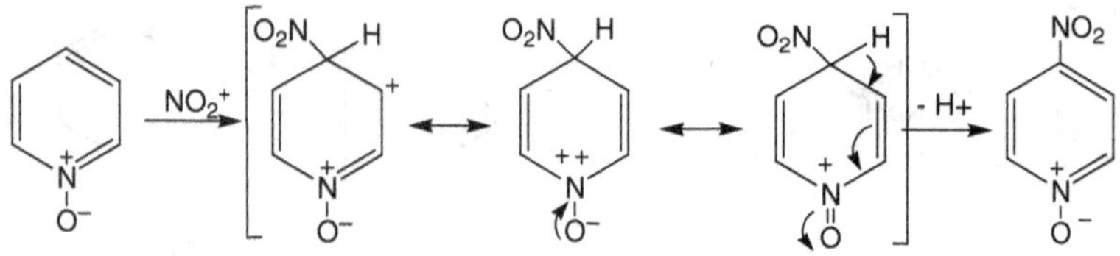

Scheme 6.23

The facile nucleophilic displacement of halogen is depicted in Scheme 6.24, using an amine as a representative nucleophile. Here, the key feature is oxygen accepting the negative charge in form **6.12**, which contributes to stabilizing the hybrid.

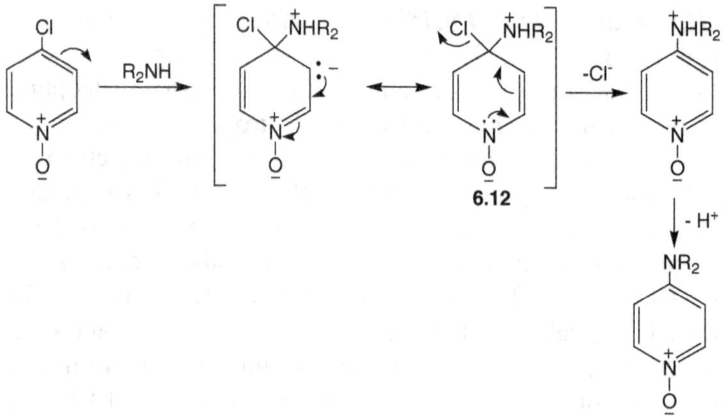

Scheme 6.24

6.3.2. Other Pi-Deficient Nitrogen Systems

6.3.2.1. Multinitrogen-Substituted Ring Systems. Rings where one or more nitrogen atom replaces carbon of pyridine exhibit the usual benzene-type of electron delocalization and have additional resonance forms arising from the electron-accepting character of each nitrogen atom. The many resonance forms for pyrimidine are shown in Scheme 6.25 to illustrate this extensive electron delocalization. As a consequence, the carbons of pyrimidine are even less reactive to electrophilic substitution than are those of pyridine, and such reactions are not practical. However, the placement on carbon of electron-releasing substituents such as amino restores reactivity to the ring and electrophilic substitutions of aminopyrimidines are well known. Other properties studied for pyridine are found in pyrimidines, such as the easy nucleophilic displacement of halogen, formation and reactions of N-oxides, stabilization of carbanions formed from C-methyl substituents, and so on.

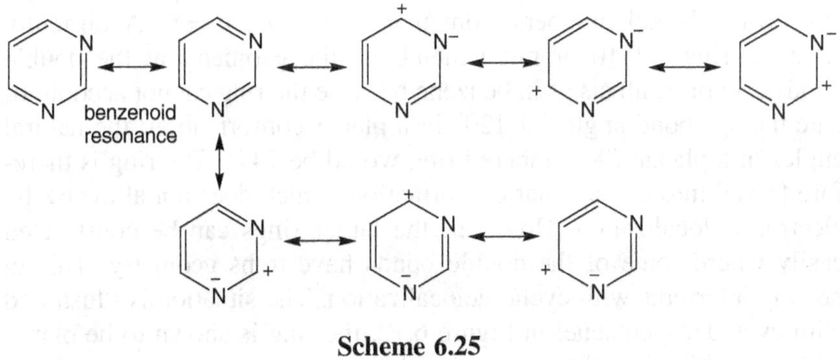

benzenoid resonance

Scheme 6.25

Similar extensive electron delocalization occurs in the other diazines, in the triazines (notably 1,3,5-), and 1,2,4,5-tetrazine; all have properties understandable from their being pi-deficient.

6.3.2.2. Benzo-fused Systems. One or more benzene rings can be fused on any of the monocyclic nitrogen cycles without changing the character of pi-deficiency and its consequences found in these rings. A notable property of such compounds is that electrophilic substitution takes place readily, but exclusively, in the benzo group; the heterocyclic ring retains its strong deactivation to such conditions. Electron delocalization is extensive in such compounds, and all ring components are involved. This is illustrated in Scheme 6.26 for quinoline.

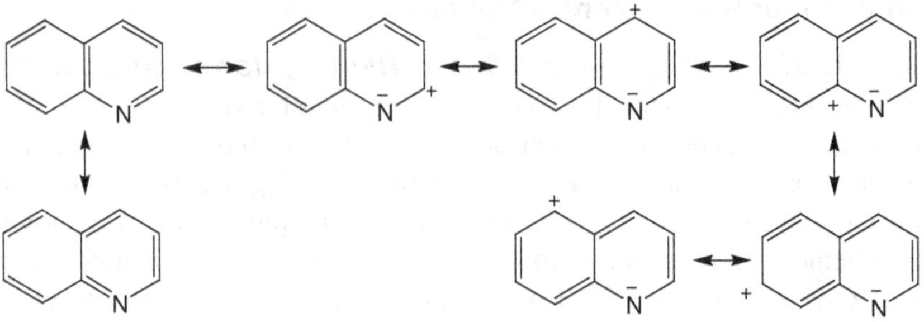

Scheme 6.26

The benzene ring of quinoline undergoes electrophilic substitution with great ease; for example, nitration can be accomplished at 0°C with HNO_3/H_2SO_4 for 30 min, giving the 5-nitro derivative. These conditions are not unlike those effective with naphthalene.

6.3.2.3. Other Hückel Systems Containing Nitrogen. In carbon chemistry, aromatic properties are well known in conjugated ring systems with Hückel numbers from $n = 0$ to about $n = 6$. A difficulty arises in rings of 10 or more members; the geometry at the double bonds cannot be all cis as in benzene because the ring cannot accommodate the sp^2 bond angles of 120° in a planar conformation; the natural angles in a planar 10-membered ring would be 144°. The ring is therefore forced into a nonplanar conformation, which does not allow cyclic electron delocalization. However, the larger rings can be constructed easily where some of the double bonds have trans geometry. This in no way interferes with cyclic delocalization. The situation is illustrated with cyclodecapentaene; in Figure 6.12, the ring is shown to be planar with a combination of three cis double bonds and two trans. Some distortion is still present, however, and this compound is not very stable. The conjugated system is better known when a methylene bridge is added (*vide infra*). Fully unsaturated large rings are known as annulenes; a numerical prefix indicates the number of CH units. Thus, the 10-membered ring is described as [10]annulene.

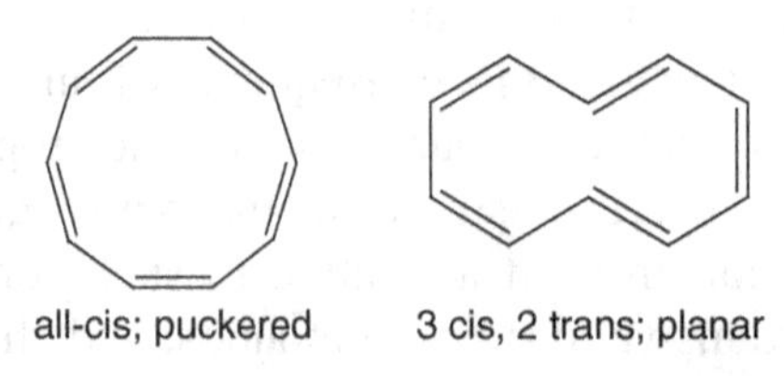

Figure 6.12. Isomers of [10]annulene.

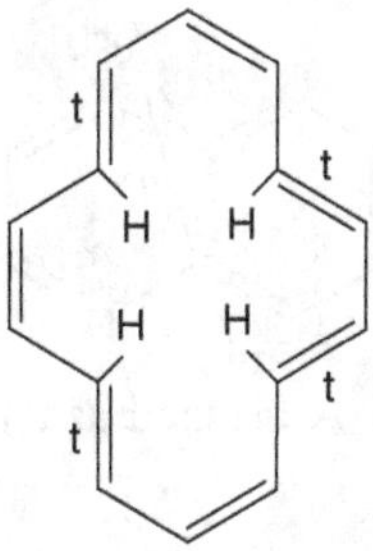

Figure 6.13. [14]Annulene (t = trans).

[14]Annulene, with n = 3, is planar, stable, and well known when there are four double bonds with trans geometry and three with cis (Figure 6.13). There is a valuable NMR consequence of this geometry; 10 protons are on the outer edge of the ring and fall in the usual deshielding area of the double bonds, and they are found at δ 8–10. There are also four protons on the *inside* of the ring, which places them in the *shielding* area of the secondary magnetic field; they resonate at δ−2 to −3. It is, therefore, easy to deduce the geometry of a large annulene from its proton NMR spectrum.

In principle, nitrogen can take the place of one or more CH units of an annulene, but such compounds are not always easily synthesized and are not well studied. Perhaps the best known compound is aza[18]annulene (Figure 6.14), which is planar and stable. Its NMR spectrum reveals its structure to have five inner H (δ−1 to −2). There are two types of outer H; 10 are found at the usual position for outer H at δ 8.8–9.1, but two are farther downfield at δ 10.1. This is the result of the additional deshielding at protons alpha to N, just as was explained for pyridine. That the two protons have identical shifts attests to the fact that they are chemically equivalent from the cyclic electron delocalization of the aromatic system.

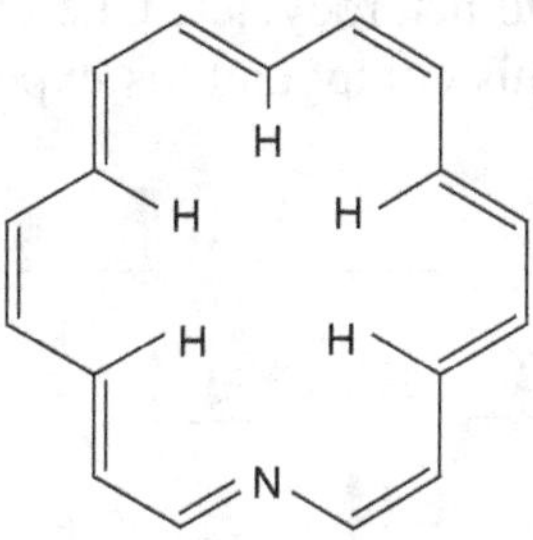

Figure 6.14. Aza[18]annulene.

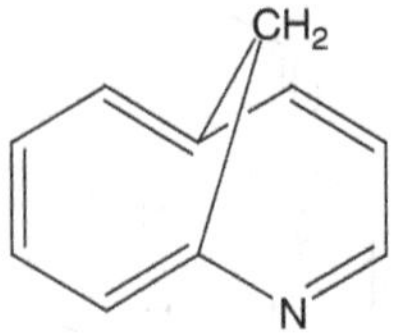

Figure 6.15. A bridged aza[10]annulene.

Aza[14]annulene is also known; its structure can be assigned from its NMR spectrum, which shows four inner H, and two of the outer H (next to N, and identical) are more strongly deshielded than the other seven.

Aza[10]annulene is best observed in the methylene-bridged structure of Figure 6.15. Several derivatives of this system are known. Here, there are two trans double bonds, with the two inner H replaced by the bridge. The bridge holds the ring in a planar conformation, and these annulenes show the NMR and X-ray properties consistent with cyclic electron delocalization.

Aza analogs of unsaturated hydrocarbons containing only 4n pi-electrons, such as cyclobutadiene ($n = 1$) and cyclooctatetraene ($n = 2$), are like the hydrocarbons in not being aromatic and are of low stability. They are referred to as antiaromatic systems. Antiaromaticity can be understood with the aid of MO theory. Using cyclobutadiene as an example, we observe in Figure 6.16 the result of placing the four electrons in the bonding MOs according to Hund's rule of maximum multiplicity, which is to achieve the maximum number of spin states. This means that for degenerate orbitals (orbitals of equal energy), each MO will have one electron, rather than one orbital with an electron pair. This leaves the two degenerate bonding MOs unfilled. The presence of unfilled bonding MOs can be taken as a definition of antiaromaticity. Aromatic systems of course have no unfilled bonding MOs. Little is known about antiaromatic heterocycles. One example that is known is azacyclobutadiene, but this compound, as expected, is unstable.

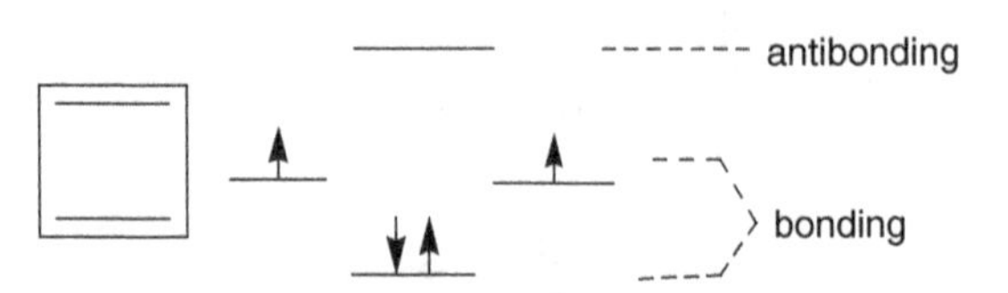

Figure 6.16. Cyclobutadiene MOs.

6.3.3. Porphyrins

The porphin ring, which was introduced in Chapter 3 as a vitally important precursor of natural products, is another example of a stable aromatic system with 18 pi-electrons (n = 4). It is unique in that some double bonds are derived from the pyrrole rings. The pyrrole N–H groups are involved in tautomeric equilibria, as shown in Scheme 6.27, but the pyrrole lone pairs do not participate in establishing the 18-pi aromatic system. The nine double bonds that do establish the planar 18-pi aromatic system in one resonance form are shown in structure **6.13**; the two double bonds not involved in this resonance are enclosed in parentheses. The net effect of the tautomerism is to make all nitrogen atoms equivalent. Furthermore, all CH units in the pyrrole rings become equivalent, as do the four =CH– units (called meso positions) linking the pyrrole rings. The proton NMR spectrum is consistent with this picture; there is only one signal for the meso hydrogens, which is well downfield (δ 10) from the aromatic ring current effect, and the protons on nitrogen are in the shielding zone of this current and resonate well upfield at δ–2. The ^{15}N NMR spectrum shows only one signal at room temperature (δ 160), but when the tautomeric proton shifts are stopped at low temperature (–60°C), the signal is split into a δ 121 component (a shift like that of pyrrole) and a δ 212 component (like a pyridine C=N shift).

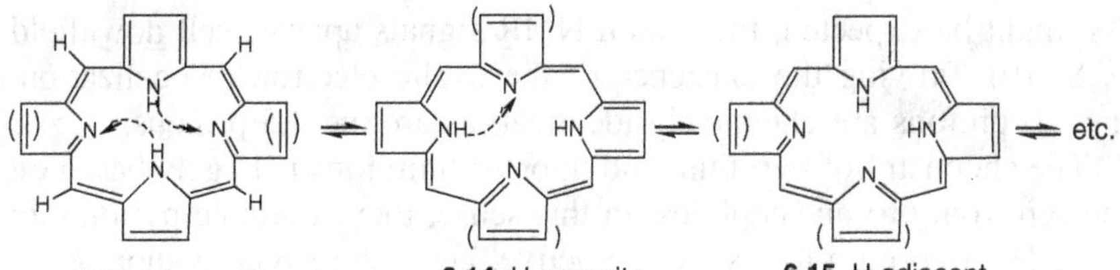

6.13, H-opposite **6.14**, H-opposite **6.15**, H-adjacent

Scheme 6.27

6.3.4. Aromatic Systems Containing Elements other than Nitrogen

The most important elements that can take the place of nitrogen in the 6-membered aromatic ring are oxygen, sulfur, and phosphorus. However, each of these ring systems has special characteristics that drastically differentiate them from the nitrogen counterparts. None of the parent rings are stable, and they are best known with multiple substituents.

This is well illustrated with the oxygen system; it is found in nature in highly substituted compounds (Chapter 3, section 3.3).

For oxygen and sulfur, it is necessary to place these atoms in the positive state (called oxonium or thioxonium ions, respectively), as found in the parent heterocycles called pyrylium (**6.16**) and thiopyrylium (**6.17**, more properly the thiinium ion).

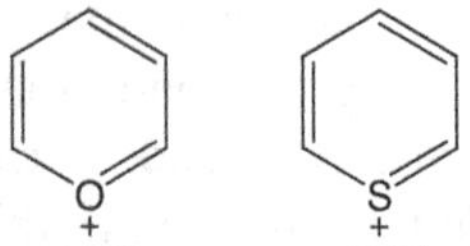

6.16 **6.17**

These systems are highly reactive to nucleophiles, and the counter anion for the parent ring is preferably non-nucleophilic, typically perchlorate (ClO_4^-) or tetrafluoborate (BF_4^-). The relation between the neutral nonaromatic pyran and the pyrylium ion is shown in Scheme 6.28.

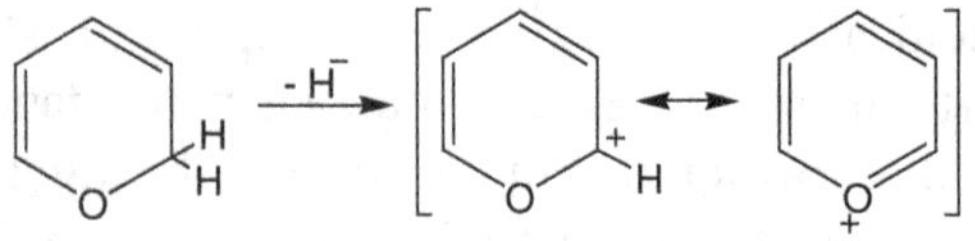

2H-pyran - H⁻ pyrylium ion

Scheme 6.28

As might be expected, the proton NMR signals appear well downfield (δ 8–10). Proving the existence of the cyclic electron delocalization, the 2,6 protons are chemically identical, as are the 3,5 protons.

The chemistry of pyrylium and thiopyrylium ions is largely based on the sensitivity to nucleophiles. In this sense, they resemble pyridinium ions. The pyrylium ion is more reactive, because oxygen is more electronegative than nitrogen. The addition of nucleophiles such as alkyl lithiums, NaCN, NaOH, and so on, occurs readily and leads initially to pyran derivatives. Usually, these adducts undergo ring opening to dienals (Scheme 6.29), which are valuable in the synthesis of other compounds.

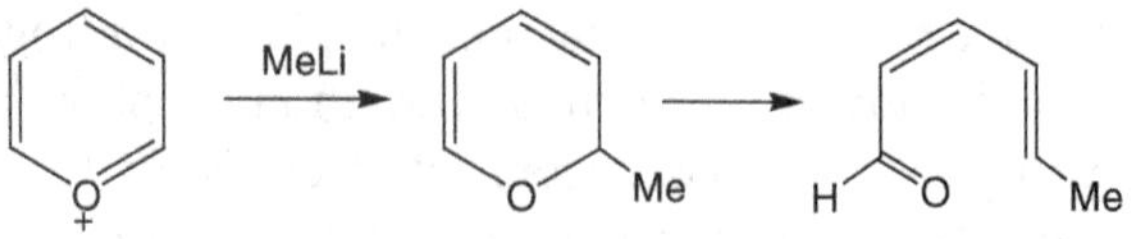

MeLi

Scheme 6.29

If the ring is substituted by a potential nucleophilic leaving group, the attack of another nucleophile may displace this group. This is illustrated with the 4-methoxy derivative **6.18** (Scheme 6.30).

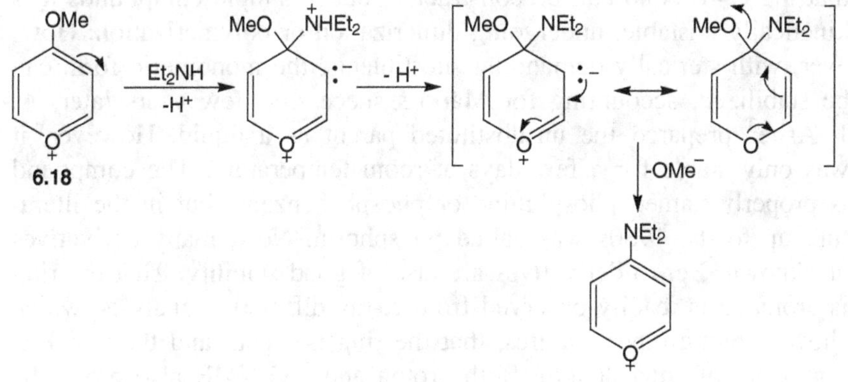

Scheme 6.30

Certain pyrylium salts react with primary amines to undergo a process of ring opening and closing, which results in the replacement of oxygen by nitrogen. This can be a useful approach to pyridinium ions, as found in Scheme 6.31.

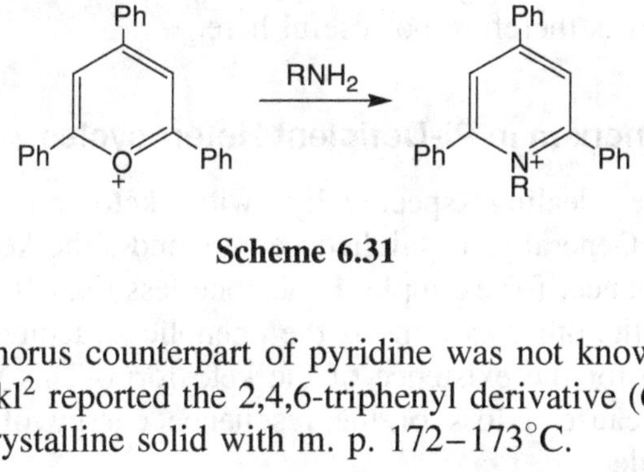

Scheme 6.31

The phosphorus counterpart of pyridine was not known until 1966, when G. Märkl[2] reported the 2,4,6-triphenyl derivative (**6.19**) as a stable, yellow crystalline solid with m. p. 172–173°C.

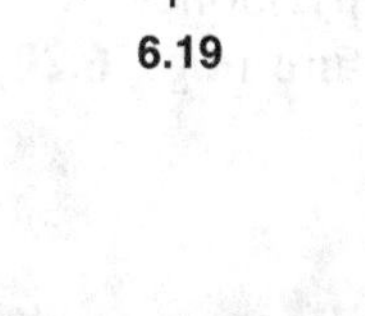

6.19

This was a major advance in phosphorus chemistry that stimulated much new research. Until that time, no example of a compound with a C-P double bond (or for that matter with a carbon doubly bonded to *any* second-row element) had ever been encountered. It is now known that the C=P bond can be constructed, but in simple compounds it is kinetically unstable, undergoing dimerization or polymerization. However, with sterically demanding substituents, the monomeric form can be stabilized, accounting for Märkl's success. A few years later, A. J. Ashe[3] prepared the unsubstituted parent as a liquid. However, it was only stable for a few days at room temperature. The compound is properly named phosphinine or phosphabenzene, but in the literature up to the 1980s was called phosphorin. Now, many derivatives are known. Benzo derivatives are also of good stability. That the ring is aromatic is readily observed from X-ray diffraction analysis, which shows, among other features, that the ring is planar and the two P-C bonds are of equal length. Both proton and ^{13}C NMR also prove the identity of the C-P bonds, which can only be the case if the system has benzene-like cyclic electron delocalization. Because the electronegativity of phosphorus (2.1) is actually less than that of carbon (2.5), the phosphinine ring will not be polarized in the way of the pyridine ring. Most of the chemical properties of phosphorus (discussed in a 2000 monograph[4]) are different from those of nitrogen, and there is no similarity between the chemistry of phosphinine and pyridine. Application of the pi-deficiency concept for the prediction of chemical properties of phosphinines is therefore not useful here.

6.3.5. Tautomerism in Pi-Deficient Heterocycles

Here, we are dealing specifically with keto-enol tautomerism (Figure 6.17). Generally, in aliphatic compounds, the keto form is in great predominance; for example, in acetone less than $10^{-5}\%$ is in the enol form. At the other extreme is the phenolic structure, where there is no evidence for the existence of the keto form. This would not be unexpected because a loss of the resonance energy of stabilization would be entailed.

The situation is different in heterocycles. When a hydroxyl group is present at the 3-position of pyridine, it is not involved in tautomerism and its chemistry is clearly that of a phenol. But when hydroxyl is placed on the 2- or 4-positions, it tautomerizes completely or nearly so to the oxo forms (called pyridones, whose synthesis was described in section 6.3.1.9); the structure (e.g., **6.20**) is analogous to that of an

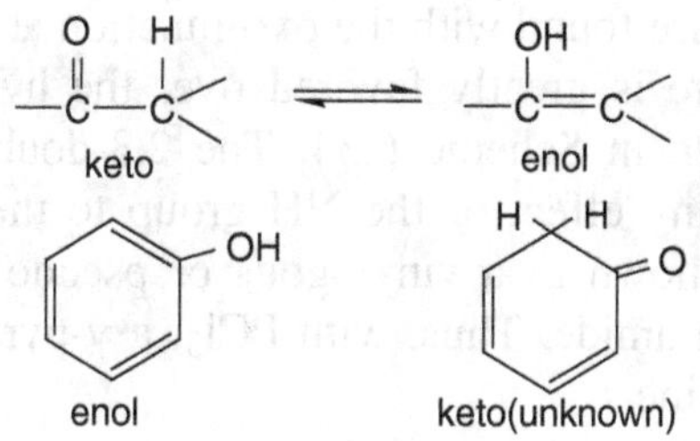

Figure 6.17. Tautomerism.

amide. The explanation for this preference lies in the fact that resonance delocalization using the lone electron pair on nitrogen is present in the oxo form and tends to restore the benzenoid resonance. This is shown in Scheme 6.32. Thus, the pyridone structure **6.20** is the only form observed in X-ray diffraction analysis, whereas in acetone solution about 8% of the compound is found as the 2-hydroxy tautomer.

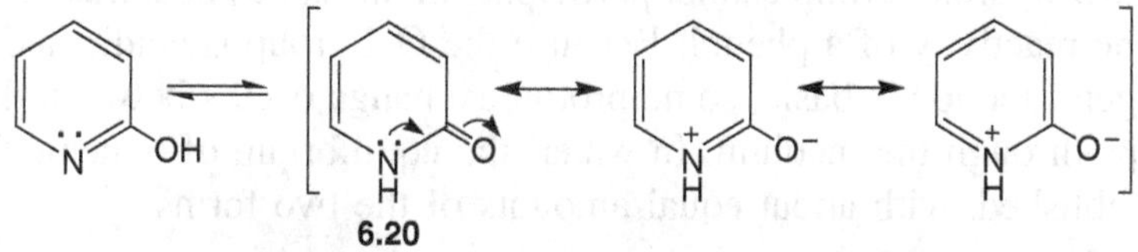

Scheme 6.32

In solution, the compound responds to reactants in either the hydroxyl or the oxo form, in the way called for by the reactant. Thus, diazomethane reacts with OH groups to form OMe groups and this occurs with 2-hydroxypyridine (Scheme 6.33). But methylation with MeI–NaOH, which is also a well-known alkylation procedure for NH groups, occurs on nitrogen.

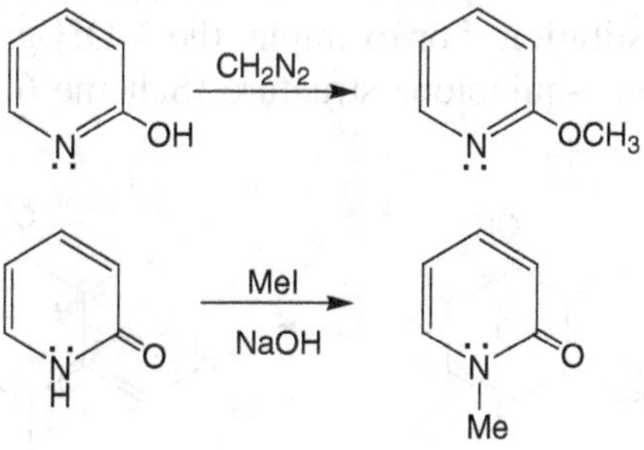

Scheme 6.33

Similar properties are found with the oxo function at the 4-position; the γ-pyridone structure is greatly favored over the hydroxyl form, with resonance as shown in Scheme 6.34. The 2,3-double bond transmits the electron-releasing effect of the NH group to the C=O group; the structural unit is known as a vinylogous or pseudo amide, and it has the reactivity of an amide. Thus, with PCl_5, a γ-pyridone is converted to a 4-chloro pyridine.

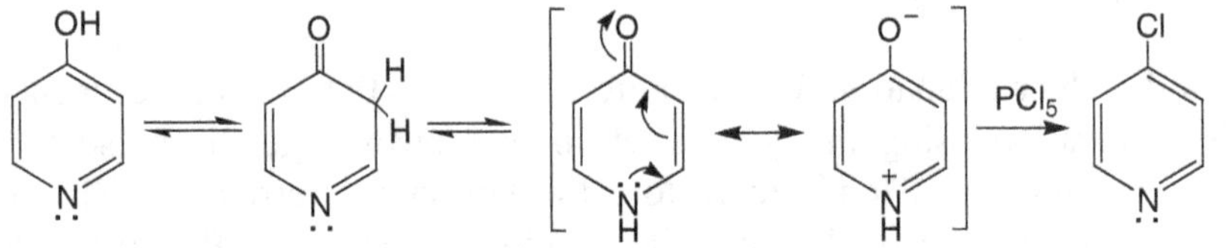

Scheme 6.34

The 3-hydroxy group cannot participate in the same phenomenon and has the reactivity of a phenol. Because the OH group is acidic and the nitrogen function is basic, some proton exchange occurs between these groups through the medium. In water, the equilibrium of Scheme 6.35 is established, with about equal amounts of the two forms.

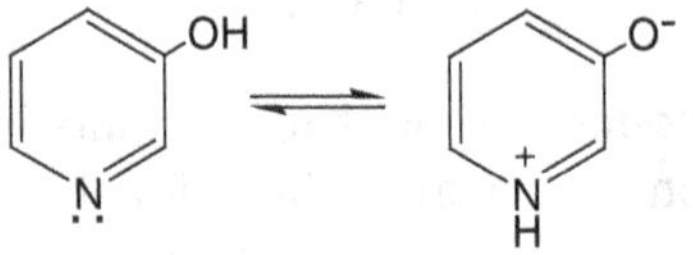

Scheme 6.35

In general, these properties of hydroxy substituted derivatives will be present in pi-deficient nitrogen heterocycles with benzo fusion or with multinitrogen substitution. For example, the 4-OH group on quinoline tautomerizes to the 4-quinolone structure (Scheme 6.36).

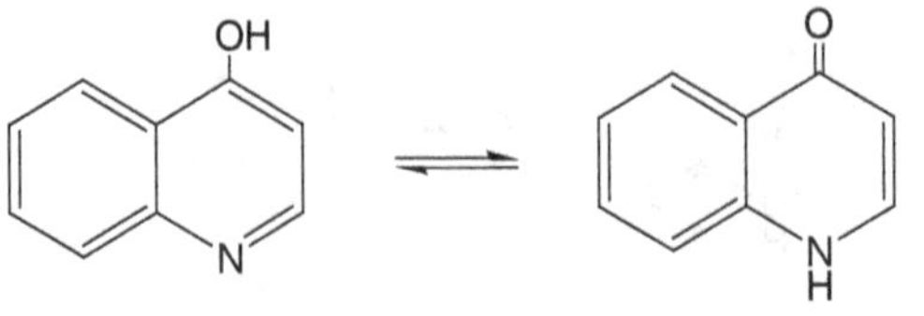

Scheme 6.36

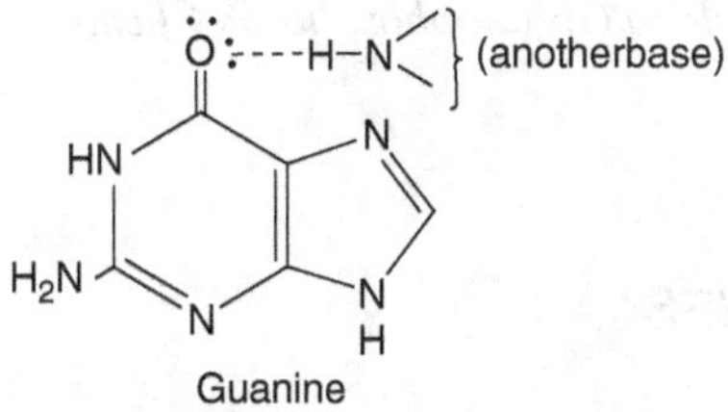

Guanine

Figure 6.18. H-Bonding in bases on nucleic acids.

Tautomerism is extremely important in pyrimidines and the related purines. For many years, the oxy function on carbon next to a nitrogen atom was written incorrectly in the OH form; in reality, the oxygen is in the carbonyl form (Scheme 6.37).

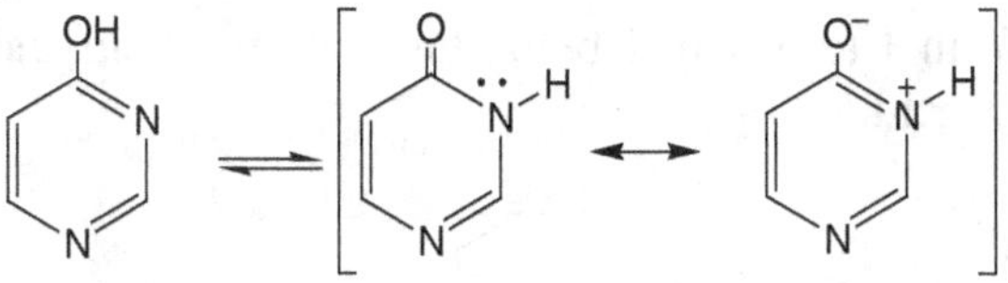

Scheme 6.37

The carbonyl oxygen can act as an acceptor in hydrogen bonding (Figure 6.18), and it was recognition of this property that aided Watson and Crick in establishing the double helix structure of nucleic acids (see Chapter 9). Here, an N–H group of another pyrimidine or purine forms a hydrogen bond with the carbonyl group, assisting in holding the two strands together. As will be discussed in Chapter 9, there is remarkable specificity in the two "bases" that will bond with each other in the double helix.

As a final note on the subject of tautomerism, the phenomenon has no significance with amino groups on pi-deficient rings; an NH_2 group remains as such and does not noticeably tautomerize to the =NH form.

REFERENCES

(1) M. K. Cyranski, T. M. Krygowski, A. R. Katritzky, and P. von R. Schleyer, *J. Org. Chem.*, **67**, 1333 (2002).

(2) G. Märkl, *Angew. Chem., Int. Ed. Engl.*, **5**, 846 (1966).

(3) A. J. Ashe, III, *J. Am. Chem. Soc.*, **93**, 3293 (1971).

(4) L. D. Quin, *A Guide to Organophosphorus Chemistry*, Wiley-Interscience, New York, 2000.

REVIEW EXERCISES

6.1. Draw 5 resonance forms for pyridine.

6.2. Rank the following from 1 to 3 with 1 being the strongest base and 3 being the weakest base.

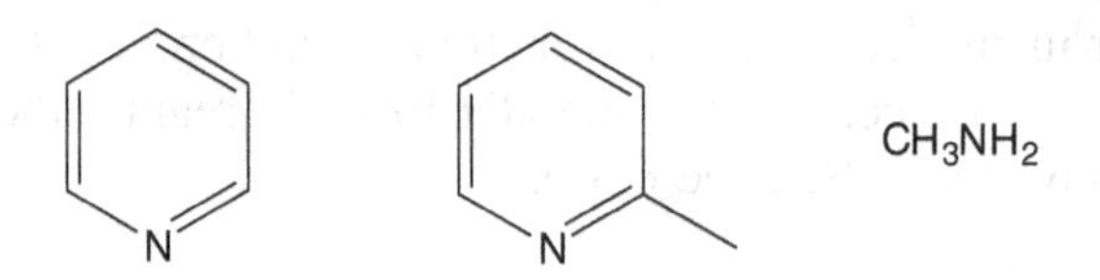

6.3. Rank from 1 to 3 with 1 being the strongest acid and 3 being the weakest acid.

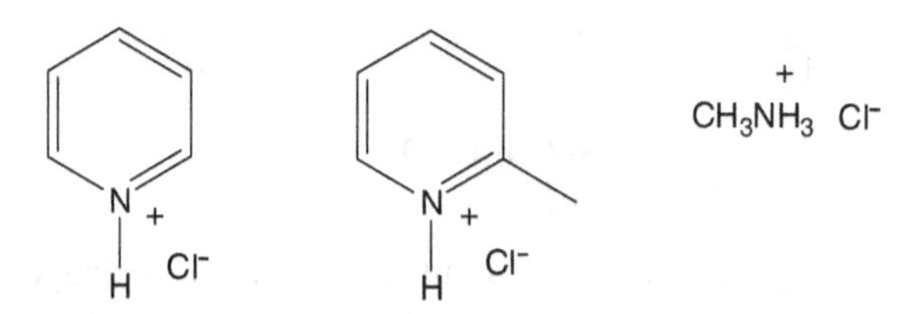

6.4. Of the following, the strongest base is?

a. Water
b. Trimethylamine
c. Pyridine
d. 2-Nitropyridine

6.5. Consider an aromatic substitution reaction to make the sulfonated compound. Rank the following from 1 (most reactive) to 4 (least reactive).

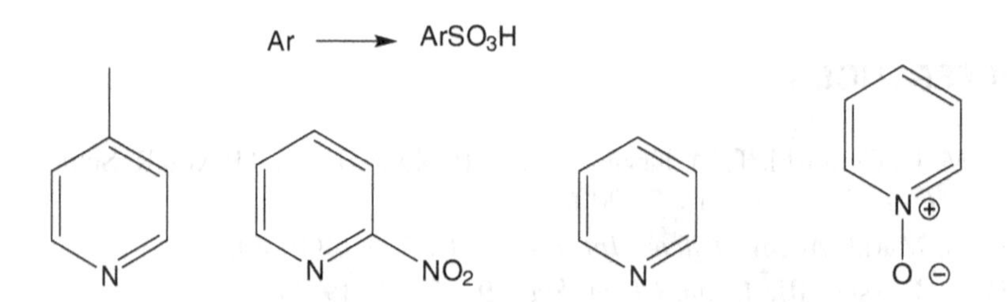

6.6. Draw the major product.

a.

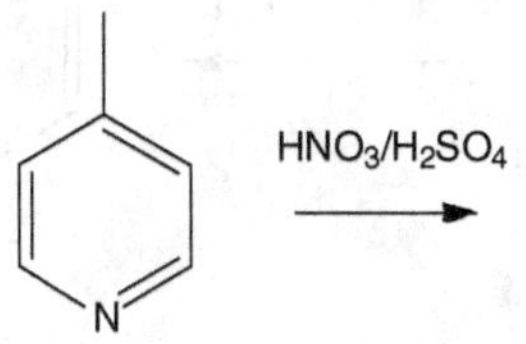

HNO_3/H_2SO_4

N

b.

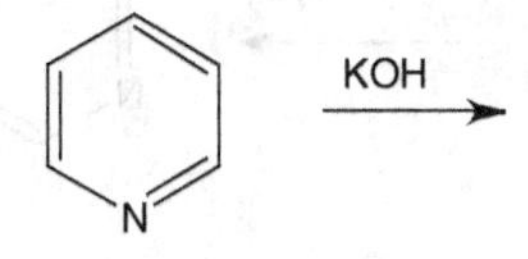

6.7. Consider an aromatic substitution reaction to make the sulfonated compound. Rank the following from 1 (most reactive) to 3 (least reactive).

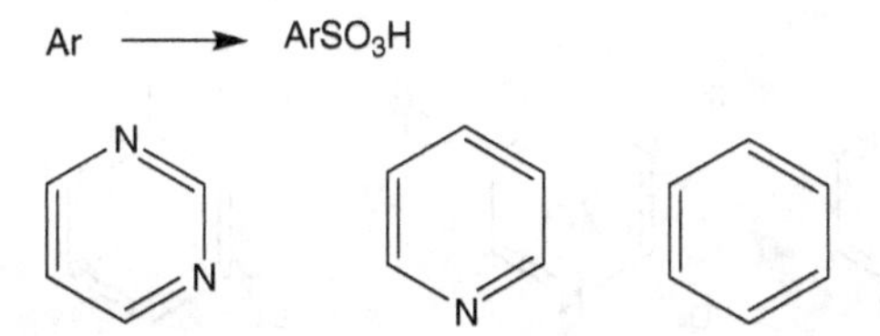

6.8. How many carbons are there in aza[18]annulene?

6.9. Consider the following reactions and choose the best one (the one that proceeds, or if more than one proceeds, the one that proceeds under the mildest conditions or with the best rate).

a.

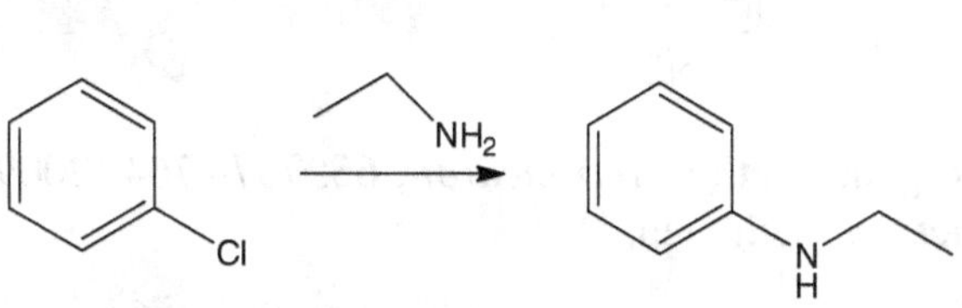

NH_2

Cl

N
H

b.

NH_2

N Cl

N N
H

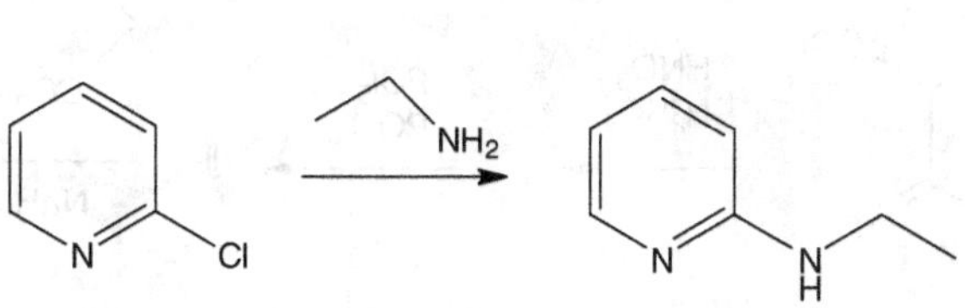

c.

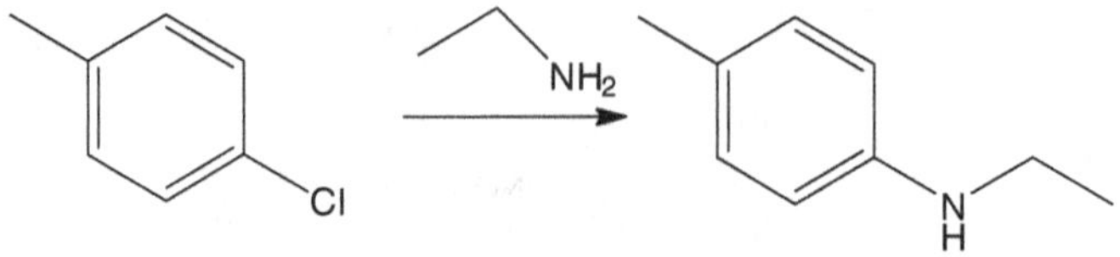

NH2

Cl

N
H

d.

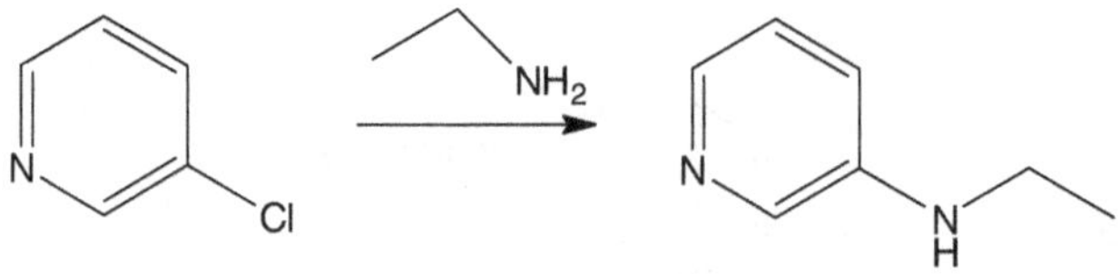

NH2

N

Cl

N

N
H

6.10. Starting with pyridine, propose a synthesis of 4-nitropyridine.

6.11. Of the following, which is the most *acidic* (most likely to lose a proton and form an anion) compound?

a. 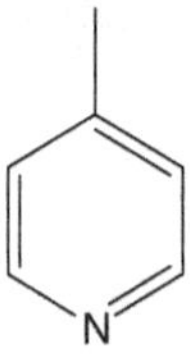 N b. 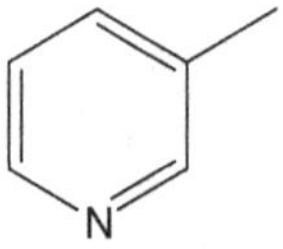 N c. 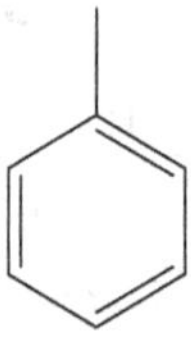 d.

6.12. The pyrazine below reacts with NaOH to give a new compound, $C_4H_4N_2O$. Write the structure of the product.

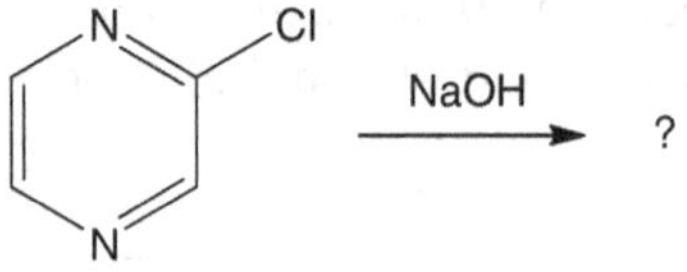

N Cl

NaOH

?

N

6.13. R. Morgentin, et al., *Tetrahedron*, **65**, 757–764 (2009). Enter the product for each step.

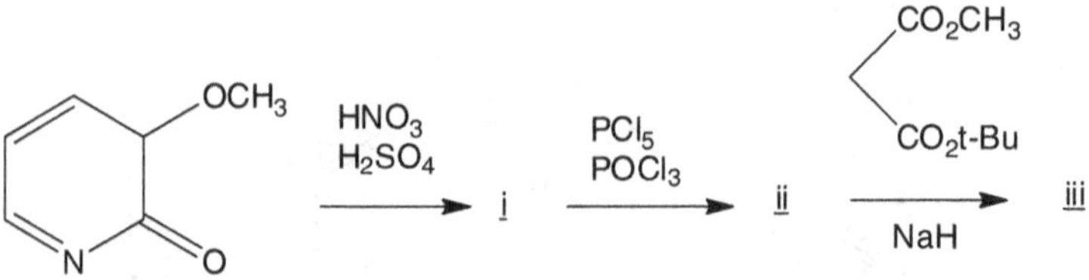

CO_2CH_3

OCH_3

HNO_3
H_2SO_4

PCl_5
$POCl_3$

CO_2t-Bu

i ii iii

NaH

N O

6.14. A. P. Krapcho, and S., Sparapani, *J. Heterocyclic Chem*., **45**, 1167–1170 (2008). Draw the expected product.

Cl
N
Cl
N
CH_3OH
19 hr
reflux

6.15. H. F. Bettinger, et al., *J. Org. Chem*., **64**, 3278–3280 (1999).

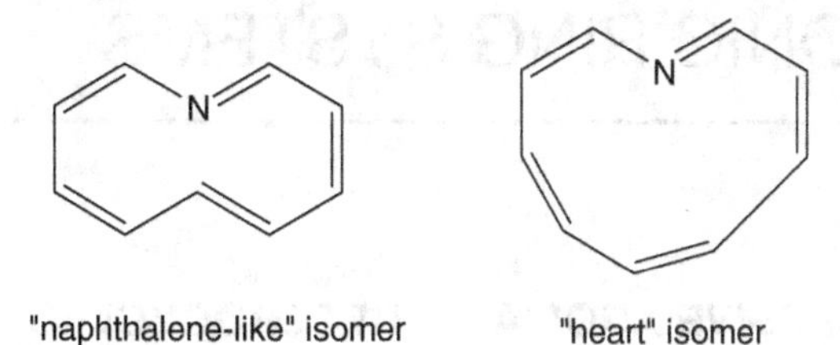

"naphthalene-like" isomer   "heart" isomer

Computations of these isomers showed that the "heart isomer" is more stable than the "naphthalene-like" isomer.

a. What is the general name for these compounds?

b. One of these isomers has a highly shielded H atom with a H at δ2.1 with respect to TMS. Which one is it?

6.16. Which is the strongest base?

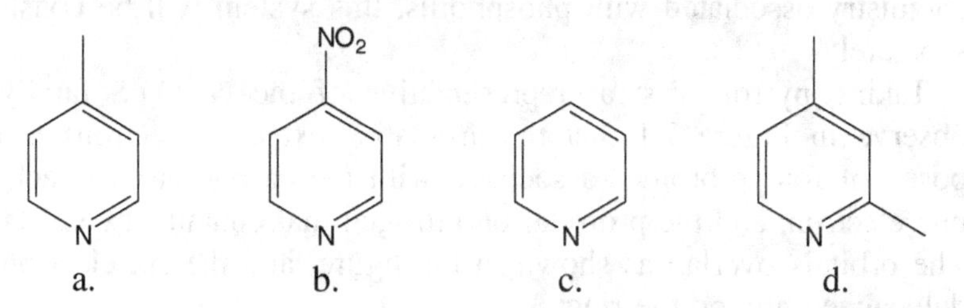

6.17. Sketch a good synthesis of 2-chloroquinoline starting from quinoline. Show key reactants and intermediates.

CHAPTER 7

AROMATICITY AND OTHER SPECIAL PROPERTIES OF HETEROCYCLES: PI-EXCESSIVE RING SYSTEMS AND MESOIONIC RING SYSTEMS

7.1. PI-EXCESSIVE AROMATIC HETEROCYCLES

7.1.1. Aromaticity of the Common 5-Membered Ring Systems

The most important ring systems have N, O, or S as the heteroatom, and the discussion here will be centered on each of these systems as a family. The ring with phosphorus as the heteroatom remained unknown until relatively recently (1959) in the long saga of heterocyclic chemistry, but it has been gaining in importance; because of the special chemistry associated with phosphorus, this system will be considered separately.

Taking pyrrole first as representative of the N, O, S family, we observe in Figure 7.1 that the aromatic sextet of electrons is composed of four p-orbitals associated with the carbon atoms, each with one electron, and the p-orbital on nitrogen that contains two electrons. The orbitals overlap as shown in the figure, and the six electrons are delocalized around the ring.

The hydrogen atom on nitrogen is known to be in the plane of the ring from X-ray diffraction studies, which means that the hybridization on nitrogen is sp^2 rather than sp^3 as is typically found in amines. The latter hybridization would put nitrogen into a tetrahedral configuration

Fundamentals of Heterocyclic Chemistry: Importance in Nature and in the Synthesis of Pharmaceuticals, By Louis D. Quin and John A. Tyrell Copyright © 2010 John Wiley & Sons, Inc.

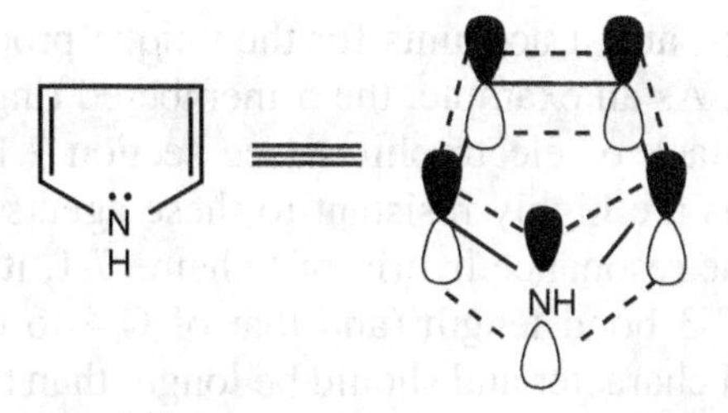

Figure 7.1. p-Orbital overlap in pyrrole.

in which the lone pair occupies one site of the tetrahedron and cannot overlap efficiently with the carbon p-orbitals.

The resonance view of the N, O, and S heterocycles is shown in Scheme 7.1 as applied to pyrrole. In all forms, the carbon atoms acquire some negativity and increased electron density, and this leads to their description as pi-excessive.

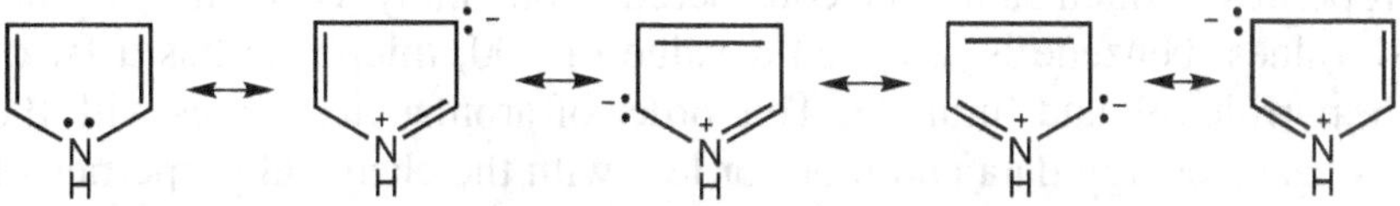

Scheme 7.1

Resonance energies have been determined for the three heterocycles; thiophene has the highest value (29.1 kcal/mol), which is similar to that for pyridine (27.9 kcal/mol), whereas pyrrole has the value 21.6 kcal/mol, and furan 16.2 kcal/mol. These values clearly are consistent with the view that these rings possess the extra stabilization associated with the presence of high electron delocalization.

Electron density calculations (shown in Figure 7.2) confirm the delocalization view; all carbons have densities above 1.0, and the heteroatom density is below 2.0, the value if the lone pair were localized on nitrogen. This is a profound difference from the densities of the

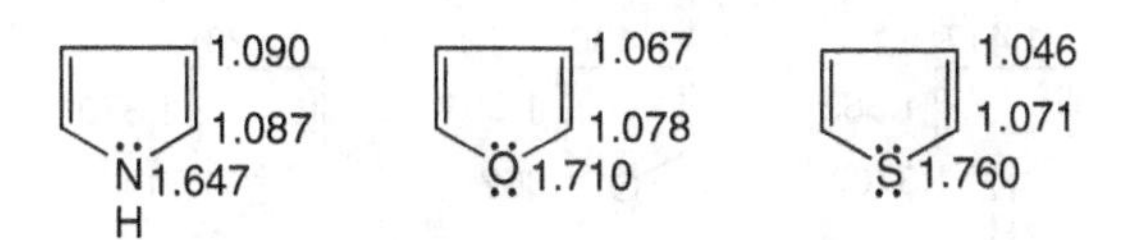

Figure 7.2. Pi-electron densities.

6-membered rings, and it accounts for the unique properties associated with each family. As an example, the 5-membered rings are easily substituted by the attack of electrophiles (see section 7.1.2), whereas the 6-membered rings are highly resistant to these agents.

By studying the resonance hybrid of Scheme 7.1, it can be observed that the C-2 to C-3 bond length (and that of C-4 to C-5) would have some single-bond character and should be longer than that of a true double bond (1.34 Å), whereas the C-3 to C-4 bond would be shorter than a true single bond (1.48 Å for sp^2 to sp^2 bonds) having acquired some double-bond character. Structural measurements[1] shown in Figure 7.3 confirm theses relations and provide experimental evidence for electron delocalization.

The deviations of the bond lengths from those of isolated bonds were first put on a quantitative basis by C. W. Bird,[2] which provided a ranking [the Bird index (BI)] of the aromaticity of various ring systems. Others have also used structural parameters in evaluating aromaticity,[3,4] but we must caution that there are many other ways to measure aromatic properties,[5] which cannot be considered in our study. Using the original BI values, benzene is assigned a value of 100, thiophene has a BI of 66, pyrrole 59, and furan 43. This order of aromaticity agrees with the resonance energy data and more or less with the chemical properties of these heterocycles. Thus furan, the heterocycle of suggested low aromaticity, shows some properties more like a diene than a significantly delocalized system (see section 7.1.3). The BI values are also low in other heterocycles where oxygen is contributing a lone pair to the aromatic sextet, as in the oxazoles. It should be noted that 6-membered heterocycles have higher BI values; thus, pyridine has the high index of 86,[6] which is consistent with its strong aromatic character. BI values for other 5-membered heterocycles are given in Figure 7.4.

Bird later developed a calculated scale [Unified Aromaticity Index (I_A)] that employed both bond lengths and resonance energies,[7] which provided a better representation for the extent of delocalization. The new scale also is said to allow a better comparison of the aromaticities of 5- and 6-membered rings. Some of these I_A values calculated by Bird

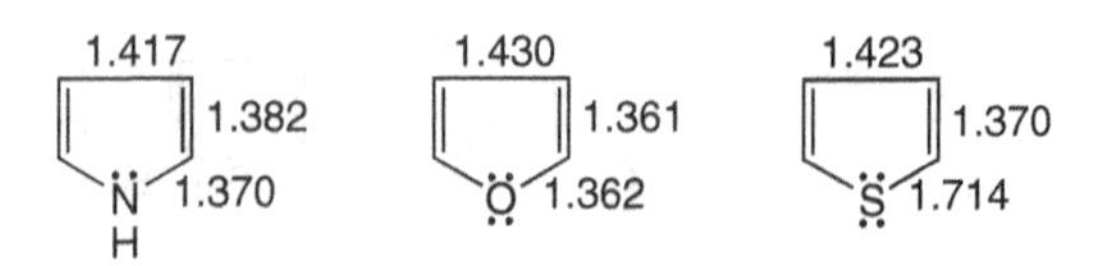

Figure 7.3. Bond lengths in angstroms.

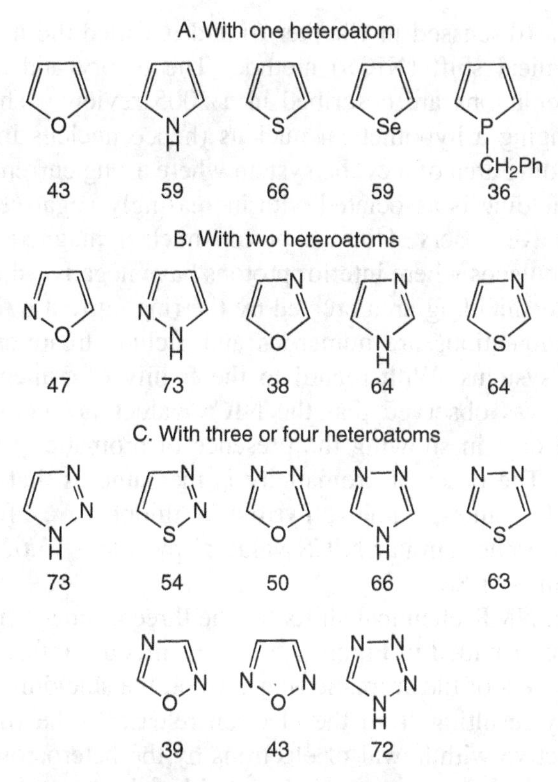

Figure 7.4. Bird indices of 5-membered rings, from Refs. 2 and 7.

are given in Appendix I at the conclusion of this book. We take note of the fact that the order of aromaticity for the 5-membered heterocycle is not in full agreement with the original BI values; pyrrole (85) and thiophene (81.5) have exchanged positions and both are close to pyridine (86). Furan remains among the least aromatic heterocycles, with an I_A of 53. We will use the I_A values when considering the aromaticity of other heterocycles.

Aromaticity has generally been recognized from an examination of experimental facts, as was shown previously. Many criteria for aromaticity have been proposed over the years; prominent in contemporary studies are criteria based on magnetic properties. The inquiry is also currently moving from experimental methods to those that employ the power of modern computational methods, especially as applied to magnetic phenomena. Prominent and widely accepted among computational methods is that introduced by Schleyer et al.[8] in 1996. This method is based on magnetic properties associated with cyclic electron delocalization (discussed in Chapter 6) and is called the nucleus independent chemical shift (NICS) method. The theory and applications of NICS calculations are described in a 2005 review.[9] The approach involves bringing a hypothetical nucleus (hence nucleus independent) into the shielding area of a cyclic system where a ring current is present. Increased shielding is associated with increasingly negative NICS values (as we have observed in the proton nuclear magnetic resonance spectra of annulenes where interior protons have negative shifts because they lie in the shielding area created by the ring current). Applications of NICS computations are numerous and include treatment of many heterocyclic systems. With regard to the family of 5-membered heterocycles, it was observed that the NICS values are consistent with experimental data in showing the presence of aromaticity of a significant degree. The order of aromaticity is the same as that established with Bird's I_A values, namely, pyrrole > thiophene > furan. However, all have rather similar NICS values:[9] pyrrole −13.62, thiophene −12.87, furan −11.88.

The proton NMR chemical shifts for the three aromatic heterocycles of interest are provided in Figure 7.5.[1] The shifts are influenced by the deshielding effect of the aromatic ring current, the shielding effect from the negativity resulting from the electron release to the ring carbons, and the inductive withdrawal of electrons by the heteroatoms. That all the values for the beta-protons are downfield of that for cyclopentene as a localized alkene might suggest that the ring current effect dominates. In furan, the inductive effect of the highly electronegative oxygen atom causes the alpha-protons to resonate at the low-field value of δ 7.29, but this cannot be the explanation for the low-field value of thiophene (δ 7.18) because sulfur has much lower electronegativity than do the other heteroatoms. This special deshielding seems to be characteristic of second-row elements and is sometimes referred to simply as the "alpha effect".

All the NMR signals show the complex coupling of the AA′BB′ or AA′MM′ type, because of the magnetic nonequivalence of chemically equivalent protons, as was discussed for the spectrum of pyridine in

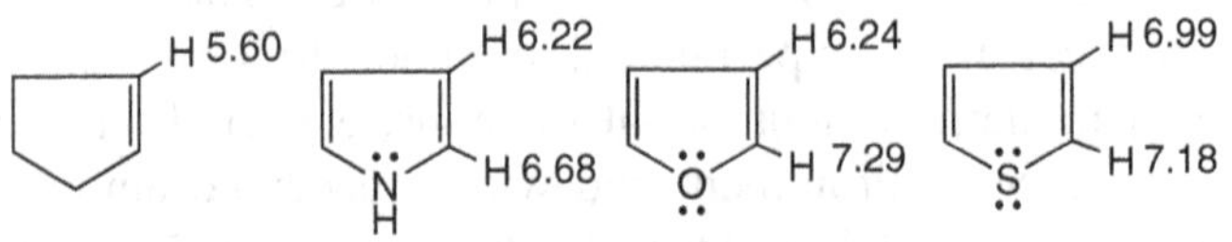

H 5.60
H 6.22
H 6.68
N
H
H 6.24
H 7.29
O
H 6.99
H 7.18
S

Figure 7.5. NMR shifts (δ) in ppm from tetramethylsilane (TMS).

Chapter 6. Thus, H-3 couples with H-2 by a different magnitude than with H-5, which makes H-2 and H-5 magnetically nonequivalent even though the chemical shifts of H-2 and H-5 are the same. Refined values for the coupling constants have been published by Katritzky, et al.[10]

7.1.2. Chemical Properties of Pyrroles

7.1.2.1. Electrophilic Substitution. A significant feature of the pi-excessive ring systems is that they are highly reactive to electrophilic species, totally unlike the pi-deficient rings. The reactivity is greater than that of benzene and is in roughly the same range as found for benzenes bearing electron releasing groups as in aniline. As a result, many useful substitution reactions are known for these heterocycles. The greater electron density in these rings accounts for this higher reactivity. The order of reactivity in aromatic substitutions is generally pyrrole > furan > thiophene. This is not the order of aromaticity as established, e.g., by the Bird I_A values (pyrrole > thiophene > furan). The I_A order reflects the thermodynamic stability of the ring systems, but the reactivity order is the result of a kinetic effect. Electron release from the heteroatom helps to stabilize the intermediate in the substitution process, thus lowering the activation energy, and nitrogen is the best at performing this effect (as revealed by the basicity of amines). The orientation of the substitutions may be explained by application of resonance theory. For all three heterocycles, electrophilic attack is favored at the alpha carbon of the ring. An attack at this position leads to an intermediate whose positive charge can be dispersed to all other ring positions; charge dispersal of course is a well-known stabilizing effect. The charge dispersal is shown in Scheme 7.2 with the use of resonance structures. Note the critical role of the heteroatom in donating electrons to the ring.

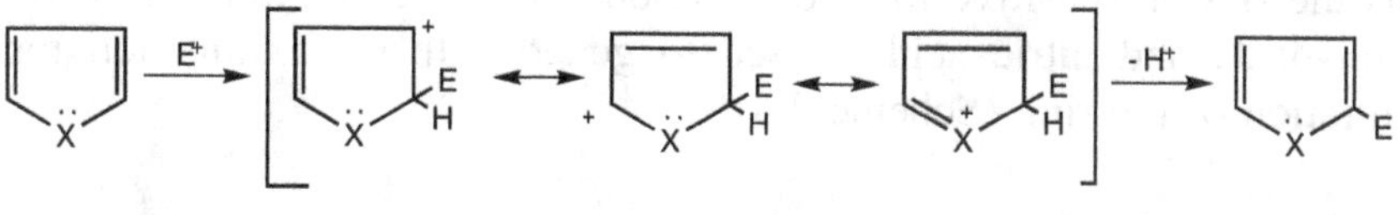

Scheme 7.2

If an attack occurred at the beta position, the charge can only be spread between the heteroatom and the alpha carbon, as shown in Scheme 7.3. Nevertheless, this still constitutes a stabilizing effect, and if the alpha positions are blocked, then substitution will occur readily at the beta positions.

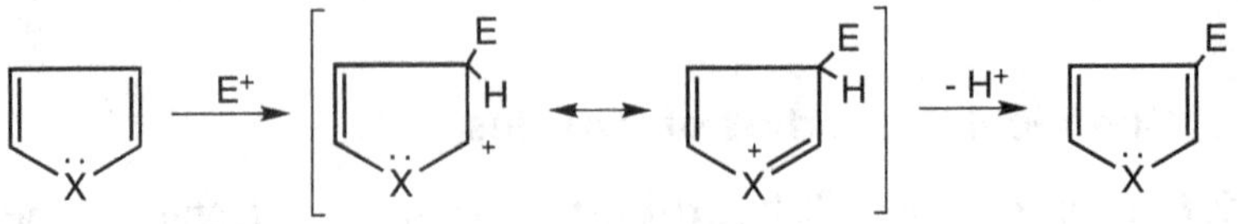

Scheme 7.3

Some typical electrophilic substitution reactions of pyrrole are described in the next section.

Acylation. It is not necessary to employ $AlCl_3$ as a catalyst, as in the Friedel–Crafts reaction of benzene chemistry, for acid chlorides to effect acylation of the pyrrole ring. The reaction (Scheme 7.4) occurs at mild temperatures, and it makes ketones readily available. Although pyrrole is formally a secondary amine, the reduced nucleophilicity of nitrogen does not allow acylation to take place at this site.

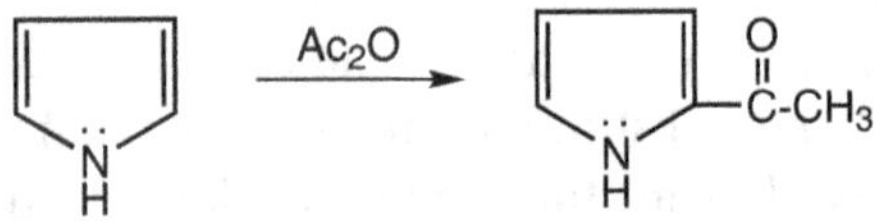

Scheme 7.4

Nitration. Nitration also occurs under mild conditions. The traditional mixture of nitric and sulfuric acids to generate the nitronium ion (NO_2^+) used for the nitration of benzene is unnecessary in pyrrole chemistry. Indeed, strong acid conditions in general are to be avoided, as the ring is sensitive to acids (section 7.1.2.2). A mixture of acetic anhydride and nitric acid is used to generate the nitronium ion for nitration of the ring (Scheme 7.5).

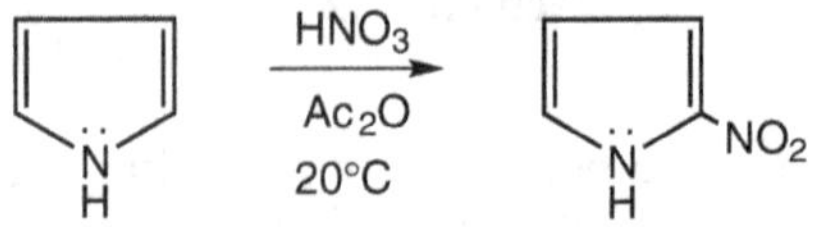

Scheme 7.5

Halogenation. Just as is true for aniline, the pyrrole ring is reactive toward bromine or chlorine, giving polysubstituted products (Scheme 7.6). No catalyst is required.

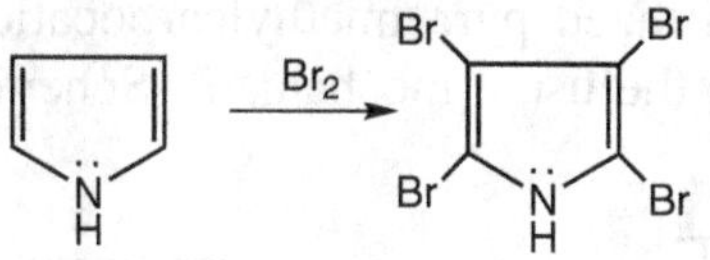
Br₂
Br
Br
Br
Br
N
H
N
H

Scheme 7.6

Vilsmeier–Haack Formylation with N,N-Dimethylformamide. In the presence of phosphorus oxychloride ($POCl_3$), the—CHO group of N,N-dimethylformamide can be attached to the pyrrole ring (Scheme 7.7). This is a highly useful process for the synthesis of pyrrole aldehydes, which are precursors of pyrrole acids by oxidation, of pyrryl carbinols by reductions with $LiAlH_4$, and of other products (Scheme 7.7).

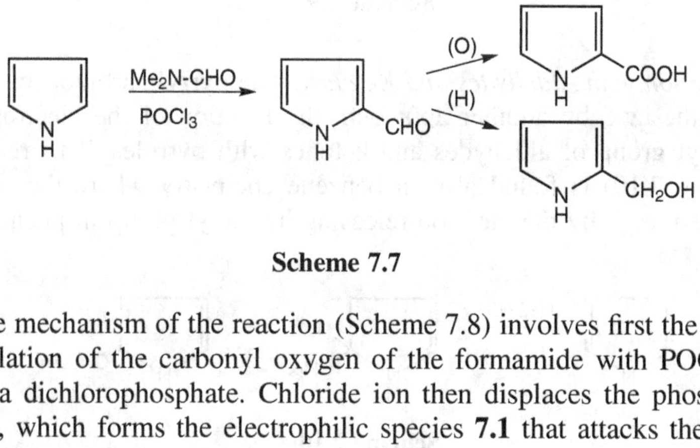
Me₂N-CHO
POCl₃
CHO
(O)
(H)
COOH
CH₂OH

Scheme 7.7

The mechanism of the reaction (Scheme 7.8) involves first the phosphorylation of the carbonyl oxygen of the formamide with $POCl_3$ to form a dichlorophosphate. Chloride ion then displaces the phosphate group, which forms the electrophilic species **7.1** that attacks the ring. Hydrolysis of the imino group restores the carbonyl group.

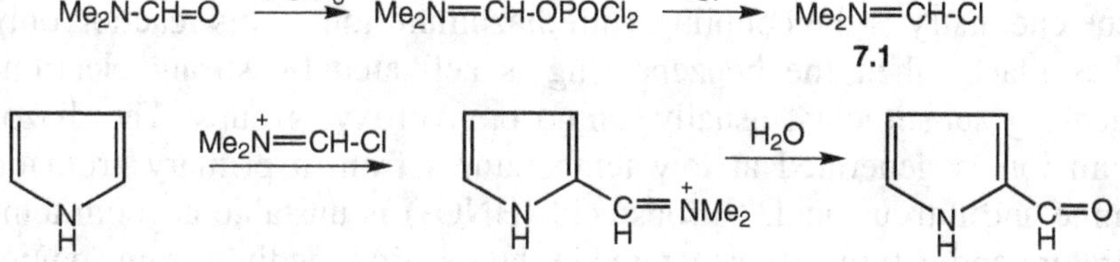
Me₂N-CH=O
POCl₃
Me₂N=CH-OPOCl₂
Cl⁻
Me₂N=CH-Cl
7.1
Me₂N=CH-Cl
C= NMe₂
H₂O
C=O

Scheme 7.8

The pyrryl carbinols occupy a special place in synthetic pyrrole chemistry; it was discovered early that with acid, they can react with another pyrrole nucleus to form the valuable dipyrrylmethane structure, which is of interest in the synthesis of porphyrins. Acid leads to the formation of the resonance stabilized pyrrylmethylcarbocation (**7.2**), which acts as an electrophile by the usual mechanism (Scheme 7.9).

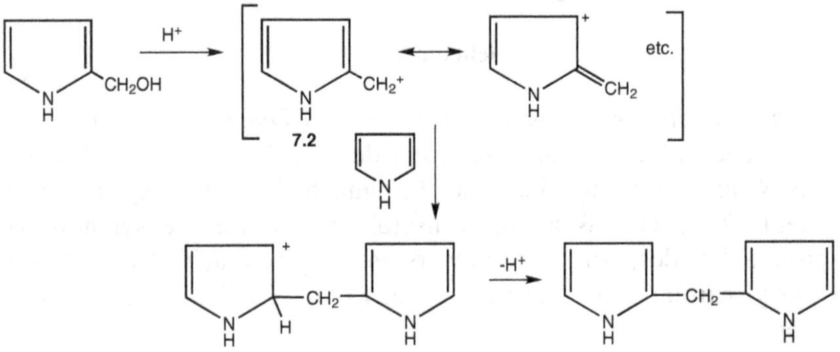

Scheme 7.9

Reaction with Aldehydes and Ketones. The pyrryl carbinols can also be synthesized by another approach: the reaction of the electrophilic carbonyl group of aldehydes and ketones with pyrroles. This reaction (Scheme 7.10) is found also in benzene chemistry where the ring is activated, e.g., by the electron-releasing hydroxyl group in phenols.

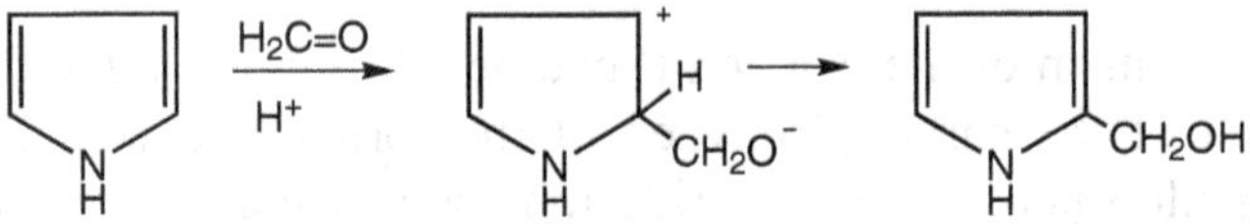

Scheme 7.10

Coupling with Diazonium Ions. Another familiar reaction from benzene chemistry is the coupling with diazonium ions. This reaction only takes place when the benzene ring is activated by strong electron-releasing substituents, usually amino or hydroxyl groups. The diazonium ion is generated at low temperatures from a primary aromatic amine and nitrous acid. Nitrous acid (HNO_2) is unstable at room temperature and is typically generated in the reaction medium from sodium nitrite and an acid (Scheme 7.11).

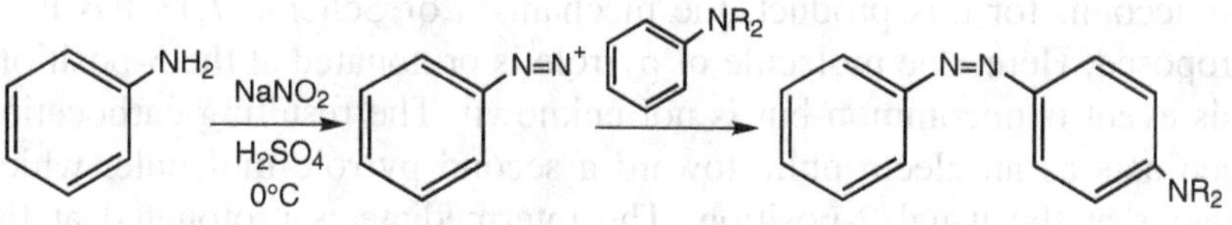

Scheme 7.11

As is consistent with the high reactivity of pyrroles toward electrophiles, the coupling reaction with diazonium ions is well known for the pyrrole ring system. The highly colored azo product, e.g., **7.3**, is useful as a precursor of aminopyrroles, which can be formed with reducing agents such as stannous chloride (Scheme 7.12).

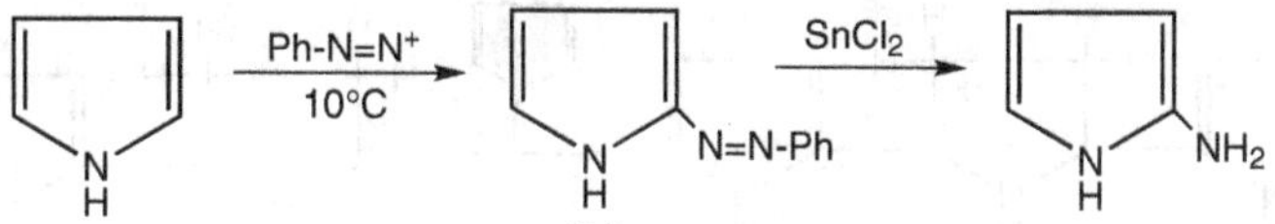

Scheme 7.12

7.1.2.2. Basic and Acidic Properties of Pyrroles. The low electron density at nitrogen in pyrroles make them weak bases (for pyrrole, K_b about 10^{-17}; cf. to noncyclic amines at about 10^{-5}), and it is not possible to make salts of pyrroles with aqueous acids. In fact, as will be shown next, protonation takes place on carbon, not on nitrogen. Also, pyrroles do not form quaternary salts with alkylating agents, or amine oxides with peroxy compounds. This is in stark contrast with pyridines. Another explanation for the unavailability of the electron pair on nitrogen in pyrroles is that the aromatic sextet (and its energy of stabilization) would be destroyed if it were used in forming a bond.

When pyrrole is heated with strong acids, a crystalline compound is formed that contains three pyrrole units. Its structure has been established as **7.4**. Strong acids can also cause the undesirable formation of polymeric products from pyrrole. These processes depend on the protonation of carbon of the ring, not of nitrogen.

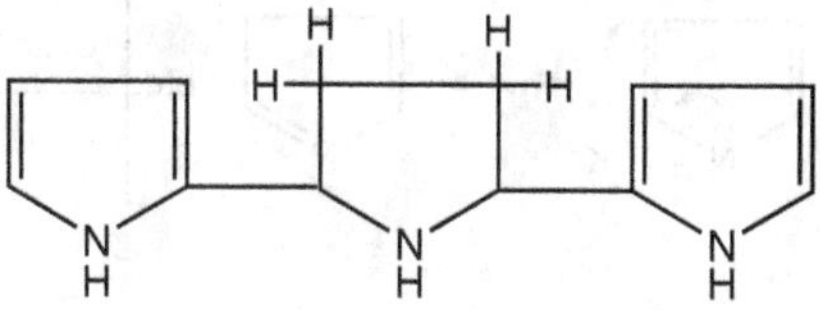

7.4

To account for this product, the mechanism of Scheme 7.13 has been proposed. Here, one molecule of pyrrole is protonated at the 3-position; this event is uncommon but is not unknown. The resulting carbocation then acts as an electrophile toward a second pyrrole molecule, which attacks at the usual 2-position. The intermediate is protonated at the 3-position of the dihydropyrrole ring, and the carbocation formed then attacks the 2-position of the third pyrrole molecule.

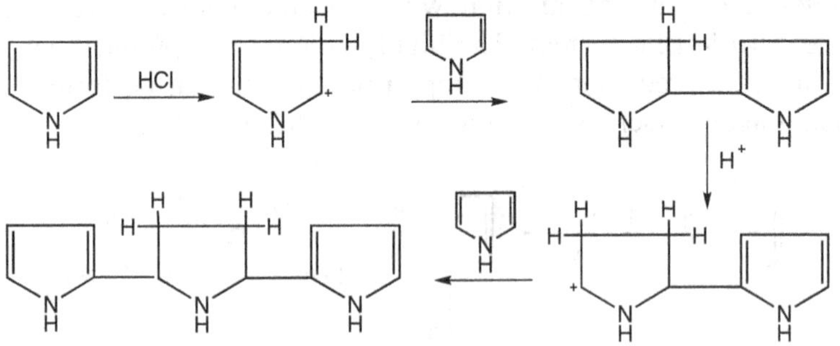

Scheme 7.13

The sensitivity of pyrroles to acids must be taken into consideration when designing syntheses, and in general, acidic conditions are to be avoided.

It is well known that the hydrogen of a secondary amine is weakly acidic and can be removed with strong bases to form anions. The hydrogen on nitrogen of pyrroles is considerably more acidic and can be removed readily with bases such as metallic amides (e.g., KNH_2) and also with active metals (Na or K). Pyrrole has an acid dissociation constant of 10^{-17}, which is about that of an alcohol. The extra stability of pyrrole anions comes from the resonance delocalization of the negative charge (Scheme 7.14). Pyrrole anions can be alkylated or acylated and can participate in Michael and other reactions typical of nitrogen anions. These properties are highly useful in the synthesis of N-substituted pyrroles. The Michael reaction is illustrated in Scheme 7.14.

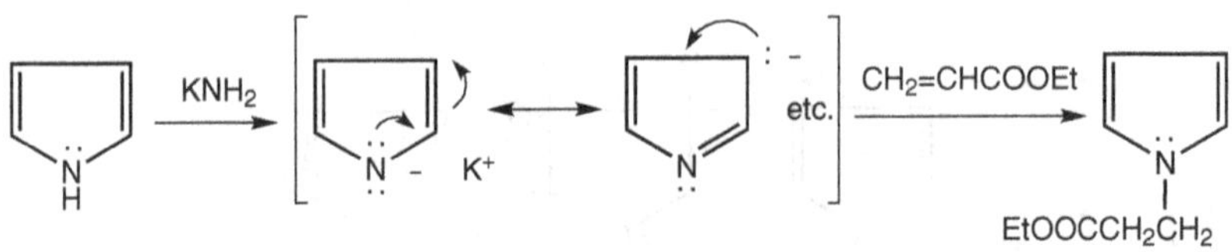

Scheme 7.14

7.1.2.3. The Diels-Alder Reaction. The aromatic ring system of benzene and pyridine prevents these compounds from entering into the Diels–Alder cycloaddition reaction as dienes except in the rarest of circumstances. So it is also with pyrrole. However, the electron delocalization into the ring can be moderated by placing electron-withdrawing groups on nitrogen, because such groups share in accepting the electron pair on N. An example is the N-carbethoxy derivative of pyrrole; with strong dienophiles, the Diels–Alder cycloaddition takes place to give a derivative of the 7-azabicyclo[2.2.1]heptadiene (7-azanorbornadiene) system (Scheme 7.15).

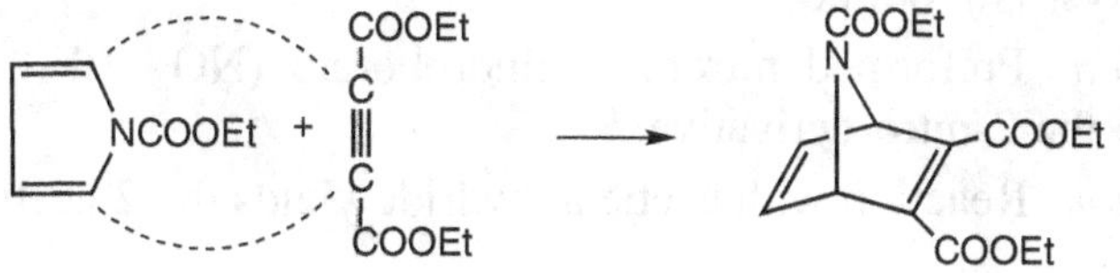

NCOOEt + COOEt–C≡C–COOEt → (N-COOEt bicyclic adduct with two COOEt)

Scheme 7.15

This method is useful for the construction of this bridged ring system.

7.1.2.4. Metallation of the 2-Position of Pyrroles. A synthetically useful property of N-alkyl pyrroles is that the proton on the 2-position can be removed with active metals (Na, Li, and K) or strong bases (butyllithium and BuLi). This behavior is different from that of pi-deficient nitrogen heterocycles, where with strong nucleophiles such as BuLi addition to a ring C=N unit takes place (see Chapter 6). The pyrrolic anions are reactive to electrophiles and give addition products with carbonyl groups, or C-alkyl derivatives with alkyl halides (Scheme 7.16), among other reactions.

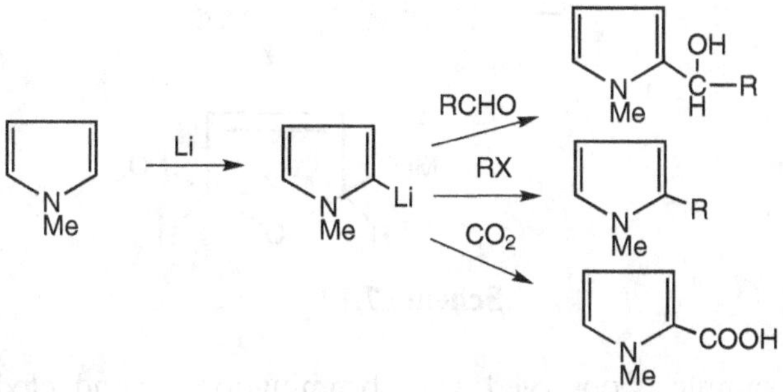

N-Me pyrrole —Li→ 2-Li-N-Me pyrrole; RCHO → 2-CH(OH)R; RX → 2-R; CO_2 → 2-COOH

Scheme 7.16

7.1.3. Reactions of Furans

Furan is the least aromatic of the N, O, S heterocycles, and special conditions are required to preserve the ring in electrophilic substitution reactions. The ring is also sensitive to acidic conditions. When substitutions are successful, it is the 2-position that is entered, for the same reasons described for pyrrole substitutions. Some processes that are useful for furan are described as follows:

Sulfonation. Sulfur trioxide in pyridine gives the 2-sulfonic acid.

Bromination. Bromine in dioxane gives the 2-bromo derivative. No catalyst is required.

Nitration. Preformed nitronium fluoroborate ($NO_2^+BF_4^-$) acts to give the 2-nitro derivative.

Acylation. Reaction with acetic anhydride yields the 2-acetyl derivative.

Under some conditions, furan undergoes addition of reactants to the diene system and thus gives nonaromatic products. This occurs by reaction with an electrophile to form the usual carbocation. which then adds a nucleophile rather than eliminating a proton in the usual way of aromatic substitution. As an example, when nitric acid in methanol solution is used as the nitration medium, the product is a 5-methoxy-2-nitro addition product (Scheme 7.17).

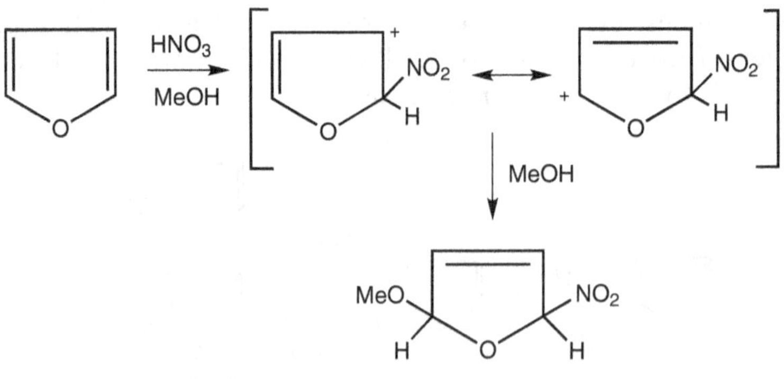

Scheme 7.17

Another example is observed when bromination is conducted in water (Scheme 7.18); here, the bromide ion acts as the nucleophile.

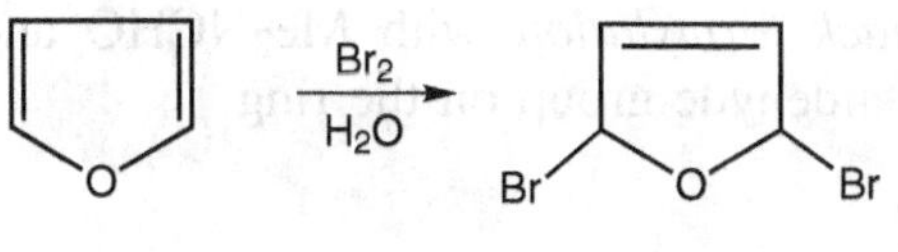

Scheme 7.18

Again in contrast to other aromatic systems, furans perform well as dienes in the Diels–Alder reaction, and many examples are known. The reaction with maleic anhydride is shown in Scheme 7.19. Here the *endo* addition product (**7.5**) is formed faster than *exo* (**7.6**), but the *exo* isomer is the more stable. Because this is an equilibrium process, ultimately *exo* is the final product. We have already shown that furan is an excellent participant in intramolecular Diels–Alder reactions (Chapter 5, section 5.1.3.7).

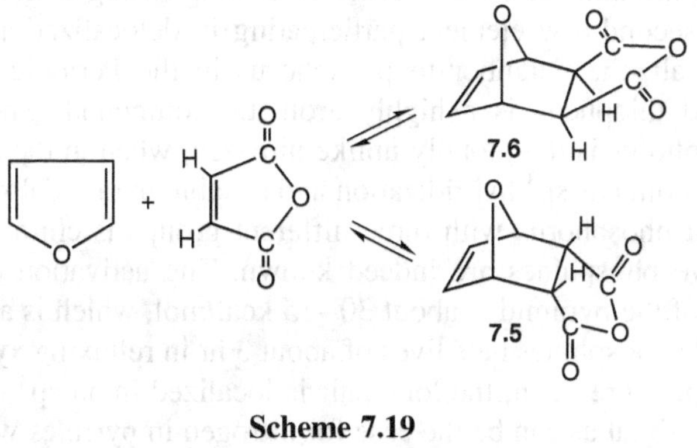

Scheme 7.19

7.1.4. Thiophene Reactivity

Thiophene undergoes a considerable number of electrophilic substitutions, some with greater ease than with benzene as might be expected from the increased pi-electron density in the ring. Substitution takes place at the 2-position. Some examples of the special conditions found useful for thiophene substitution are listed below.

Sulfonation occurs with concentrated sulfuric acid at 30°C.

Nitration is accomplished with nitric acid-acetic anhydride.

Chlorination in the dark, or with SO_2Cl_2, gives the mono-chloro derivative.

Friedel–Crafts acylation with acetyl chloride requires the milder catalyst $SnCl_4$ rather than the usual $AlCl_3$.

Vilsmeier–Haack formylation with Me_2NCHO and $POCl_3$ readily installs the aldehyde group on the ring.

7.1.5. The Phosphole Ring System

In principle, the ring system where PH replaces NH should be capable of having aromatic character because the lone electron pair on P could be delocalized into the ring as in pyrrole. The parent phosphole molecule, however, is unstable and has only been detected spectroscopically at $-90°C$,[11] but the P-methyl derivative is stable and distillable, and many other substituted phospholes are known. The field is growing regularly and deserves brief consideration here. The literature has been reviewed extensively.[12,13]

There has been much discussion about the reality of phospholes possessing aromatic character. There is nothing wrong with the general idea of a second-row element participating in delocalization phenomena; after all, the neighbor to phosphorus in the Periodic System is sulfur, and thiophene is a highly aromatic compound. The problem with phosphorus is that, totally unlike nitrogen, when in the tricovalent state, this atom has sp^3 hybridization and a stable pyramidal shape. This means that phosphorus with three different groups is chiral, and optically active phosphines are indeed known. The activation energy for inversion of the pyramid is about 30–35 kcal/mol, which is adequate to give simple phosphines half-lives of about 3 hr in refluxing xylene.[14] At room temperature, then, the lone pair is localized in an sp^3 orbital and not in a p-orbital as can be the case for nitrogen in pyrroles with its low barrier to pyramidal inversion. The sp^3 orbital is oriented away from the plane of the ring and thus cannot overlap effectively with the carbon p-orbitals as happens with nitrogen. The view has become prevalent that phospholes are of low aromaticity. It should be noted, however, that the proton NMR spectra do show the downfield shifting found in aromatic system; for 1-methylphosphole, both the alpha and beta protons are at δ 6.5–7.5 in a complex AA′BB′X spectrum ($X = {}^{31}P$). Simple phospholes are generally liquids, but 1-benzylphosphole was found to be a crystalline solid at around room temperature. This allowed the structure to be determined by X-ray diffraction analysis, which clearly showed the pyramidal shape of phosphorus.[15] That some electron delocalization was present, however, was suggested from determination of the bond lengths in the ring; this gave a low BI of 35.5,[2,7,16] which is below that of furan (43) but nevertheless points to a measure of delocalization. The NICS value is also well below that of furan.[9] It was later reasoned and justified theoretically that the delocalization could be increased if the phosphorus pyramid could be partially flattened by increasing steric hindrance between the P-substituent and the ring atoms. This was accomplished experimentally by placing the large 2,4,6-tri(*tert*-butyl)phenyl substituent on P. X-ray diffraction analysis indeed showed significant flattening of the P pyramid and marked changes in bond lengths in the ring.[17] The Bird index was 56.5, which is close to the value for pyrrole (59). Consistent with this increased aromaticity, the compound underwent the Friedel–Crafts reaction with acetyl chloride and $AlCl_3$, which is the first phosphole to undergo any electrophilic substitution on carbon (Scheme 7.20). The major product had structure **7.7**; acylation also occurred at positions 4 and 5. Other properties of phospholes are unique to phosphorus and for the most part are unlike pyrrole properties.

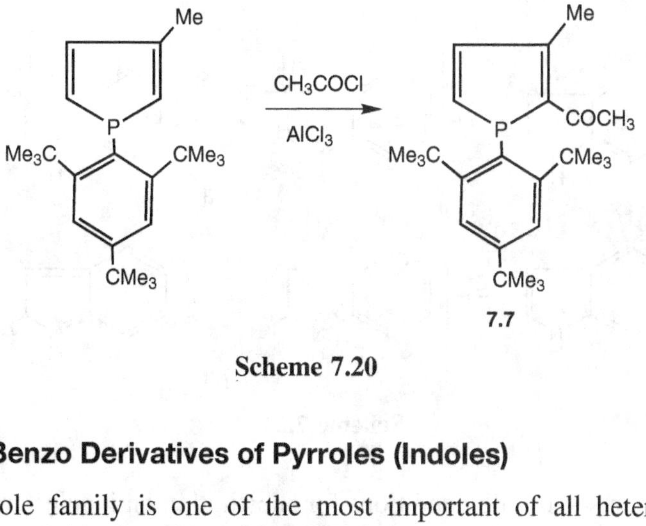

Me P Me_3C CMe_3 CMe_3 $\xrightarrow[AlCl_3]{CH_3COCl}$ Me $COCH_3$ P Me_3C CMe_3 CMe_3 **7.7**

Scheme 7.20

7.1.6. Benzo Derivatives of Pyrroles (Indoles)

The indole family is one of the most important of all heterocyclic families, and the chemistry of this system is vast. Many natural products and synthetic medicinals contain this nucleus. We are restricted here to matters of aromaticity, and we find that we are dealing with a stable ring system that exhibits all the properties expected from being pi-excessive. Calculated electron densities are shown in Figure 7.5, where it is found that all carbons have densities greater than 1.00.

Electrophilic substitutions occur readily with an attack on the electron-rich pyrrole moiety rather than the benzene ring. The 3-position is entered in preference to the 2-position, but if the 3-position is blocked, substitution occurs at the 2-position, thus again

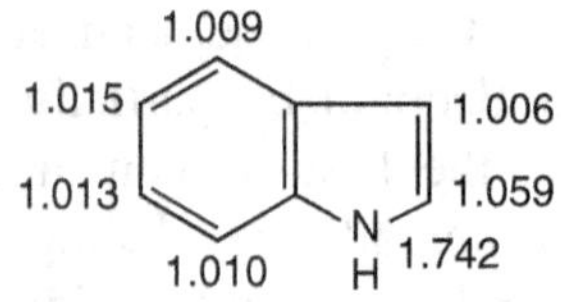

Figure 7.6. Pi-electron densities for indole.

in preference to attack on the benzene ring. The preference for the 3-position can be explained by considering the resonance possibilities from attack of a positive species (Scheme 7.21). The intermediate **7.8** is stabilized effectively by electron release from nitrogen, whereas stabilization of the positive intermediate from attack at the 2-position requires disruption of the aromatic system of benzene (**7.9**). Although it is still a stabilizing effect, it is less so than that of direct electron release from nitrogen as in **7.8**.

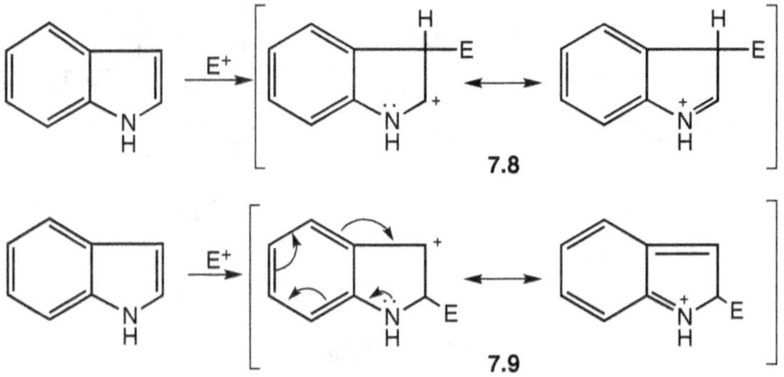

Scheme 7.21

The reagents and mild conditions for substitution on indoles are similar to those employed for pyrroles. It should be added that protonation, not surprisingly, occurs at the 3-position rather than on nitrogen.

7.1.7. Large-Ring Pi-Excessive Heterocycles

A Hückel aromatic system with ten electrons (n = 2) can be established with the 9-membered azonine ring system involving the eight pi-electrons of the double bonds and the lone pair on nitrogen. The all-cis ring (**7.10**) is known but not stable because of the angle strain as described in the aza-annulenes (Chapter 6). Its proton NMR spectrum clearly shows the deshielding expected from the operation of a ring current. With one trans double bond, the strain is relieved and compound **7.11** is of reasonable stability.

δ 6.9–7.0
H δ 6.0
H δ 7.07

7.10 **7.11**

Derivatives of the 7-membered azepine ring are well known, although the ring is not aromatic with eight pi-electrons. It is, in fact, classed as antiaromatic from its MO description of having unfilled bonding MOs.

Large-ring oxygen and sulfur heterocycles are also known and those possessing 4n+2 pi-electrons show manifestations of aromaticity.

7.1.8. Aromatic Ring Systems with Mixed Heteroatom Bondings

Here, we will consider the situation where in a 5-membered ring there is a saturated heteroatom (e.g., NH, O, or S) that acts as an electron donor and a pyridine-like C=N that acts as acceptor. This situation leads to strong delocalization and to rings of great stability and common occurrence.

The imidazole system provides a suitable example with which to consider this situation. With the NH group, however, tautomerism is present and the proton shifts from one nitrogen to the other (Scheme 7.22).

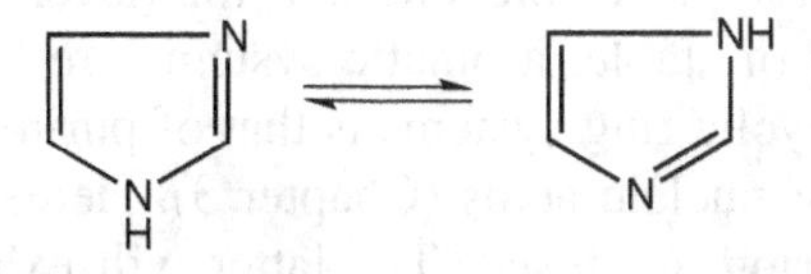

Scheme 7.22

To circumvent this problem, we will consider the N-methyl derivative of imidazole, where no tautomerism is possible. The resonance hybrid for this compound is shown in Scheme 7.23.

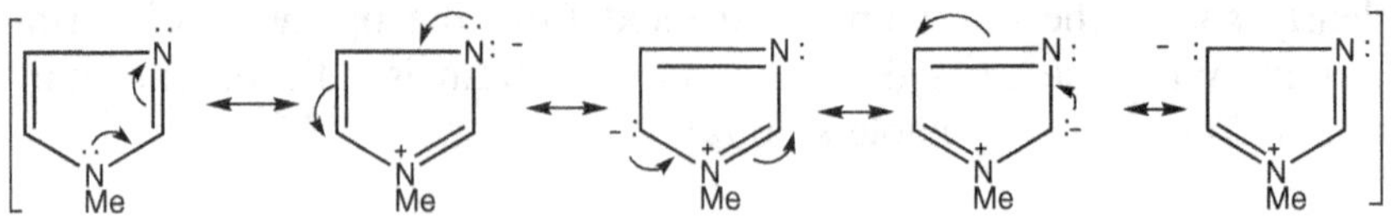

7.12

Scheme 7.23

Contributor **7.12** illustrates the direct interaction of donor N-Me and acceptor C=N, which is a major contributor to the hybrid. The negative charge on N is acceptable and of course is involved in pyridine-type delocalization. As a result, the I_A of 79 for imidazole is close to that of pyrrole (85) and thiophene (81.5).

The resonance hybrid for donation of oxygen electrons in 1,3-oxazoles and of sulfur electrons in 1,3-thiazoles is of the same form. In thiazole, the Bird I_A value is 79. These compounds have all the characteristics of aromatic systems and readily undergo electrophilic substitutions.

Similarly, pronounced delocalization is present in the triazoles **7.13** and **7.14**. The Bird I_A values are even higher (90 and 100, respectively).

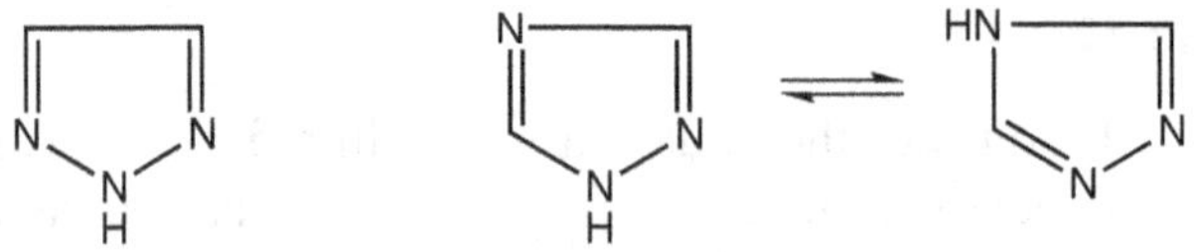

7.13, 1H-1,2,5-triazole **7.14,** 1H-1,2,4-triazole

A considerable variety of heterocyclic systems from various mixtures of the heteroatoms O, N, and S is possible, and where the donor-acceptor situation is present, the aromaticity can be expected to be pronounced.

These principles apply in bicyclic and multicyclic systems as well, and large numbers of stable, aromatic systems are known. One of the most important bicyclic ring systems is that of purine, which is present among the bases of nucleic acids (Chapter 3). Here, a pyrimidine ring and an imidazole ring are fused. The latter will exhibit its usual tautomeric behavior, but now the two forms are not equivalent, and in purine itself form **7.15** is known to predominate over **7.16**. In the nucleoside component of the nucleic acids (and in other compounds), the bond to the sugar moiety is located at the NH of form **7.15** (incorrectly but historically labeled the 9-position).

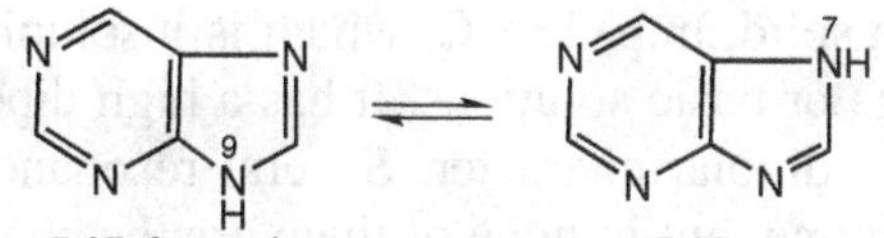

7.15, favored 7.16

Scheme 7.24

Several resonance forms can be written for purine where the NH electrons are donated to C=N units, either in the pyrimidine ring or the imidazole ring. One such form is shown as **7.17**.

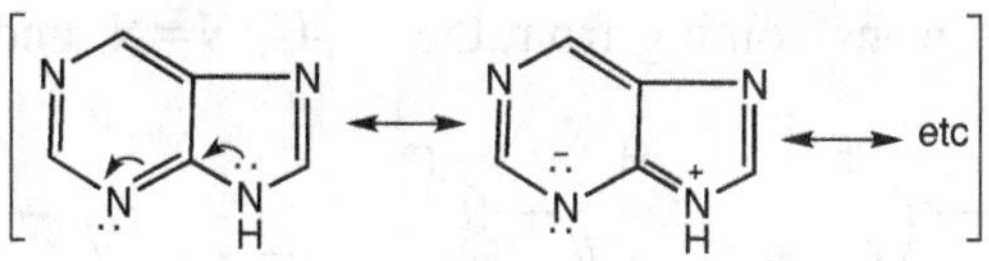

7.17

7.2. MESOIONIC HETEROCYCLES

Certain types of 5-membered heterocyclic rings have the unique feature of possessing dipolar structure with no resonance form that is neutral and fully covalent. Thus, they are cyclic zwitterions. There are more than 50 dipolar compounds (including some that have 6-membered rings) that have been synthesized; they have become known as mesoionic compounds. Some of their properties suggest that they can be classified as aromatic substances.

The 1,2,3-oxadiazole derivative **7.19** was the first mesoionic compound to be recognized as such. This occurred at the University of Sydney, Australia, by Earl and Mackney in 1935,[18] and the name "sydnone" was coined for it. Its rather simple synthesis is shown in Scheme 7.25. The starting material is the nitroso derivative (**7.18**) of N-phenylglycine, which is converted to a mixed anhydride with acetic anhydride. The oxygen of the nitroso group performs nucleophilic displacement of the acetate ion to close the ring. This leaves nitrogen with a positive charge, which is lost with expulsion of a proton.

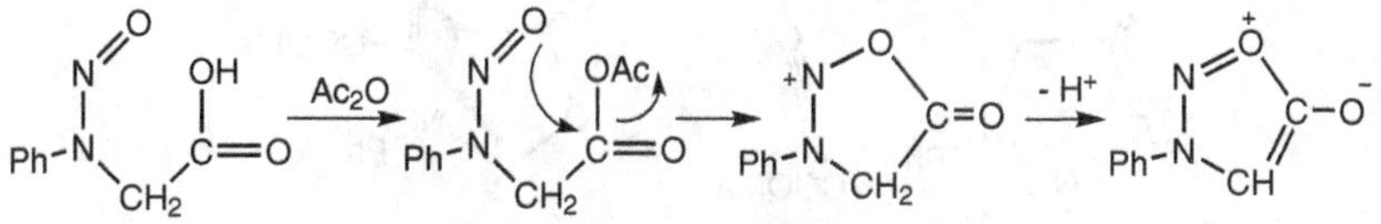

7.18 7.19

Scheme 7.25

Sydnone **7.19** is a solid, m.p. 135°C, which is insoluble in water but is not stable in acidic or basic solutions. It has a high dipole moment of 7 D in keeping with dipolar character. Several resonance structures can be written for sydnone, but in none of them are the charges neutralized. In the resonance structures **7.20** and **7.21** shown in Scheme 7.26, the positive charge is delocalized in the ring, with oxygen carrying the negative charge outside the ring. This leads to a commonly used general structure **7.22** for expressing these charge properties and accounting for the high dipole moment. It also conveys the notion that the ring system is delocalized with six pi-electrons, and thus it is an aromatic system, with the six electrons coming from the C=C, N=N, and O units.

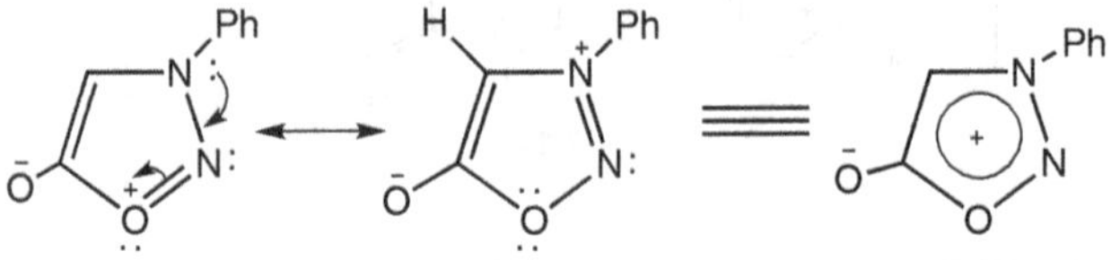

7.20 7.21 7.22

Scheme 7.26

However, other resonance structures (**7.23** and **7.24**) can be written where both charges are in the ring.

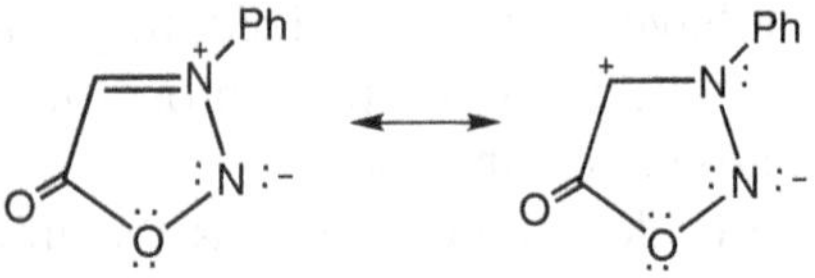

7.23 7.24

Structure **7.24** is of special importance because the charges are separated by a single atom (N) bearing a lone pair, and thus the structure can be classified as a 1,3-dipole. Indeed, sydnone is well known to function in cycloadditions as a 1,3-dipole, and it gives some interesting products; one such reaction is shown in Scheme 7.27.

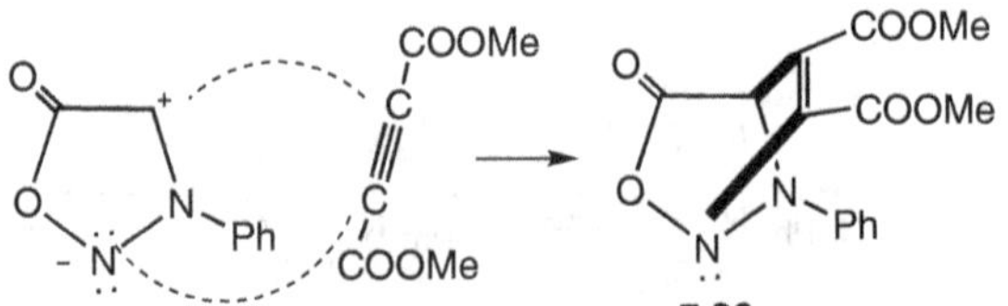

7.25 7.26

Scheme 7.27

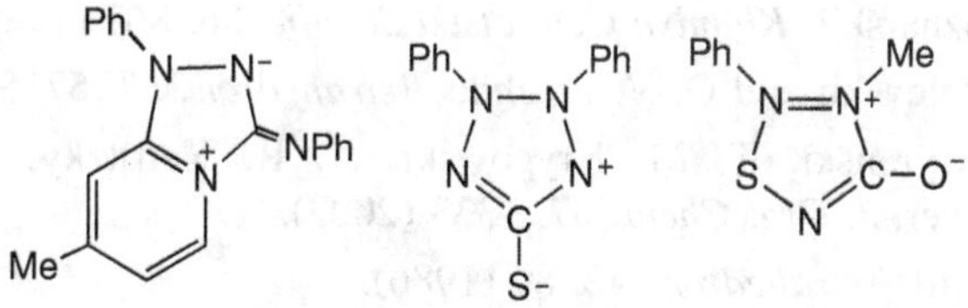

Figure 7.7. Some mesoionic compounds.

Another striking property of sydnones that is consistent with the view of aromaticity is that they undergo electrophilic substitution reactions, as with bromine in Scheme 7.28.

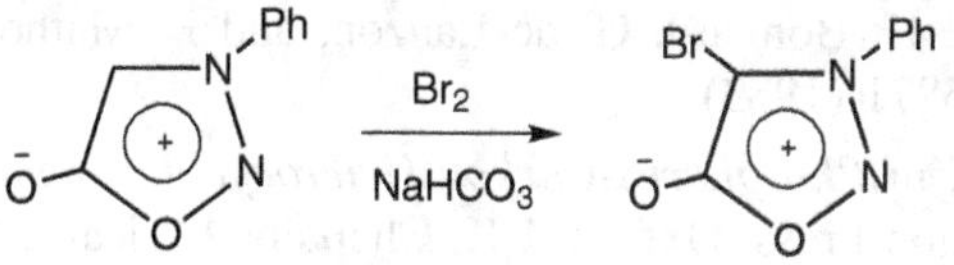

Scheme 7.28

Munchone is another mesoionic compound formed by a reaction similar to that for sydnone (R. Huisgen, University of Munich, hence its name[19]). Here, a C=O unit takes the place of N=O on nitrogen of a glycine derivative (Scheme 7.29). The ring formed has the dipolar 1,3-oxazole structure (**7.27**).

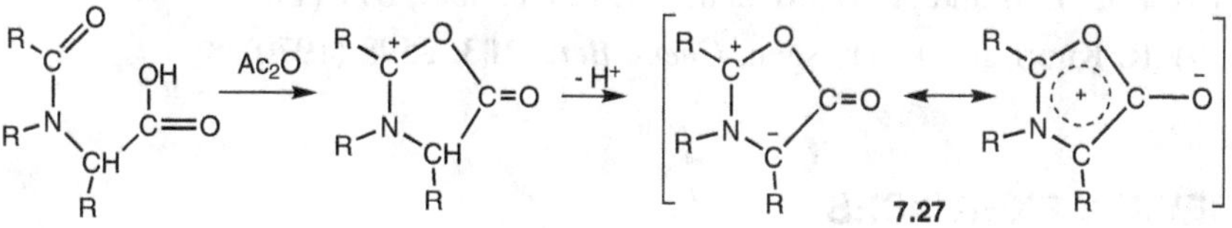

Scheme 7.29

A few other mesoionics are shown in Figure 7.7.

REFERENCES

(1) A. R. Katritzky, *Handbook of Heterocyclic Chemistry*, Pergamon Press, Oxford, UK, 1985, pp. 58, 571.

(2) C. W. Bird, *Tetrahedron*, **41**, 1409 (1985).

(3) A. F. Pozharskii, *Khimiya Geterotsikl Soedin*, **56**, 867 (1985).

(4) S. I. Kotelevskii and O. V. Prezhdo, *Tetrahedron*, **57**, 5715 (2001).

(5) M. K. Cyranski, T. M. Krygowski, A. R. Katritzky, and P. von R. Schleyer, *J. Org. Chem*. **67**, 1333 (2002).

(6) C. W. Bird, *Tetrahedron*, **42**, 89 (1986).

(7) C. W. Bird, *Tetrahedron*, **48**, 335 (1992).

(8) P. v. R. Schleyer, C. Maerker, A. Dransfeld, H. Jiao, and N. J. R. v. E. Hommes, *J. Am. Chem. Soc*., **118**, 6317 (1996).

(9) Z. Chen, C. S. Wannere, C. Cominboeuf, R. Puchta, and P. v. R. Schleyer, *Chem. Rev*., **105**, 3842 (2005).

(10) A. R. Katritzky, N. G. Akhmedov, J. Doskocz, P. B. Mohapatra, C. D. Hall, and A. Güven, *Magn. Reson. Chem*., **45**, 532 (2007).

(11) C. Charrier, H. Bonnard, G. de Lauzon, and F. Mathey, *J. Am. Chem. Soc*., **105**, 6871 (1983).

(12) L. D. Quin, in *Phosphorus-Carbon Heterocyclic Chemistry*, F. Mathey, Ed., Pergamon Press, Oxford, UK, Chapters 4.2.1 and 4.2.2.

(13) L. D. Quin, *Current Org. Chem*., **10**, 43 (2006).

(14) L. Horner, H. Winkler, A. Rapp, A. Mentrup, H. Hoffmann, and P. Beck, *Tetrahedron Lett*., 161 (1961).

(15) P. Coggon, J. F. Engel, A. T. McPhail, and L. D. Quin, *J. Am. Chem. Soc*., **92**, 5779 (1970).

(16) C. W. Bird, *Tetrahedron*, **41**, 5097 (1985).

(17) Gy. Keglevich, Zs. Böcskei, Gy. Keserü, K. Újszászy, and L. D. Quin, *J. Am. Chem. Soc*., **119**, 5095 (1997).

(18) J. C. Earl and A. W. Mackney, *J. Chem. Soc*., 899 (1955).

(19) R. Knorr and R. Huisgen, *Chem. Ber*., **103**, 2598 (1970).

REVIEW EXERCISES

7.1. Rank from 1 (most aromatic; highest Bird index) to 3 (least aromatic; lowest Bird index).

Benzene

Furan

Thiophene

7.2. Which of the following has excessive pi-electron density in the ring?

Furan Pyrrole Thiophene

7.3. Draw the expected product.

Li

H

O

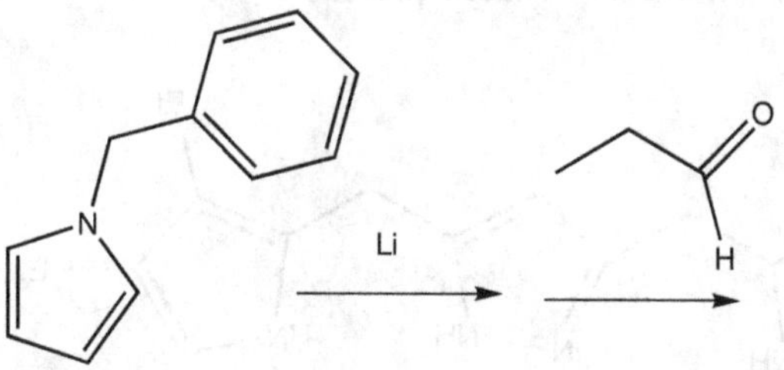

7.4. True or false: Compared with pyridine, pyrrole is less basic and more acidic.

7.5. Draw the expected product.

S

HNO_3

O

O O

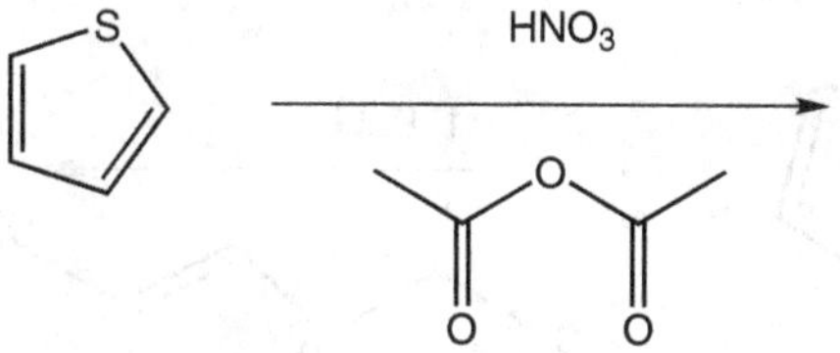

7.6. V. I. Terenin, et al., *Chem. Heterocycl. Comp.*, **44**, 200 (2008). Draw the expected product.

Cl

O

CH_3

N

N

$AlCl_3$

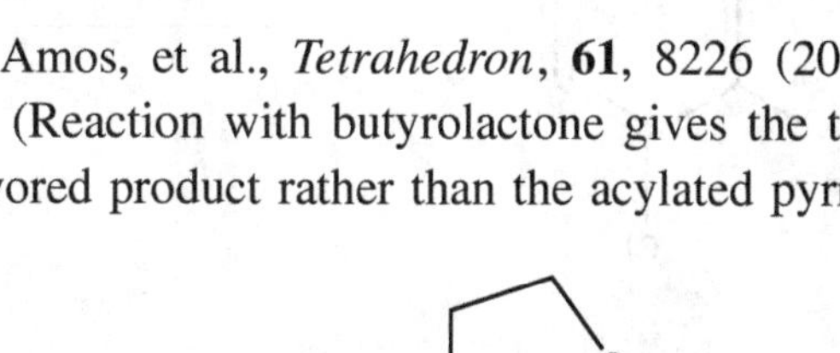

7.7. R. I. J. Amos, et al., *Tetrahedron*, **61**, 8226 (2005). Draw the product. (Reaction with butyrolactone gives the thermodynamically favored product rather than the acylated pyrrole.)

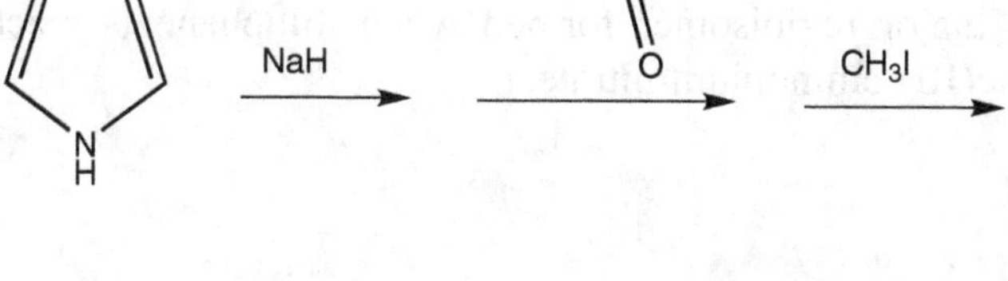

O

NaH

O

CH_3I

N
H

7.8. S. Katsiaouni et al., *Chem. Eur. J.*, **14**, 4823 (2008). Draw the expected Vilsmeier–Haack product.

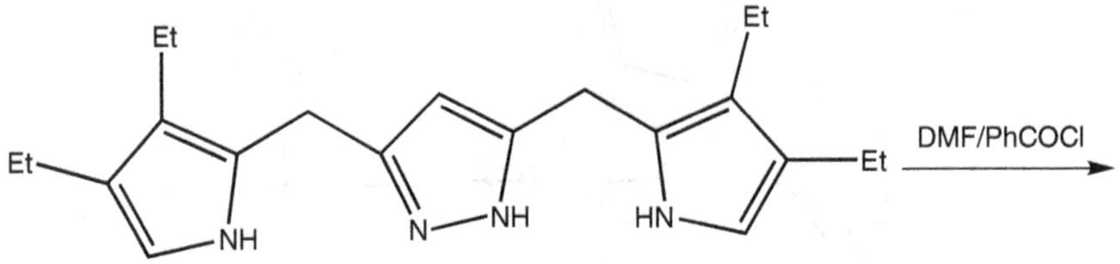

Et Et NH N NH HN Et Et DMF/PhCOCl

7.9. S. Katsiaouni, et al., *Chem. Eur. J.*, **14**, 4823 (2008). Draw the expected product formed by the lithiation of pyrrole followed by reaction with the pyrazole.

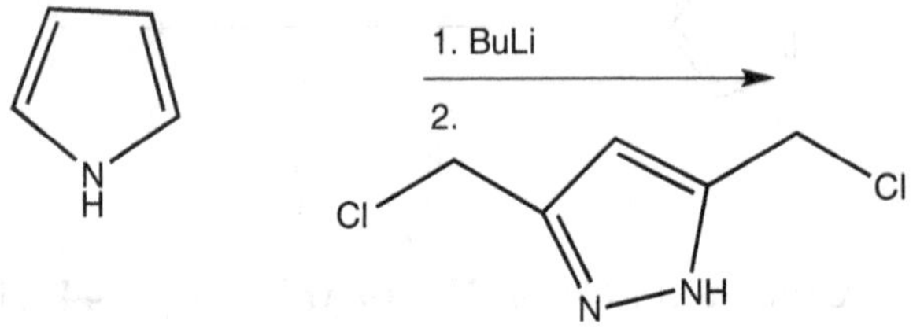

N H 1. BuLi 2. Cl N NH Cl

7.10. H. Kikuchi, et al., *Org. Lett.*, **11**, 1693 (2009). Draw the product.

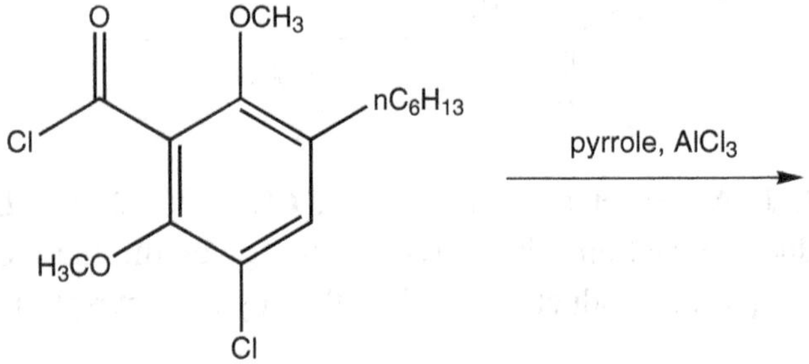

O OCH_3 nC_6H_{13} Cl H_3CO Cl pyrrole, $AlCl_3$

7.11. T. Kiyoshi et al., *J. Chem. Res. Synop.*, **8**, 497 (2003). Ceric(III) ammonium nitrate is used as a mild nitrating agent. Draw the major regioisomer formed when thiophene is reacted with ceric(III) ammonium nitrate.

7.12. J. S. Yadav, et al., *Tetrahedron Lett*., **48**, 5573 (2007). Draw the expected alkylation product.

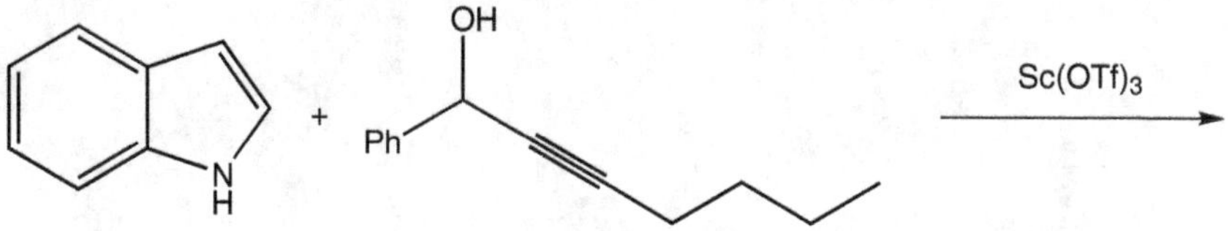

Indole + Ph–CH(OH)–C≡C–CH₂CH₂CH₃ —$Sc(OTf)_3$→

7.13. The ^{15}N NMR of imidazole has how many peaks? Explain.

7.14. The ^{15}N NMR of N-methylimidazole has how many peaks? Explain.

CHAPTER 8

THE IMPORTANCE OF HETEROCYCLES IN MEDICINE

8.1. GENERAL

Most pharmaceuticals are based on heterocycles. An inspection of the structures of the top-selling brand-name drugs in 2007[1] reveals that 8 of the top 10 and 71 of the top 100 drugs contain heterocycles. This is not new. Heterocycles have dominated medicinal chemistry from the beginning. Consistent with their importance, many U.S. patents by pharmaceutical companies involve heterocyclic compounds. For example, a search of the patent literature from 1976 to September 2008 reveals that 1729 patents issued to Pfizer, as a representative company, contain the word "pyridine." Merck has 3504 U.S. patents containing the word pyridine. This is not peculiar to pyridine as shown in Table 8.1 where other heterocycles were searched for three major companies.

There is every reason to expect this trend to continue. All the major pharmaceutical companies have significant research efforts involving heterocycles. For example, Pfizer had 12 U.S. patents issue in August of 2009 and 9 involved heterocycles.

Because of the magnitude of involvement of heterocycles used in medicine, a comprehensive discussion, even if possible, would be beyond the scope of this book. Rather, this chapter will discuss

Fundamentals of Heterocyclic Chemistry: Importance in Nature and in the Synthesis of Pharmaceuticals,
By Louis D. Quin and John A. Tyrell Copyright © 2010 John Wiley & Sons, Inc.

Table 8.1. Number of U.S. Patents Containing Certain Words describing Heterocycles

Word Searched	Pfizer	Merck	Abbott
Pyridine	1729	3504	711
Indole	424	623	158
Pyrrole	267	423	154
Imidazole	655	1382	377
Quinoline	441	699	290

briefly a little of the history of heterocycles in medicine, followed by examples of pyridines, indoles, quinolines, azepines, and pyrimidines in pharmaceutically active ingredients. Chapter 9 has other examples of these ring systems including their synthesis. Selection of these five groups is arbitrary and ignores several other types of heterocycles, but it is meant to give examples of the use of heterocycles in medicine. This classification is also an oversimplification. Many pharmaceutical compounds contain more than one type of ring system. For example, the first four exemplified pyridine compounds (section 8.3) used as proton pump inhibitors also contain a benzimidazole structure. Dimebon is discussed in the section on pyridines, but it also contains the indole ring. This, too, is arbitrary and not meant to imply that the pyridine structure is more important for Alzheimer's treatment than the indole structure. The United States Adopted Names Council serves health professionals by selecting nonproprietary names for new drugs based on pharmacological and/or chemical relationships. One monograph[2] organizes these new drugs by chemical structure, and much of the text involves heterocycles. There are chapters on 5-membered heterocycles, 6-membered heterocycles, 5-membered heterocycles fused to one benzene ring, 6-membered heterocycles fused to one benzene ring, bicyclic-fused heterocycles, and polycyclic-fused heterocycles. The reader is directed to this and preceding volumes for more examples of heterocycles in medicine.

8.2. HISTORICAL

Heterocycles have been used medicinally since the beginning of written records. Shen Nung, a Chinese scholar-emperor who lived in 2735 B.C., wrote of the herb Ch'ang Shan, as being helpful in treating fevers.[3] Ch'ang Shan was later found to contain dichroins,[4] for example, beta–dichroine.[5]

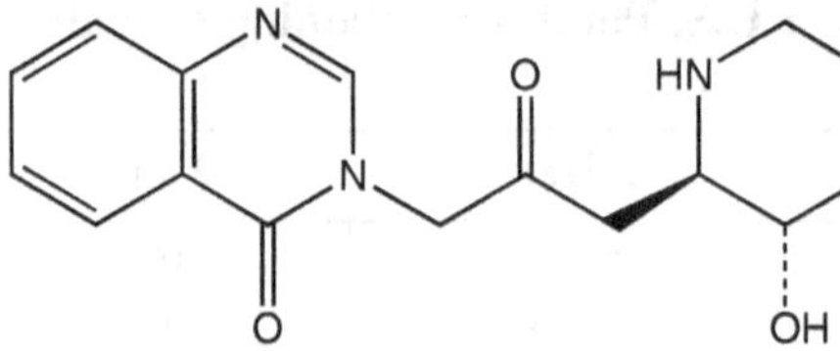

beta-dichroine

Another example of ancient usage of a heterocyclic compound is opium. Opium contains several alkaloids including morphine and was imported from Greece by the Egyptians before the war of Troy, which was waged in approximately 1200 B.C.[6] Even before the ancient Greeks, the Sumerians (Babylonians) carved tablets with pictures of the opium poppy.[3] Some of the first animal studies of drugs were done with opium. For example, in the 1700s, Robert Whytt used frogs to study the effect of opium on the heart.[7]

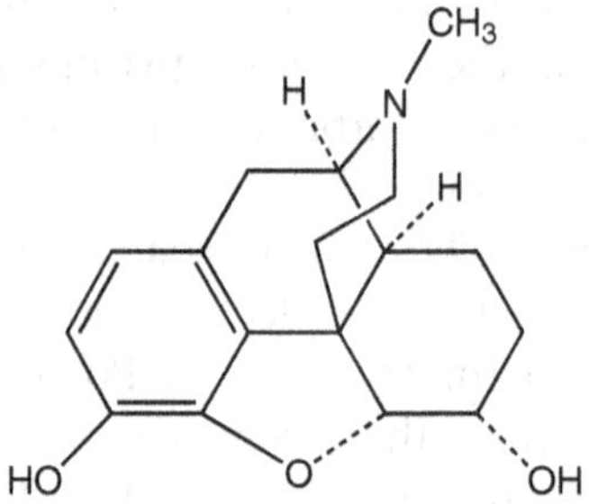

morphine

The first synthetic heterocyclic pharmaceutical seems to be antipyrine. Antipyrine is a pyrazole analgesic and an antipyretic, like aspirin. Ludwig Knorr used Emil Fischer's discovery of phenylhydrazine to synthesize antipyrine, and in 1883, Knorr was granted a patent on the synthesis. In 1885, one year after market introduction, almost 6 metric tons were sold, and in 1899, sales had grown to almost 800 metric tons.[8] More recently, antipyrine has been used in a solution with benzocaine to relieve ear pain and swelling caused by middle ear infections. Knorr's synthesis is shown in Scheme 8.1.[9,10] Details of the Knorr synthesis are discussed in Chapter 4.

Another class of early drugs is based on malonylurea, which was discovered in 1864 by von Baeyer. Legend has it that after making the

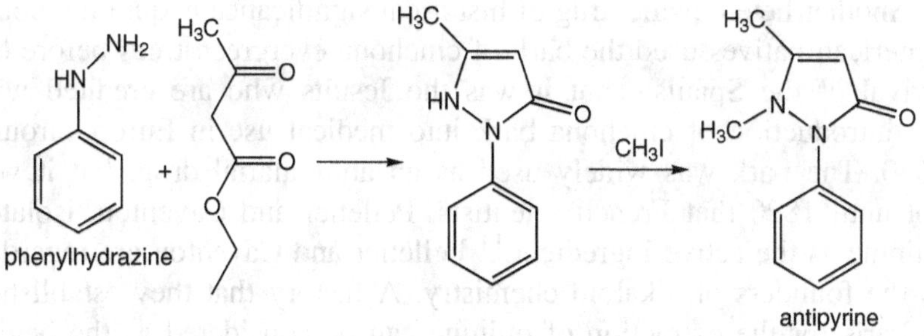

phenylhydrazine
+
CH3I
antipyrine

Scheme 8.1

compound from urea and malonic acid, von Baeyer was in a pub and attended a celebration of St. Barbara, who is patron saint of artillerymen. This prompted him to give malonylurea the name barbituric acid, which is a combination of the words Barbara and urea.[11]

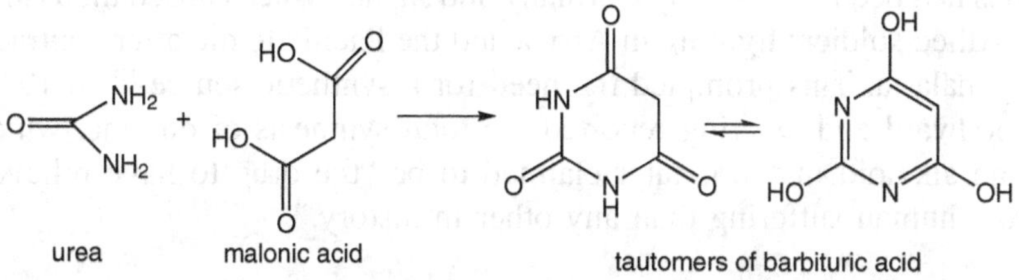

urea
malonic acid
tautomers of barbituric acid

Scheme 8.2

There are many derivatives of barbituric acid. The first to the market was diethylbarbituric acid, which is also known as barbital, malonal, or gardenal.[12] Phenobarbital was introduced by Bayer Pharmaceuticals in 1912[11] and is used currently for the treatment of epilepsy. In 1926, the effect of phenobarbital on cerebral circulation was studied.[13] During the twentieth century, more than 2500 barbiturates were synthesized, 50 of which were eventually employed clinically.[12]

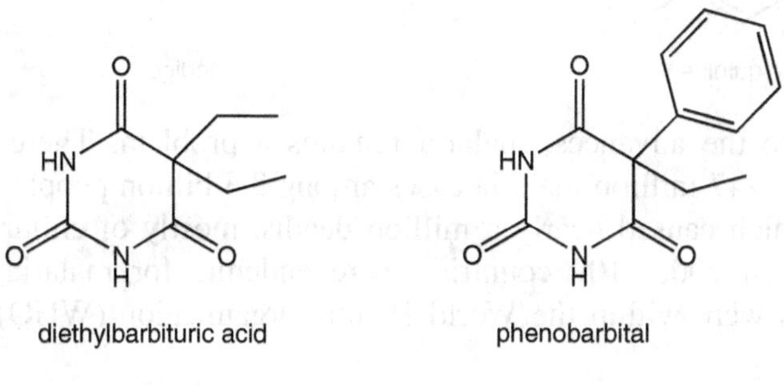

diethylbarbituric acid
phenobarbital

Another heterocyclic drug of historical significance is quinine. South American natives used the bark of cinchona evergreen trees before the arrival of the Spanish, but it was the Jesuits who are credited with the introduction of cinchona bark into medical use in Europe around 1640. The bark was widely used as an antimalarial drug, but it was not until 1820 that French scientists, Pelletier and Caventou, isolated quinine as the active ingredient.[14] Pelletier and Caventou are regarded as the founders of alkaloid chemistry. A factory that they established in Paris for the extraction of quinine can be considered as the beginning of the modern pharmaceutical industry. The synthesis of quinine remained elusive for more than 100 years after its isolation. In 1856, Perkin synthesized mauveine, which is an indigo dye. This was the first synthetic dye and was an offshoot of his unsuccessful efforts to synthesize quinine. Today, the Perkin Medal is widely acknowledged as the highest honor in American industrial chemistry.

The cinchona bark was in scarce supply in World War II. The plantations had been captured by Germany and Japan, which caused thousands of Allied soldiers fighting in Africa and the Pacific to die after contracting malaria. This prompted the need for a synthetic source.[15] In 1944, Woodward and Doering reported the total synthesis of quinine, which is an alkaloid that was later claimed to be "the drug to have relieved more human suffering than any other in history."

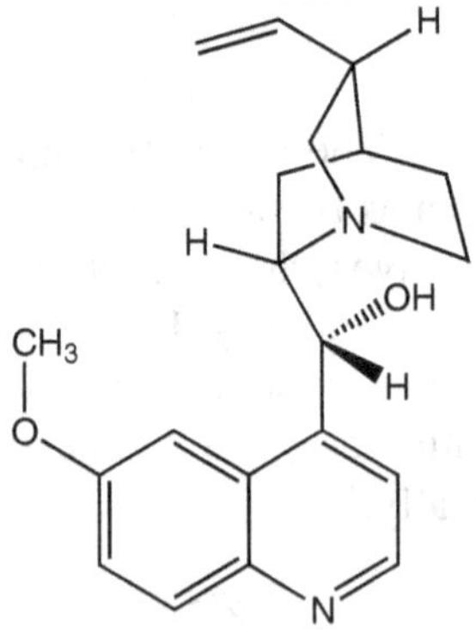

quinine

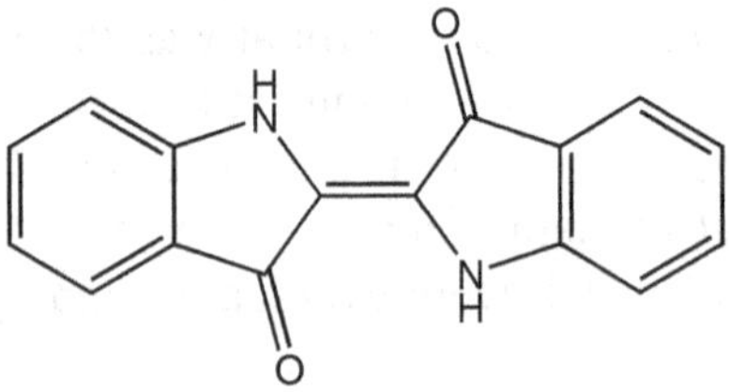

indigo

Despite the advances, malaria remains a problem. There were an estimated 247 million malaria cases among 3.3 billion people at risk in 2006, which caused nearly a million deaths, mostly of children under 5 years. In 2008, 109 countries were endemic for malaria, and 45 countries were within the World Health Organization (WHO) African region.[16] Quinine is still used as are the structurally simpler chloroquine and primaquine. Chloroquine is a purely synthetic drug discovered to solve the problem of quinine shortage during the war. Also prescribed today are the arteminisin-based combination therapies (ACTs).[17]

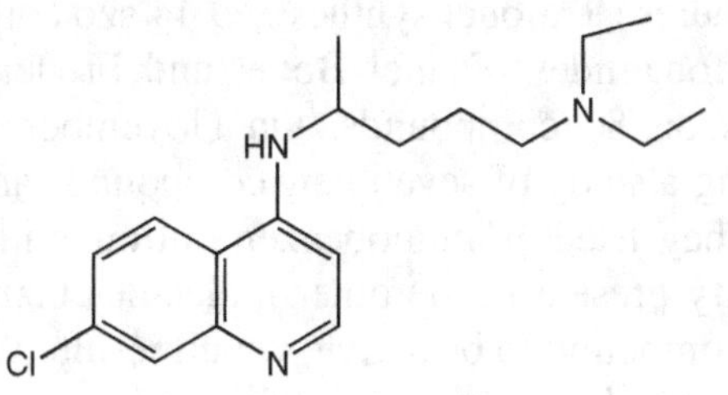

chloroquine

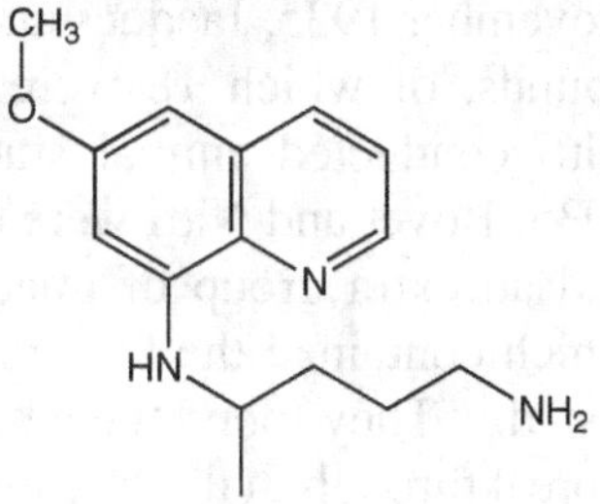

primaquine

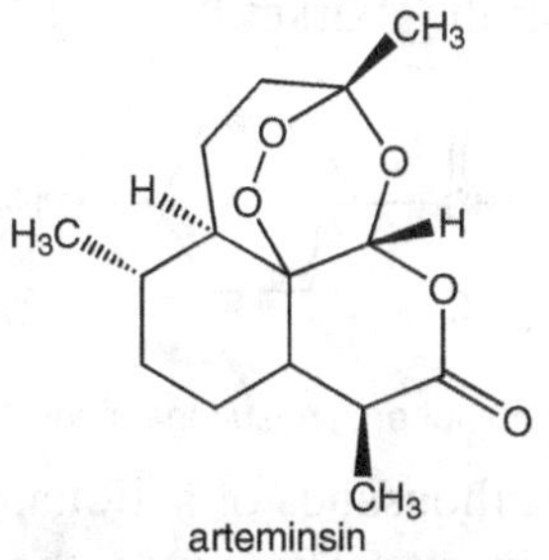

arteminsin

In 1932, Gerhard Domagk, who was working for I.G. Farbenindustrie, tested a red dye, 4-[(2,4-diaminophenyl)azo] benzenesulfonamide hydrochloride on mice that had been injected with streptococci. All the controls died and all the treated mice survived. The dye was later found to have curative powers in humans and went on the market in 1935 as Prontosil.[11]

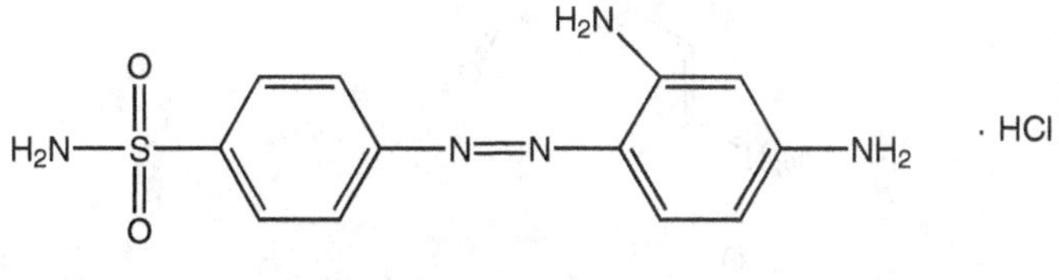

4-[(2,4-diaminophenyl)azo] benzenesulfonamide hydrochloride

In 1939, the Nobel Prize in Physiology or Medicine was awarded to Domagk. Domagk was cited for his work with Prontosil. At the time, the Germans were forbidden to accept the Nobel prize and Domagk was twice arrested and temporarily jailed by the Gestapo.[18] By 1939, Prontosil was recognized as the first of a new class of antibacterial drugs called the sulfa drugs.

The discovery of the antibacterial action of Prontosil prompted a flurry of research activity. At the Pasteur Institute, between July and November 1935, Jacques and Therese Trefouel synthesized 44 azo compounds, of which 18 were sulfonamides; Daniel Bovet and Frederic Nitti conducted animal studies on 80 compounds. On November 6, 1935, Bovet and Nitti were doing a study of seven new compounds and had an extra group of mice. They tried *p*-aminobenzenesulfonamide, which contained the functionality present in the other molecules being tested.[18] They found the white compound to be active, which resulted in a breakthrough in the research that had previously been focused on dyes. In 1937, *p*-aminobenzenesulfonamide, by now called sulfanilamide in the United States, went on the market.[18]

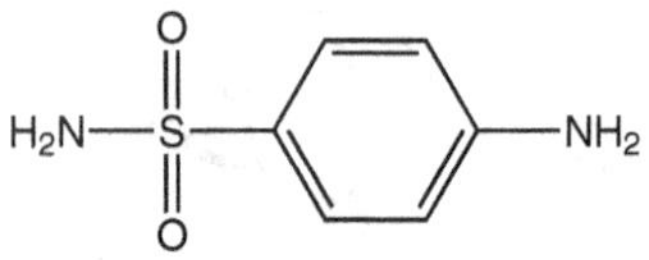

O
H_2N—S—NH$_2$
O

p-aminobenzenesulfonamide

During the next decade, thousands of sulfonamides were synthesized and tested as antibacterial agents. These were the first structure–activity studies. Also, this was one of the first examples where new lead compounds for other diseases were revealed from observed side effects.[19]

Many sulfonamides are still in use. More than a dozen antibiotics are listed in the U.S. Pharmacopeia,[20] some of which are shown next with varying substituents on the sulfonamide nitrogen.

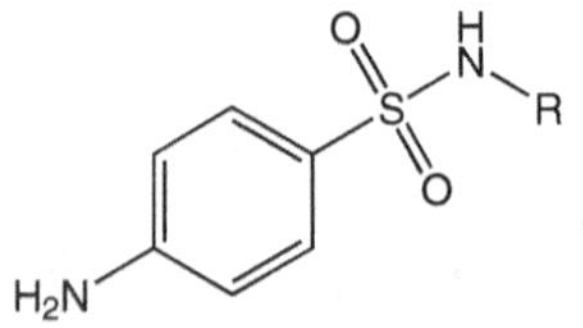

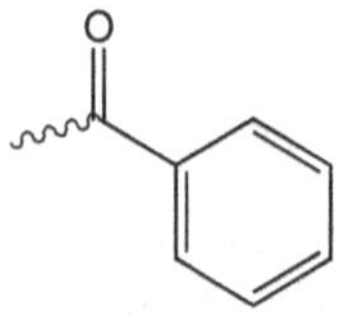

O H N R S O H_2N

O | O CH$_3$ | N N Cl | N N

sulfabenzamide 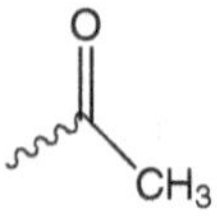 | sulfacetamide 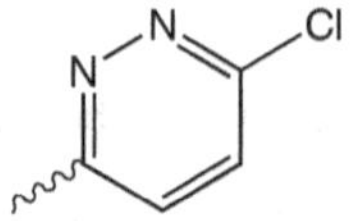 | sulfachloropyridazine 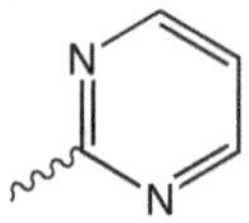 | sulfadiazine sulfadoxine 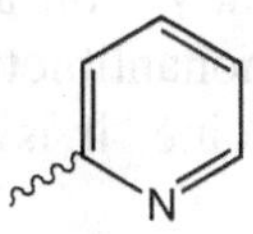 sulfapyridine 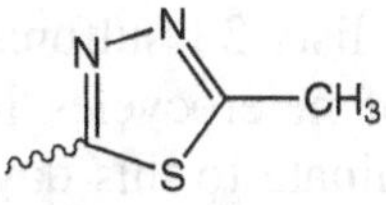 sulfamethizole 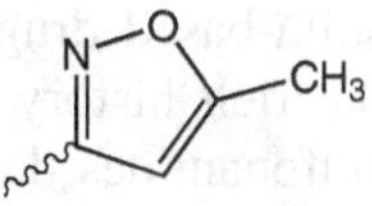 sulfamethoxazole

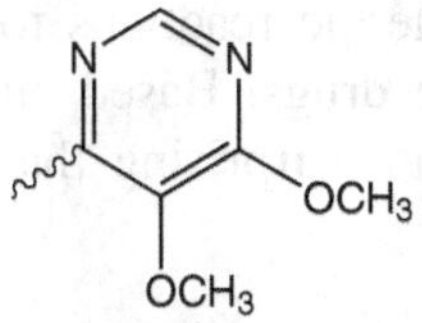

Another common drug based on sulfonamide chemistry is hydrochlorothiazide, which is sometimes abbreviated HCT or HCTZ. Hydrochlorothiazide is a diuretic that inhibits the kidney's ability to retain water. This reduces sodium levels, and hydrochlorothiazide is used for hypertension and other treatments such as the prevention of kidney stones. Hydrochlorothiazide is used in combination with various angiotension II antagonists in brand-name drugs such as DIOVAN HCT (Novartis Pharmaceuticals Corporation), HYZAAR (Merck & Co. Inc.), BENICAR HCT (Daiichi Sankyo Inc.), AVALIDE (Bristol-Myers Squibb Sanofi-Synthelabo Partnership), and MICARDIS HCT (Boehringer Ingelheim Pharmaceuticals, Inc.).

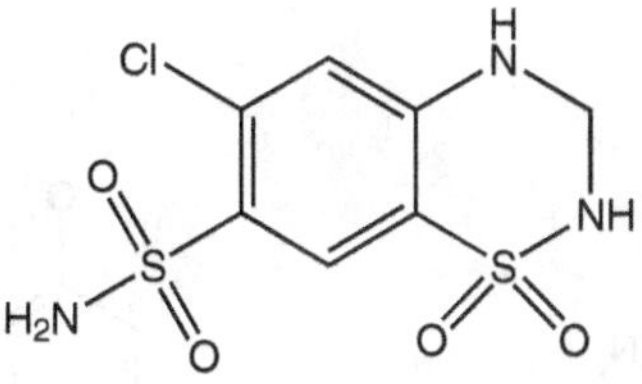

hydrochlorothiazide

Hydrochlorothiazide is one of several sulfonamide nonantibiotic drugs. Other examples are dorzolamide and brinzolamide, which are used for the treatment of glaucoma.

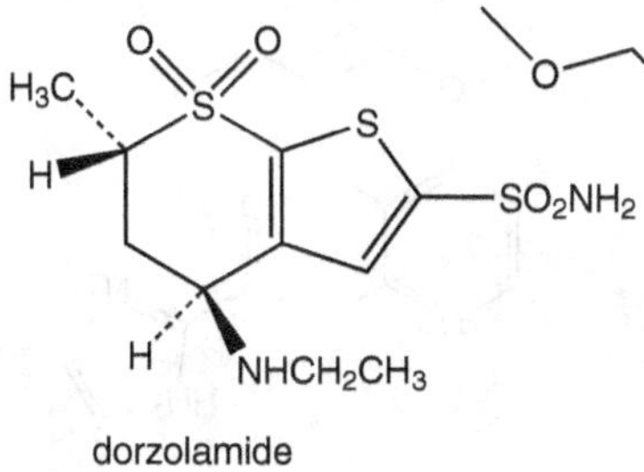

dorzolamide 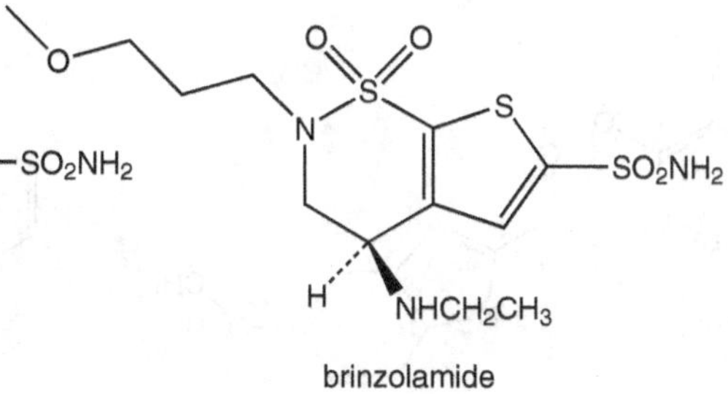 brinzolamide

There are many other examples. A study[21] on allergic reactions to sulfa-based drugs lists 38 sulfonamide nonantibiotic drugs. Based on the rich history of heterocycles in medicine, it is not surprising that sulfonamides dominate to this day.

8.3. PYRIDINES

The pyridine ring is found in many current pharmaceuticals. It is present in some proton pump inhibitors used for reducing the amount of acid produced by the stomach. These drugs can be used to treat reflux disease, ulcers, or heartburn. Omeprazole is sold by AstraZeneca Pharmaceuticals LP as the magnesium salt in the racemic form as PRILOSEC and as the S enantiomer as NEXIUM. Lansoprazole is sold by TAP Pharmaceuticals Inc. as PREVACID. Pantoprazole is sold by Wyeth Pharmaceuticals Inc. as the sodium salt under the name PROTONIX. Rabeprazole was developed by Eisai Co. Ltd., who sells it as the sodium salt under the name ACIPHEX. Each of these also contains a benzimidazole ring.

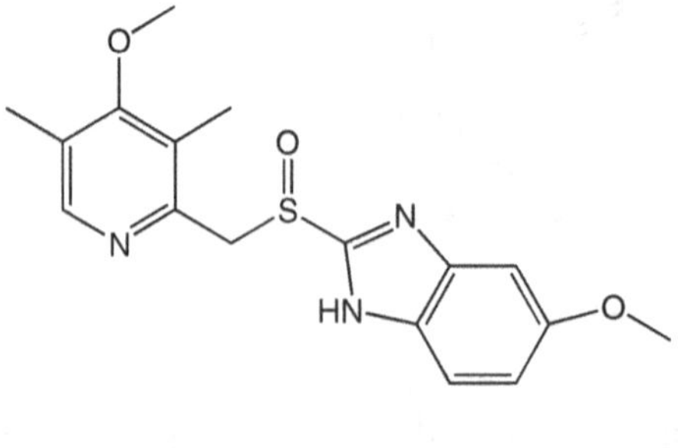

omeprazole

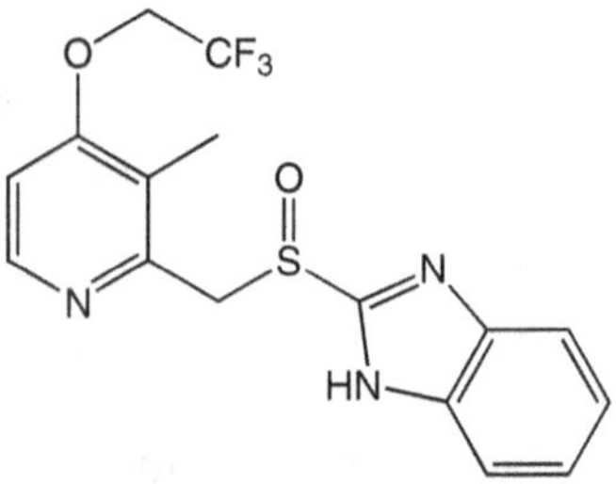

lansoprazole

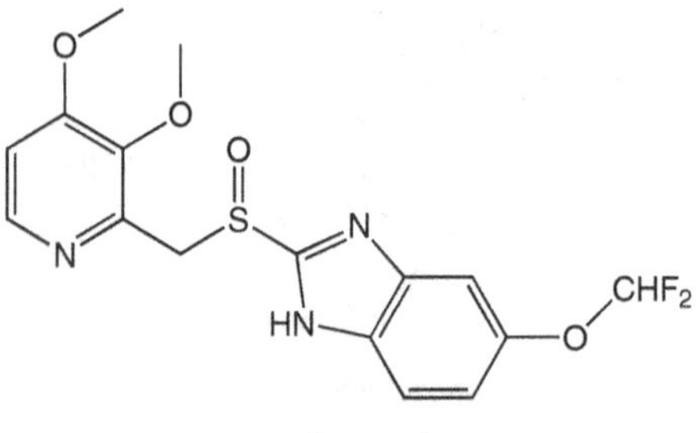

pantoprazole

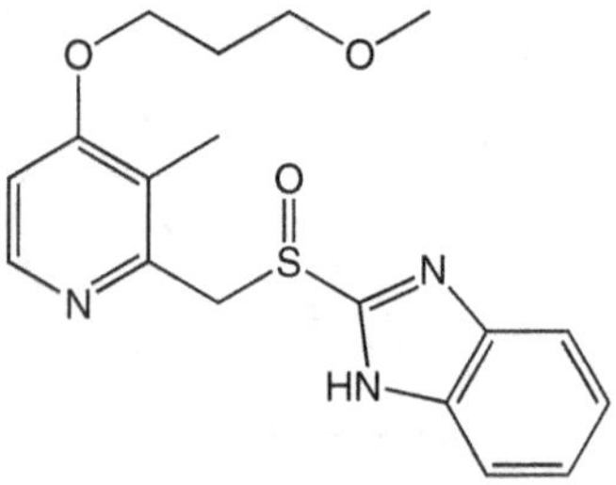

rabeprazole

Two thiazolidinedione compounds that contain the pyridine ring and that are used for diabetes are pioglitazone (the hydrochloride salt is sold by Takeda Pharmaceuticals Inc. as ACTOS) and rosiglitazone (sold as the maleate by GlaxoSmithKline as AVANDIA).

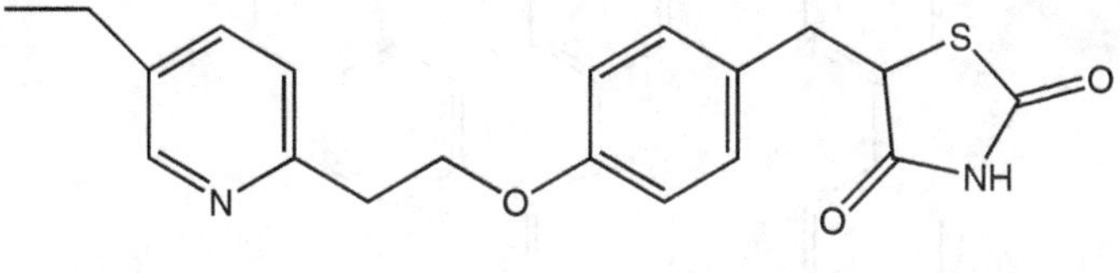

pioglitazone

rosiglitazone

Eszopiclone is marketed for insomnia as LUNESTA by Sepracor Inc. Eszopiclone contains a pyridine ring as well as pyrrolopyrazine and piperazine rings.

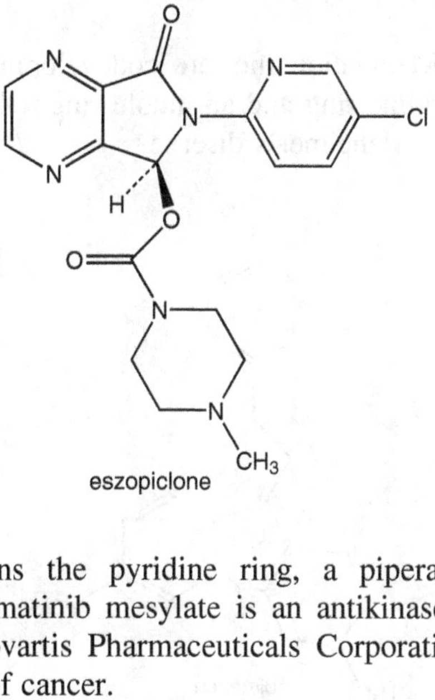

eszopiclone

Imatinib contains the pyridine ring, a piperazine ring, and a pyrimidine ring. Imatinib mesylate is an antikinase inhibitor sold as GLEEVEC by Novartis Pharmaceuticals Corporation as a treatment for certain forms of cancer.

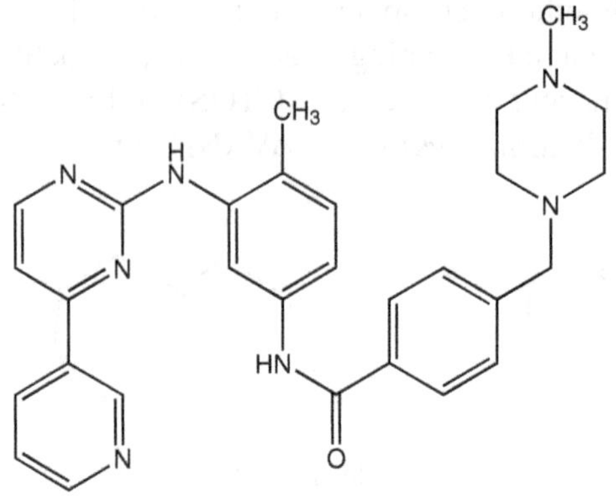

imatinib

Another pyridine-containing pharmaceutical is niacin, also known as nicotinic acid or vitamin B_3. It is sold in a controlled-release form as NIASPAN by Abbott Laboratories for improving cholesterol levels.

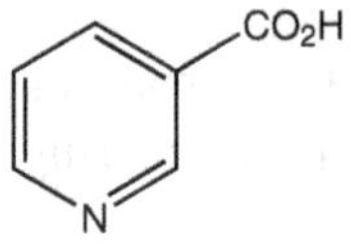

niacin

Pfizer Inc. and Medivation Inc. are codeveloping dimebon, which contains both a pyridine ring and an indole ring. Dimebon is in Phase III clinical trials for Alzheimer's disease.[22]

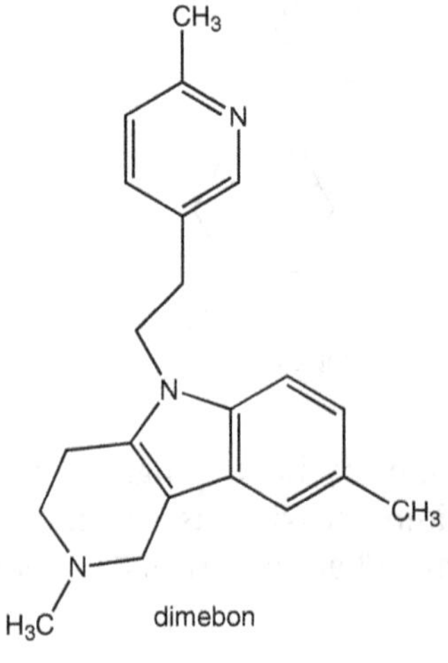

dimebon

8.4. INDOLES

Serotonin, which is an indole, occurs naturally in the body. In most cases of migraines, serotonin levels decrease. Many migraine medications are based on the indole structure. Sumatriptan (sold as the succinate salt by GlaxoSmithKline as IMITREX), rizatriptan (sold as the benzoate salt by Merck & Co. Inc. as MAXALT), and eletriptan (sold as the hydrobromide salt by Pfizer Inc. as RELPAX) are all prescribed for migraines.

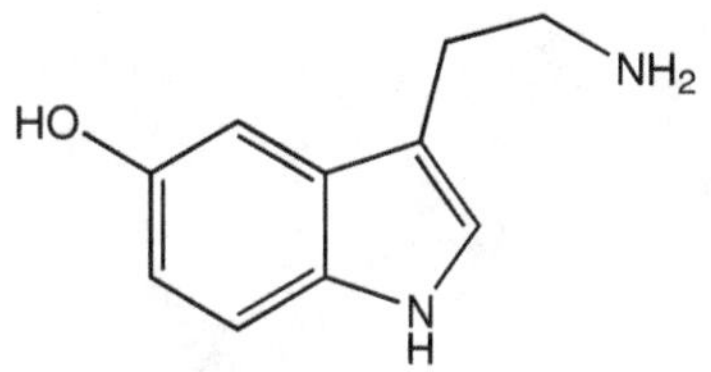

serotonin

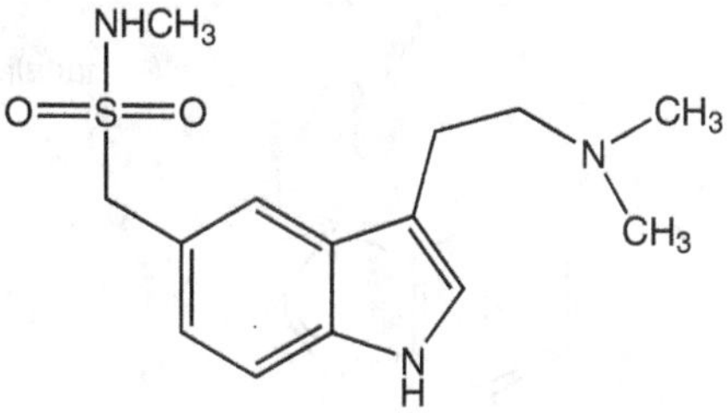

sumatriptan

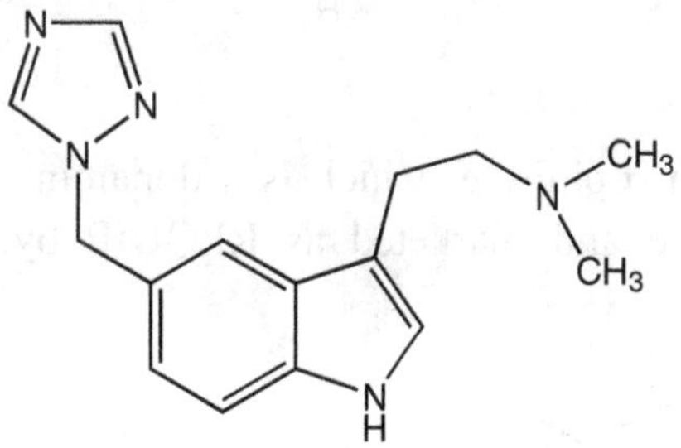

rizatriptan

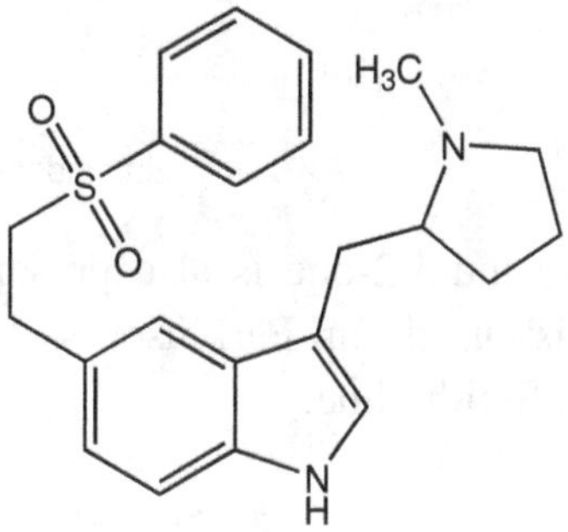

eletriptan

Tadalafil, an erectile dysfunction drug (marketed by Lilly as CIALIS), contains an indole ring, a pyrazine ring, and a dioxole.

The indole ring structure (as the indol-2-one) is shown in ziprasidone (sold as the hydrochloride salt by Pfizer Inc. as GEODON and prescribed for bipolar disorder).

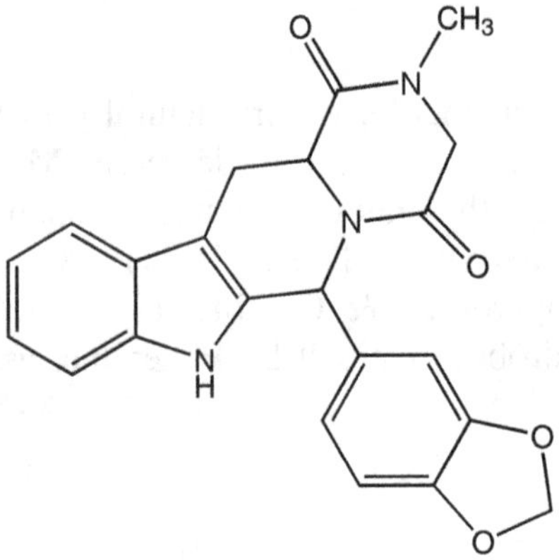

tadalafil

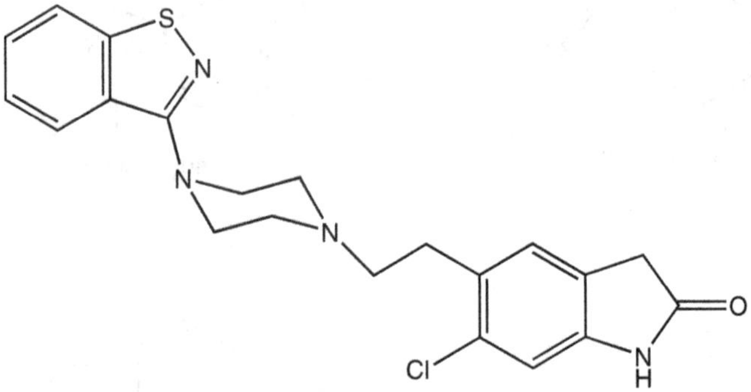

ziprasidone

The indol-2-one is also present in ropinirole, which is a dopamine agonist used for Parkinson's disease and marketed as REQUIP by GlaxoSmithKline.

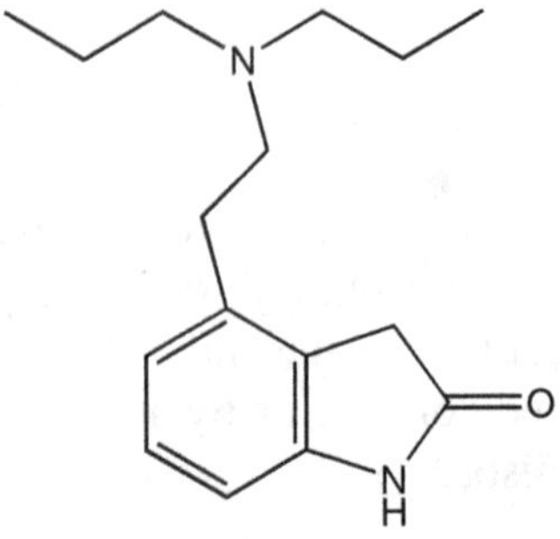

ropinirole

8.5. QUINOLINES

A quinoline compound called montelukast sodium is the active ingredient in SINGULAIR, which is the Merck drug for asthma.

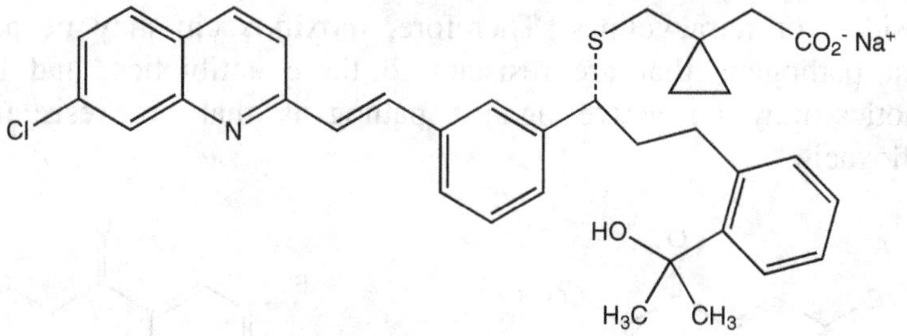

montelukast sodium

One class of drugs containing the quinoline ring is the quinolone antibiotics, especially the fluoroquinolone antibiotics. The first quinolone antibacterial was discovered serendipitously in the early 1960s.[23] Chemists at the Sterling-Winthrop laboratories in Rensselaer, NY, isolated a by-product in their synthesis of chloroquine.

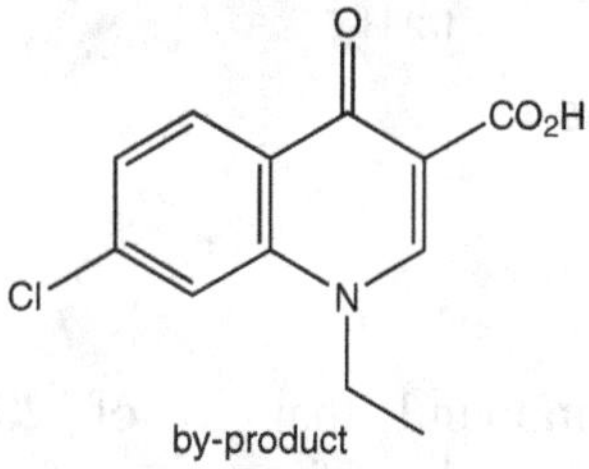

by-product

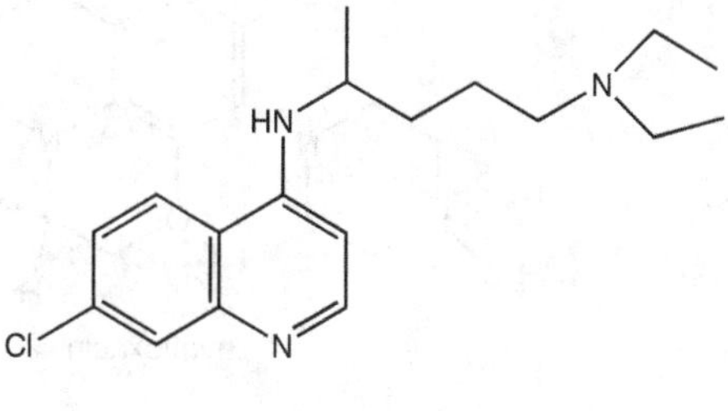

chloroquine

The by-product was found to exhibit antibacterial activity. Since the 1960s, more than 10,000 structurally related agents have been described in many hundreds of patents and journal articles.[23]

The fluoroquinolones are second-generation antibacterials. Ciprofloxacin (CIPRO, Bayer AG), moxifloxacin (AVELOX, developed by Bayer AG and marketed in the United States by Schering-Plough Corp.), and levofloxacin (LEVAQUIN, Ortho-McNeil-Janssen Pharmaceuticals, Inc.) are examples of this group. All are antibiotics used to treat or prevent certain infections caused by bacteria. Ciprofloxacin is also used to treat or prevent anthrax in people who may have been exposed to anthrax germs in the air. Similarly, moxifloxacin kills sensitive bacteria by stopping the production of essential proteins needed by the bacteria to survive. Moxifloxacin is used in a sterile ophthalmic solution as VIGIMOX (moxifloxacin HCl ophthalmic solution). The mechanism of action for quinolones, including moxifloxacin, is different from that of macrolides, aminoglycosides, or tetracyclines. Therefore, moxifloxacin may be active against pathogens that are resistant to these antibiotics, and these antibiotics may be active against pathogens that are resistant to moxifloxacin.

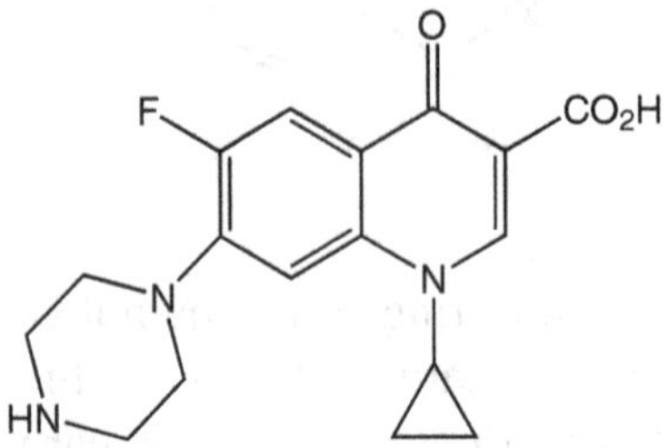

ciprofloxacin

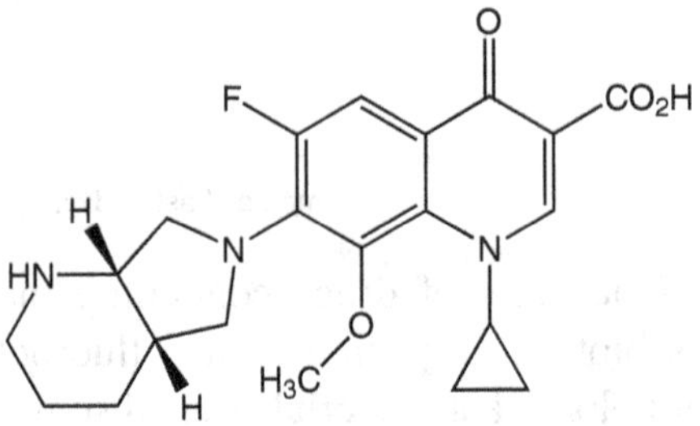

moxifloxacin

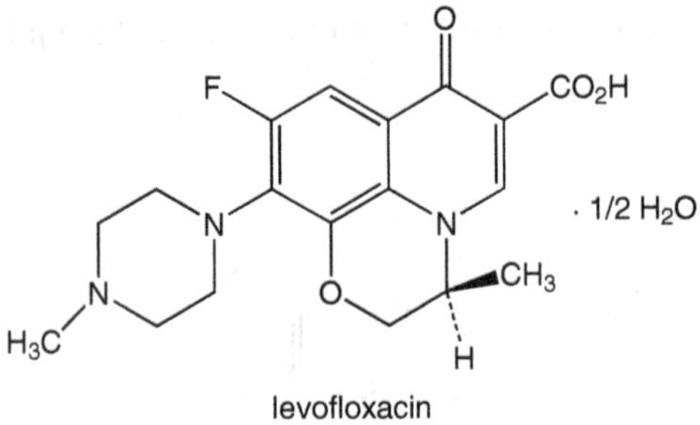

levofloxacin

PF-2545920 is a quinoline-based compound that as of 2008, Pfizer had entered into phase II clinical trials for treatment of schizophrenia.[24,25]

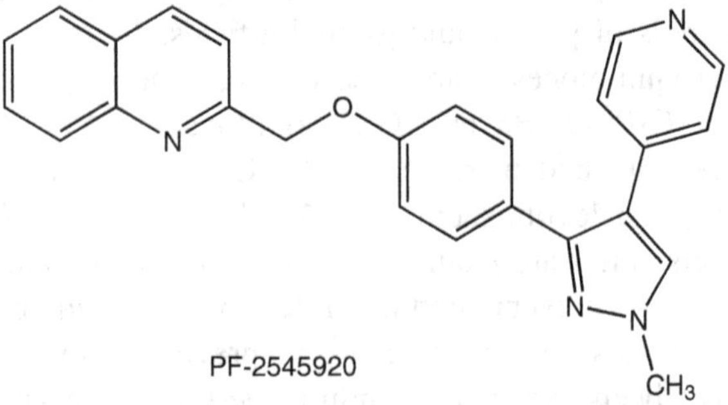

PF-2545920

8.6. AZEPINES

Perhaps the most common drugs based on 7-membered rings are the benzodiazepines. Different benzodiazepines have been used for the treatment of seizures, insomnia, depression, or anxiety. Examples of benzodiazepines include alprazolam (XANAX, Pfizer, Inc.), chlordiazepoxide (LIBRIUM, Hoffman-LaRoche, Inc.), diazepam (VALIUM, Roche Laboratories), and lorazepam (ATIVAN, Biovail Pharmaceuticals, Inc.).

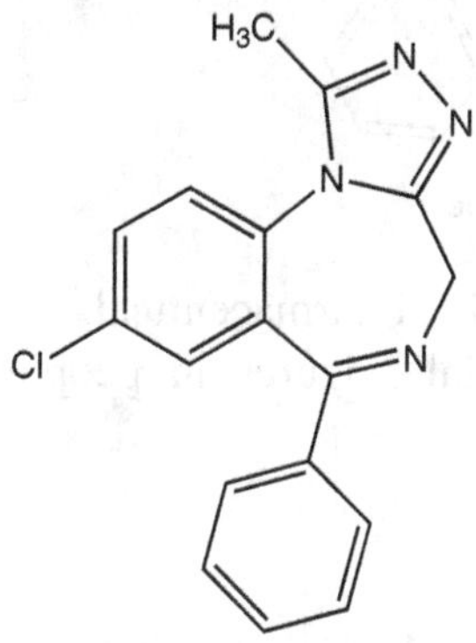

alprazolam

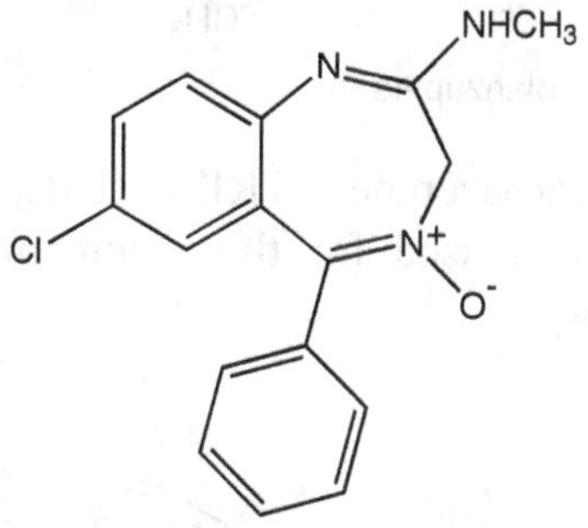

chlordiazepoxide

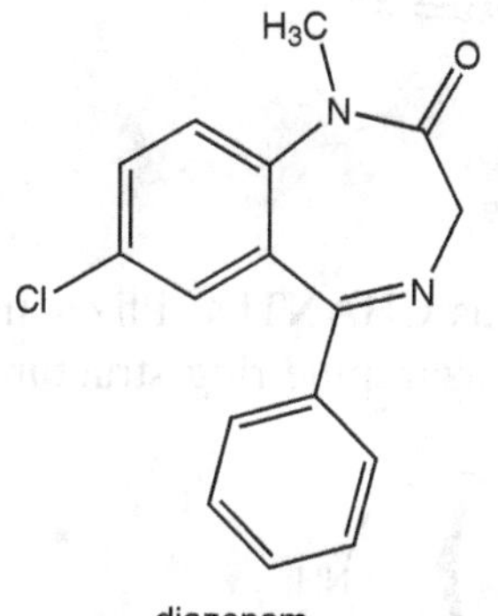

diazepam

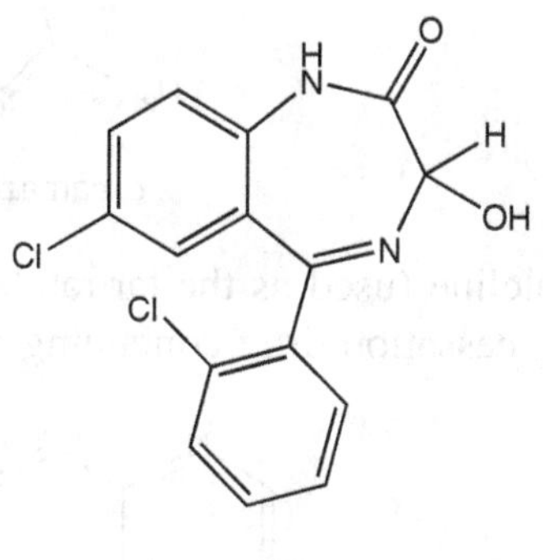

lorazepam

Olanzapine is a psychotropic agent that belongs to the thienobenzodiazepine class. Olanzapine (ZYPREXA; Eli Lilly and Company) is approved by the U.S. Food and Drug Administration (FDA) for treating the symptoms of schizophrenia and acute mixed, manic episodes, or maintenance treatment of bipolar disorder. Quetiapine

(SEROQUEL, AstraZeneca Pharmaceuticals LP), a dibenzothiazepine, is a mood-stabilizing medication approved by the FDA to treat both the highs and lows of bipolar disorder.

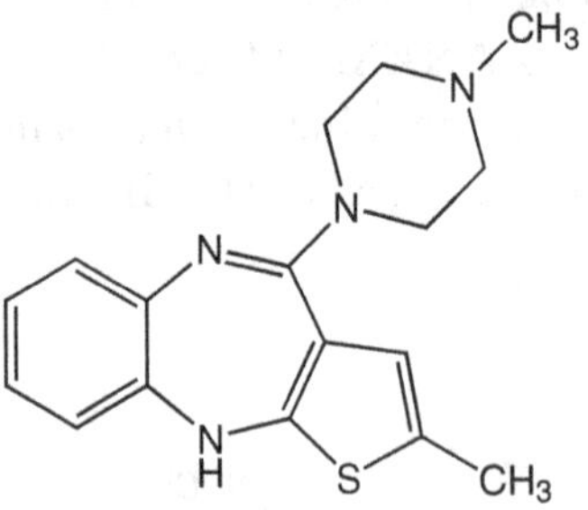

olanzapine 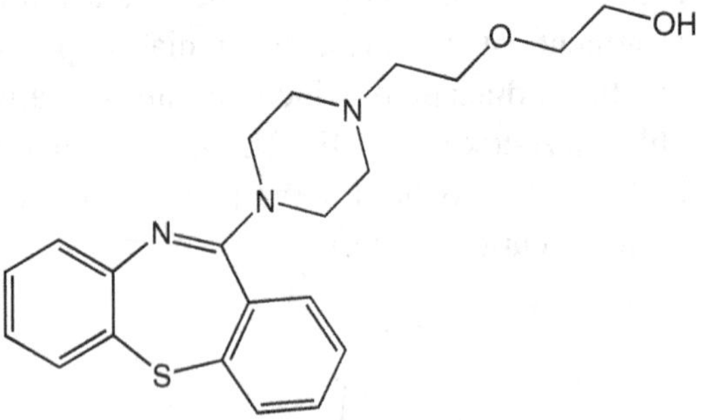 quetiapine

Oxcarbazepine (TRILEPTAL, Novartis Pharmaceuticals Corporation) is used for the treatment of partial seizures in people with epilepsy.

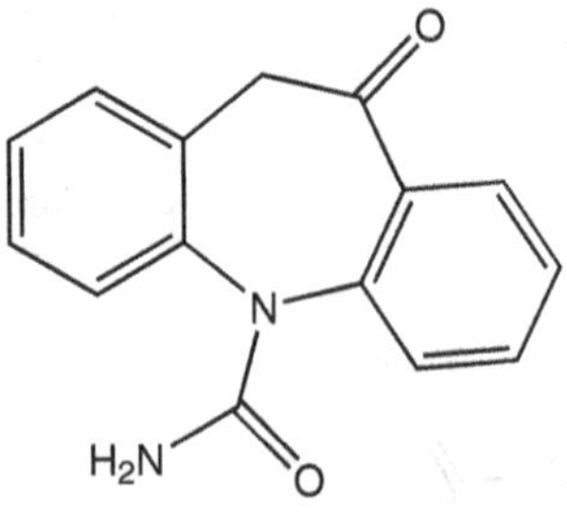

oxcarbazepine

Varenicline (used as the tartrate salt as CHANTIX, Pfizer Inc.) is a smoking cessation drug containing a benzazepine ring structure.

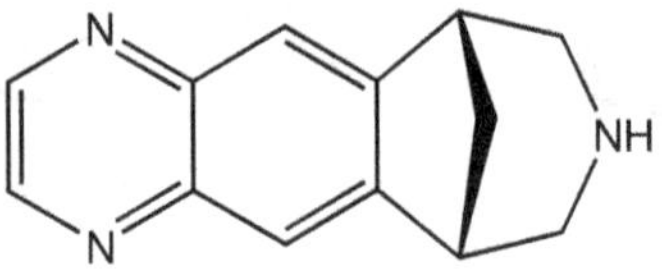

verenicline

Azelastine (hydrochloride salt is ASTELIN, Meda Pharmaceuticals Inc.) is an antihistamine that is used as a nasal spray and provides relief for seasonal allergies.

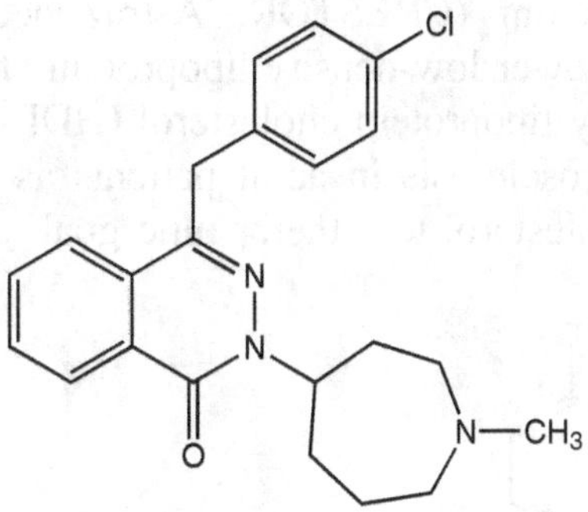

azelastine

8.7. PYRIMIDINES

As discussed in Chapter 3, the nucleic acid bases cytosine, thymine, and uracil contain a pyrimidine ring. Adenine and guanine are based on the purine ring, which includes a pyrimidine ring. Because the five nucleic acid bases contain the pyrimidine ring, perhaps we should not be surprised that pyrimidines are prominent in the pharmaceutically active ingredients used in a variety of therapies including antipsychotic, cholesterol reduction, cancer, erectile dysfunction, antivirals, and human immunodeficiency virus (HIV).

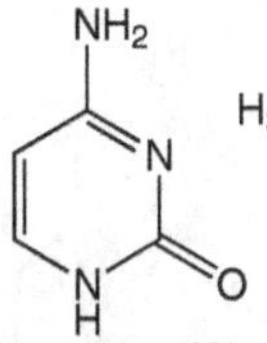

cytosine (C) 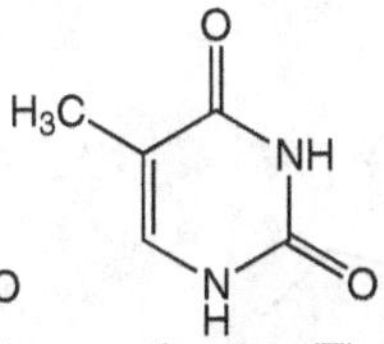 thymine (T) 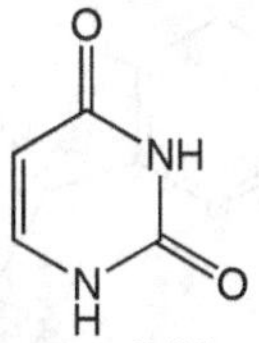 uracil (U) 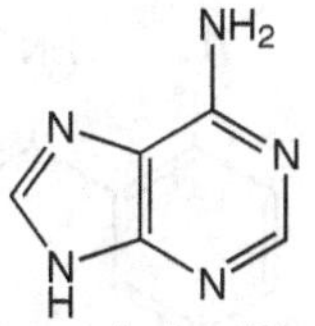 adenine (A) 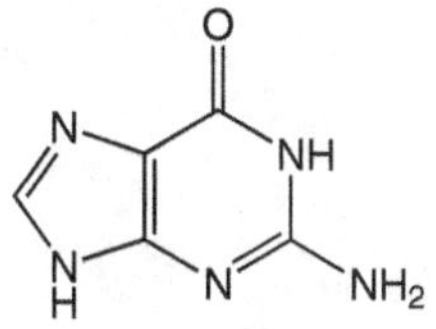 guanine (G)

Risperidone (RISPERDAL, Ortho-McNeil-Janssen Pharmaceuticals, Inc.) is used for the treatment of irritability associated with autistic disorder in children ages 5–17, the treatment of schizophrenia in adults, and the treatment of bipolar mania.

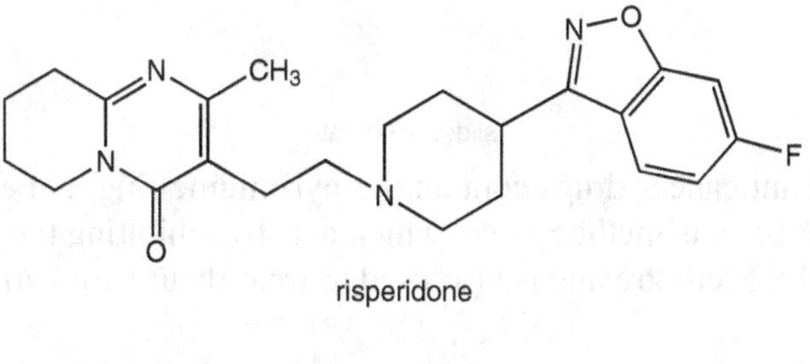

risperidone

Rosuvastatin calcium (CRESTOR, AstraZeneca Pharmaceuticals LP) is indicated to lower low-density lipoprotein cholesterol (LDL-C), increase high-density lipoprotein cholesterol (HDL-C), and to slow the progression of atherosclerosis in adult patients as part of a treatment plan to decrease cholesterol to a therapeutic goal.

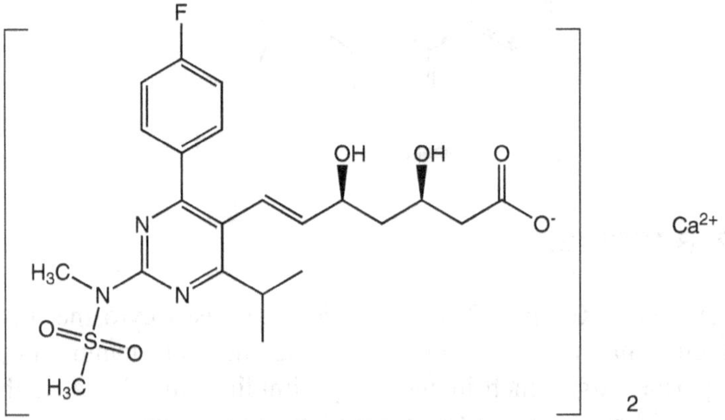

rosuvastatin calcium

Sildenafil citrate (VIAGRA, Pfizer Inc.) is an oral therapy for erectile dysfunction that contains the pyrimidine group.

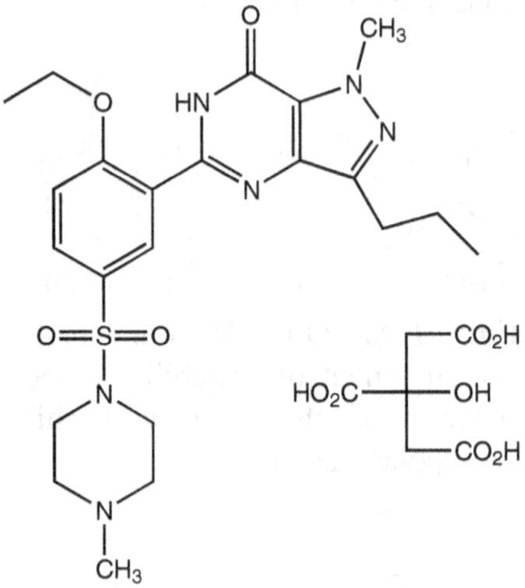

sildenafil citrate

Several anticancer drugs contain the pyrimidine ring. An early drug, still in use today is methotrexate, which acts by inhibiting the formation of folic acid. Methotrexate is also used to treat rheumatoid arthritis. For the synthetic chemist, methotrexate is particularly interesting because it can be prepared in a one-step "shotgun" reaction from tetraaminopyrimidine, *p*-(N-methylamino)-benzoyl glutamic acid, and a three-carbon synthon such as dibromopropionaldehyde or 1,1,3-tribromoacetone.[26]

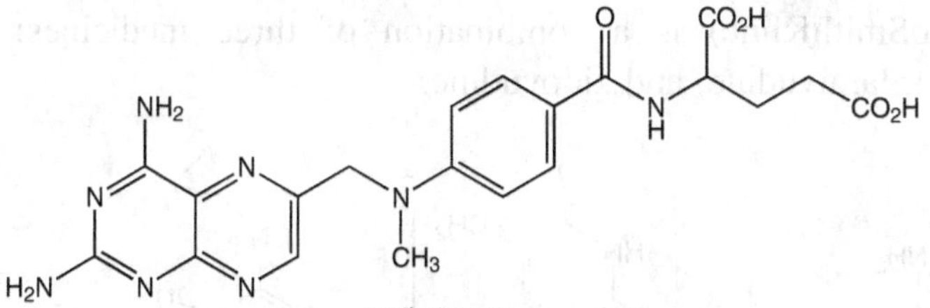

methotrexate

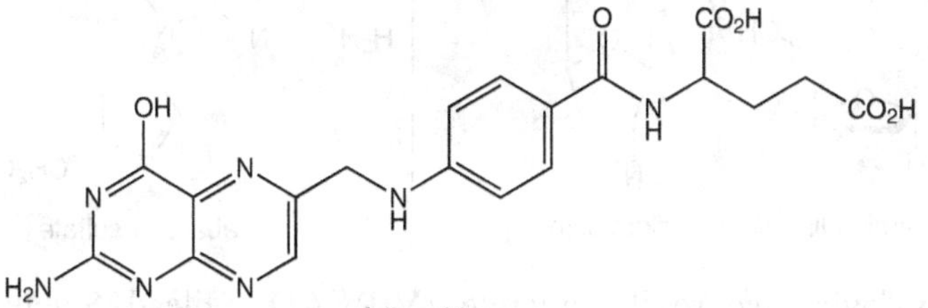

folic acid

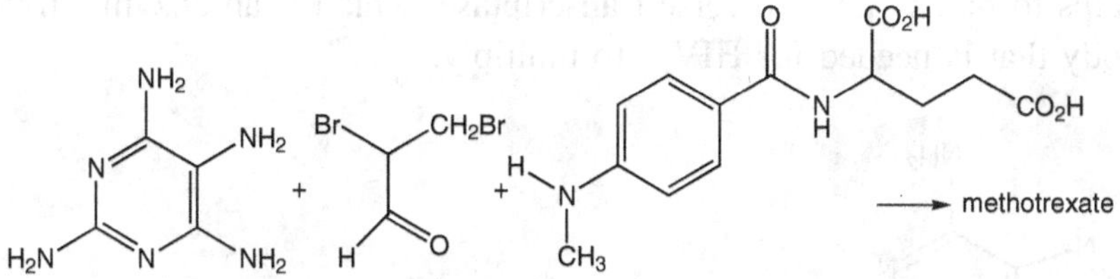

Imatinib mesylate (discussed in Section 8.3 under pyridines) contains a pyrimidine ring and is sold as GLEEVEC by Novartis Pharmaceuticals Corporation as a treatment for certain forms of cancer. Capecitabine (XELODA, Roche Laboratories Inc.) is prescribed for breast and colorectal cancer.

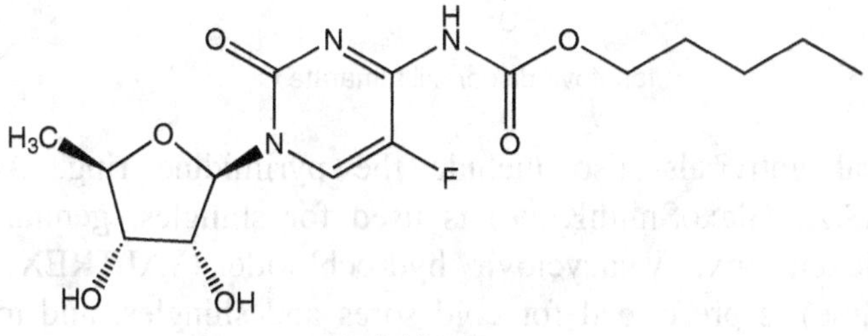

capecitabine

Many pharmaceuticals used to treat HIV infection contain the pyrimidine ring. Among those are COMBIVIR tablets (GlaxoSmithKline), which are combination tablets that contain lamivudine and zidovudine. EPZICOM (GlaxoSmithKline) combines abacavir sulfate and lamivudine (two HIV medicines) in one tablet. TRIZIVIR (GlaxoSmithKline) is a combination of three medicines: abacavir sulfate, lamivudine, and zidovudine.

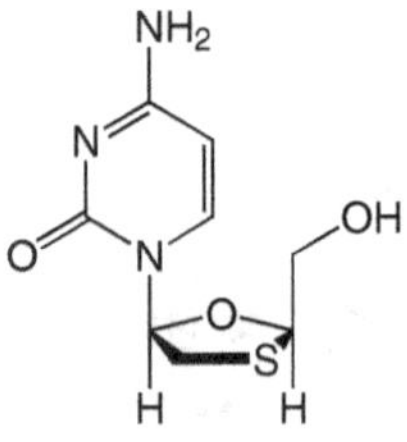

lamivudine 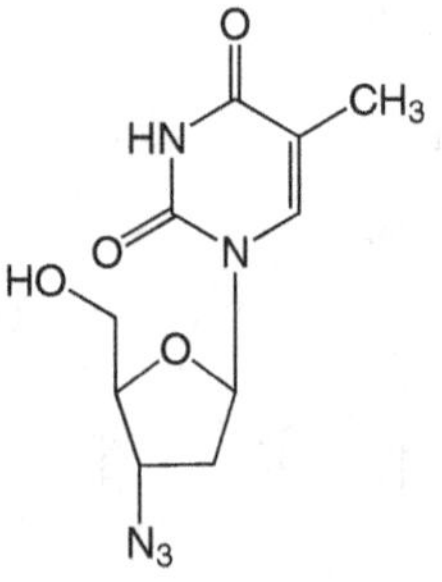 zidovudine 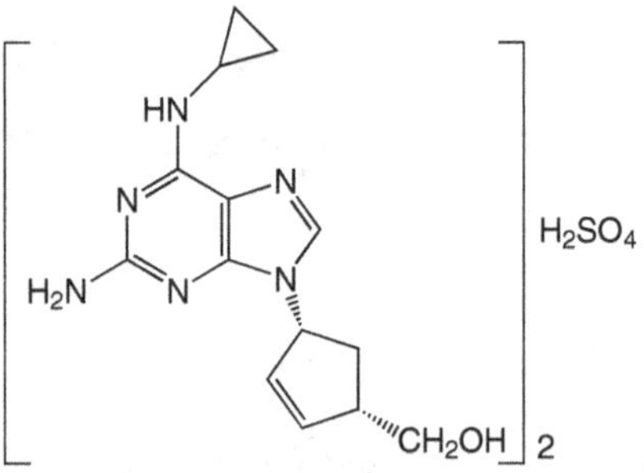 abacavir sulfate

Tenofovir disoproxil fumarate (VIREAD, Gilead Sciences, Inc.) helps to block HIV-1 reverse transcriptase, which is an enzyme in the body that is needed for HIV-1 to multiply.

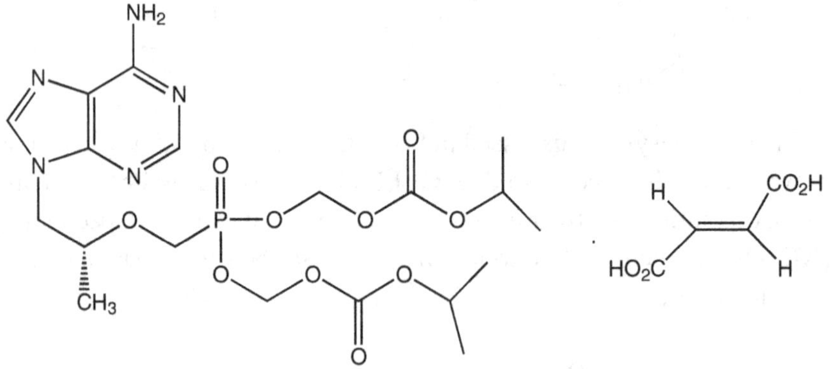

tenofovir disoproxil fumarate

Several antivirals also include the pyrimidine ring. Acyclovir (ZOVIRAX, GlaxoSmithKline) is used for shingles, genital herpes, and chicken pox. Valacyclovir hydrochloride (VALTREX, GlaxoSmithKline) is prescribed for cold sores and shingles, and to reduce outbreaks of genital herpes. After oral administration, valacyclovir hydrochloride is rapidly absorbed from the gastrointestinal tract and nearly completely converted to acyclovir and *L*-valine by first-pass intestinal and/or hepatic metabolism. Famciclovir (FAMVIR, Novartis Pharmaceuticals Corporation) is structurally similar and is also similarly prescribed.

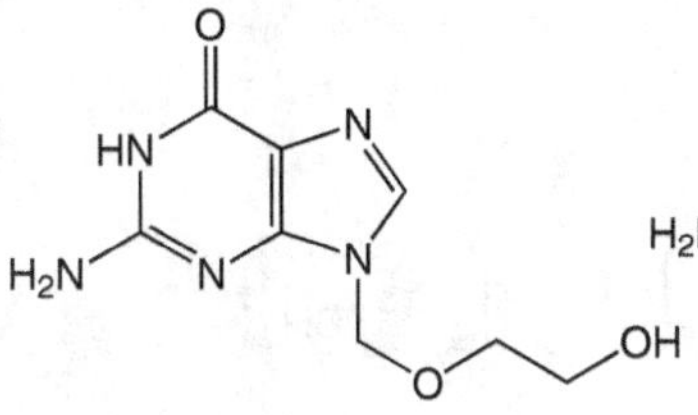

acyclovir

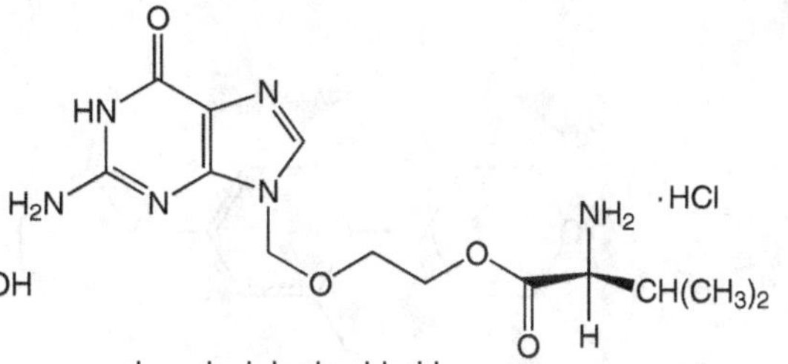

valacyclovir hydrochloride

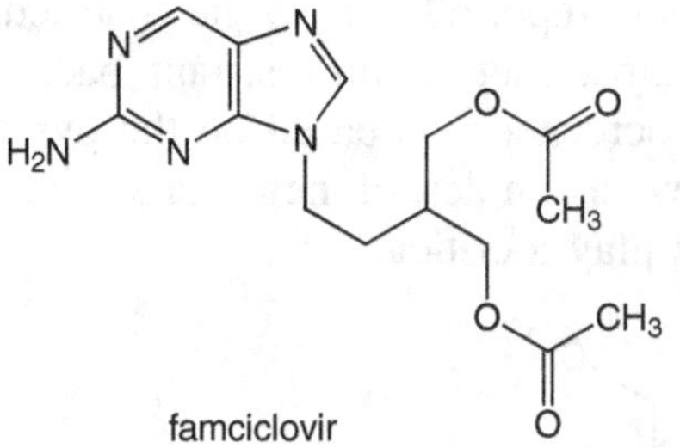

famciclovir

8.8. CONCLUDING REMARKS

There are a myriad of other examples of heterocycles used in medicine and based on these five ring structures. Other rings systems such as imidazoles, triazoles, piperazines, etc. are also prevalent. The preceding examples are not meant to be comprehensive but rather to demonstrate the pervasiveness of heterocycles in medicine. This will undoubtedly continue. For example, bacteria become resistant to commonly prescribed antibiotics, and the need for new antibiotics is ongoing. Vancomycin, which is derived from the word "vanquished," is a glycopeptide antibiotic traditionally prescribed after treatment with other antibiotics had failed. However, vancomycin-resistant organisms have developed and other compounds are needed. Linezolid (ZYVOX, Pfizer Inc.) is a synthetic antibacterial agent of a new class of antibiotics, the oxazolidinones. Linezolid inhibits bacterial protein synthesis through a mechanism of action different from that of other antibacterial agents; therefore, cross-resistance between linezolid and other classes of antibiotics is unlikely. In 2000, linezolid was approved by the FDA, but already linezolid-resistant bacteria have been found.

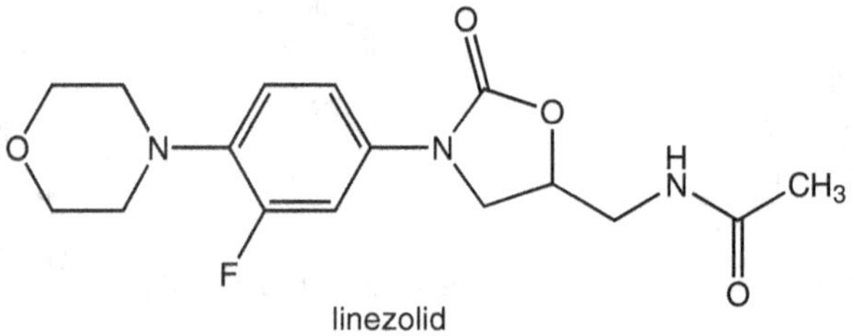

linezolid

Emmacin has been reported[27] as a new structural subclass that exhibits activity against methicillin-resistant bacteria. Another class of antibiotics being screened[28] is based on the pyridopyrimidine ring. These are just a few examples of new classes of antibiotics. As is shown, heterocycles play a critical role.

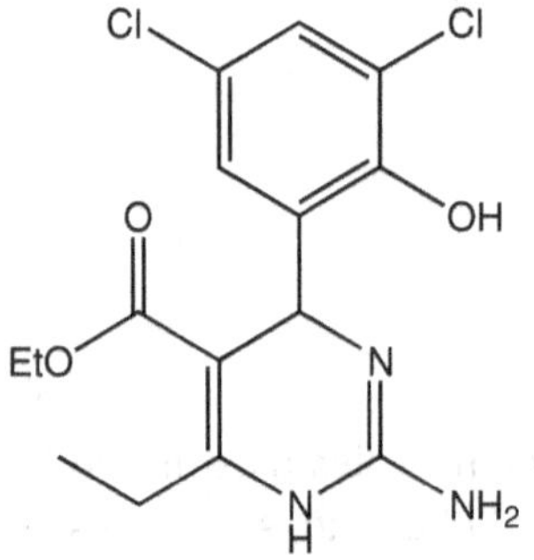

emmacin

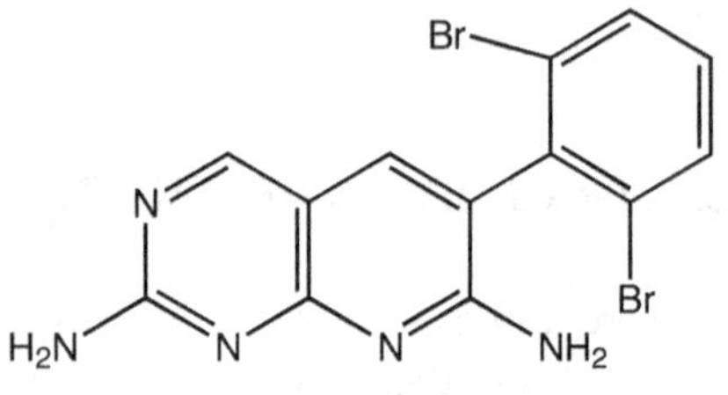

pyridopyrimidine antibiotic

Other molecules are being built with multiple functions to target cancerous cells. The compound shown next[29] uses the crown ether functionality to detect sodium cations and the pyridine for protons. When both Na^+ and H^+ are at high levels, as in cancerous cells, the molecule acts as a photosensitizer to form singlet oxygen to kill cancer cells.

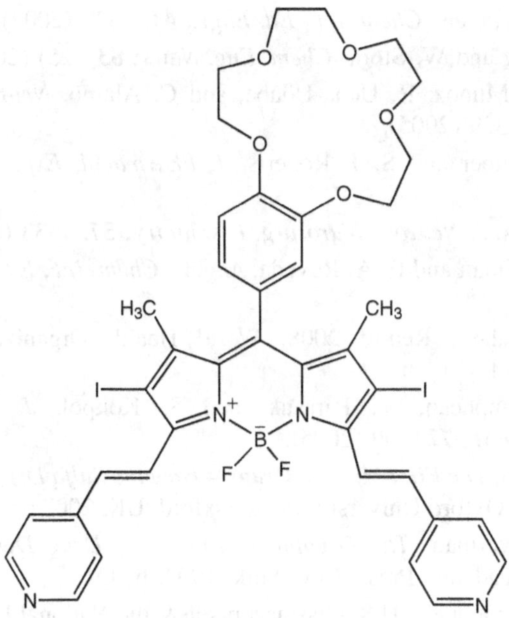

There is every reason to believe that most newly discovered pharmaceutically active compounds will continue to be based on heterocycles. As researchers develop new leads, an understanding of the physical properties and syntheses of heterocycles will be even more critical in the future.

REFERENCES

(1) *Drug Topics*, March 10, 2008.

(2) D. Lednicer, *The Organic Chemistry of Drug Synthesis*, Vol. 7, Wiley, New York, 2008.

(3) A. Burger, *Understanding Medications; What the Label Doesn't Tell You* Vol. 4, American Chemical Society, Washington, DC, 1995.

(4) T. Q. Chou, F. Y. Fu, and Y. S. Kao, *J. Am. Chem. Soc.*, **70**, 1765 (1948).

(5) Y. Deng, R. Xu, and Y. Yang, *J. Chin. Pharm. Sci.*, **9**, 116 (2000).

(6) P. Prioreschi, *History of Medicine*, Horatius Press, Omaha, NE, 1996.

(7) M. P. Earles, *Ann. Sci.*, **19**, 241 (1963).

(8) K. Brune, *Acute Pain*, **1**, 33 (1997).

(9) L. Knorr, *Ber. Dtsch. Chem. Ges.*, **16**, 2597 (1883).

(10) H. Kunz, *Angew. Chem. Int., Ed. Engl.*, **41**, 4439 (2002).
(11) R. Pepling and W. Stork, *Chem. Eng. News*, **83**, (25) (2005).
(12) F. Lopez-Munoz, R. Ucha-Udabe, and C. Alamo, *Neuropsychiat. Dis. Treat.*, **1**, 329 (2005).
(13) C. M. Gruber and S. J. Roberts, *J. Pharmacol. Exp. Ther.*, **27**, 349 (1926).
(14) L. F. Haas, *J. Neurol. Neurosurg. Psychiatry*, **57**, 1333 (1994).
(15) T. S. Kaufman and E. A. Ruveda, *Angew. Chem. Int., Ed. Engl.*, **44**, 854 (2005).
(16) World Malaria Report 2008, World Health Organization, Geneva, Switzerland.
(17) S. Tonmunphean, V. Parasuk, and S. Kokpol, *J. Molec. Struct. THEOCHEM*, **724**, 99 (2005).
(18) J. E. Lesch, *The First Miracle Drugs—How the Sulfa Drugs Transformed Medicine*, Oxford University Press, Oxford, UK, 2007 pp. 100, 126, 131.
(19) R. B. Silverman, *The Organic Chemistry of Drug Design and Drug Action*, Academic Press, New York, 1992, p. 155.
(20) U.S. Pharmacopeia, U.S. Pharmacopoeia & the National Formulary: USP 23 NF18, Basel, Switzerland, 1995.
(21) B. L. Strom, R. Schinnar, A. J. Apter, D. J. Margolis, E. Lautenbach, S. Hennessy, W. B. Bilker, and D. Pettitt, *N. Engl. J. Med.*, **349**, 1628 (2003).
(22) *Chem. Eng. News*, **86**, 22 (2008).
(23) V. Andriole, *The Quinolones*, third edition Academic Press, New York, 2000, pp. 34, 36.
(24) *Chem. Eng. News*, **86**, 38 (2008).
(25) P. R. Verhoest, C. J. Helal, D. J. Hoover, and J. M. Humphrey, *U.S. Pat. No. 7,429,665* (2008).
(26) B. Singh and F. C. Schaefer, *U.S. Pat. No. 4,374,987* (1983).
(27) E. E. Wyatt, W. R. J. D. Galloway, G. L. Gemma, M. Welch, O. Loiseleur, A. T. Plowright, and D. R. Spring, *Chem. Commun.*, 4962 (2008).
(28) *Chem. Eng. News*, **87**, 31 (2009).
(29) S. Ozlem and E. U. Akkaya, *J. Am. Chem. Soc.*, **131**, 48 (2009).

CHAPTER 9

SYNTHETIC METHODS FOR SOME PROMINENT HETEROCYCLIC FAMILIES: EXAMPLES OF PHARMACEUTICALS SYNTHESIS

9.1. SCOPE OF THE CHAPTER

The various approaches to forming heterocyclic ring systems by intramolecular cyclizations were discussed in Chapter 4 and by cycloaddition reactions in Chapter 5. In Chapters 6 and 7, numerous reactions of heterocyclics were described. We will now reexamine some of this material from a different standpoint, taking a particular heterocyclic family and considering some of the methods that are useful for the construction of the family. In addition, some information on the synthesis of important heterocycles not yet considered will be presented. As was discussed in Chapter 8, great numbers of new heterocyclic compounds have been, and continue to be, synthesized in the research laboratories of the pharmaceutical industry in their search for new materials of value in medicine, and the many accomplishments from this work provide a multitude of examples that can be used to illustrate the practical application of heterocyclic chemistry. However, the presentation must be limited to a small number of examples; for more in-depth discussions of important syntheses, two works by

Fundamentals of Heterocyclic Chemistry: Importance in Nature and in the Synthesis of Pharmaceuticals, By Louis D. Quin and John A. Tyrell Copyright © 2010 John Wiley & Sons, Inc.

Dr. Daniel Lednicer are particularly valuable: the series *The Organic Chemistry of Drug Synthesis*[1] and *Strategies for Organic Drug Synthesis and Design*.[2] We are most appreciative that Dr. Lednicer compiled such valuable summaries of synthetic pharmaceuticals, and we acknowledge our liberal use of the references cited in his works.

9.2. PYRROLES

Because of its high reactivity, pyrrole is a useful starting material for many syntheses. It is available commercially by the reactions of Scheme 9.1.

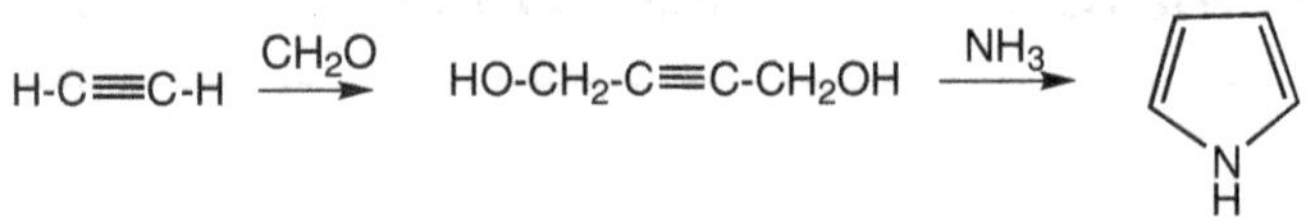

$$H\text{-}C{\equiv}C\text{-}H \xrightarrow{CH_2O} HO\text{-}CH_2\text{-}C{\equiv}C\text{-}CH_2OH \xrightarrow{NH_3} \text{pyrrole (N-H)}$$

Scheme 9.1

An early pyrrole synthesis, which was originally useful on the laboratory scale, involves the pyrolysis of the ammonium salt of the glucose-derived mucic acid (**9.1**, Scheme 9.2).

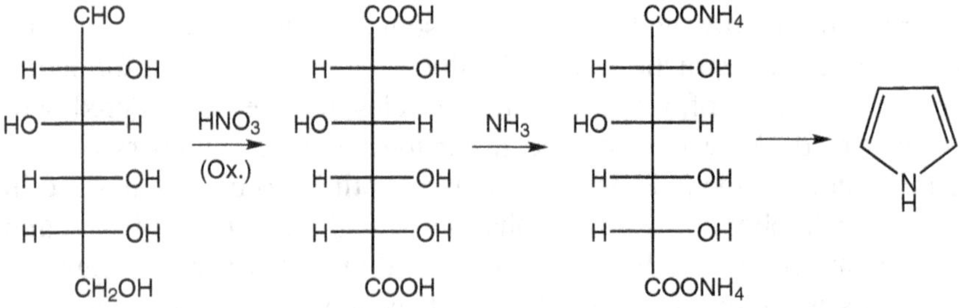

CHO; H–OH; HO–H; H–OH; H–OH; CH_2OH $\xrightarrow[\text{(Ox.)}]{HNO_3}$ COOH; H–OH; HO–H; H–OH; H–OH; COOH (**9.1**) $\xrightarrow{NH_3}$ $COONH_4$; H–OH; HO–H; H–OH; H–OH; $COONH_4$ $\longrightarrow$ pyrrole (N-H)

Scheme 9.2

Chapter 4 discussed two of the traditional methods of pyrrole synthesis, the Paal-Knorr synthesis and the Knorr synthesis. The basic reactions are repeated here as Scheme 9.3. Details of the mechanisms were given in sections 4.2.1 and 4.2.3, respectively.

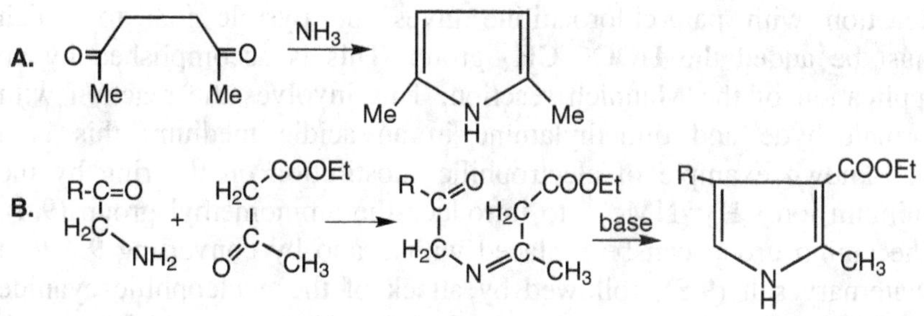

Scheme 9.3

Another classic method is that known as the Hantzsch pyrrole synthesis (Scheme 9.4). The nitrogen starting material is an enamine (**9.2**), which is prepared from a beta-keto ester and ammonia. The beta-position of the enamine is electron-rich and is alkylated with an alpha-haloketone. The amino and the carbonyl groups interact in the familiar way to close the ring.

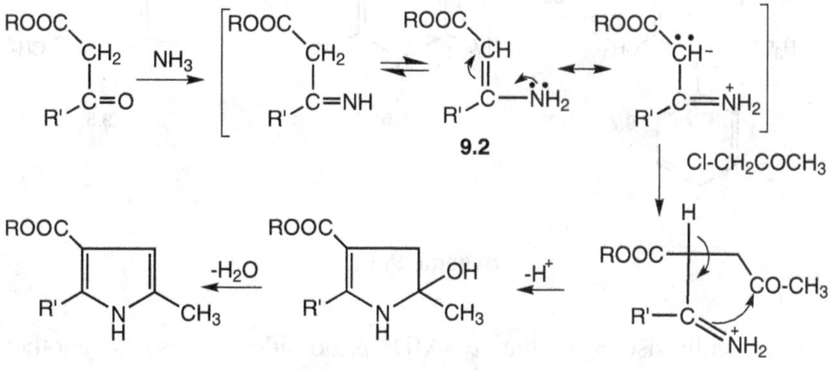

Scheme 9.4

The pyrrole ring appears in some pharmaceuticals, but multicyclic compounds, especially based on indole, are of more common occurrence. Pyrrolidines also are of importance. An application of the Paal-Knorr method in the pharmaceutical field is found in the synthesis of clopirac, which is a nonsteroidal anti-inflammatory drug (a group of agents[2] known as NSAIDs). The starting material (Scheme 9.5) is hexane-2,5-dione (actually introduced as the diketal 2,5-dimethoxytetrahydrofuran, which hydrolyzes to the diketone).

Reaction with para-chloroaniline gives the pyrrole **9.3**, to which must be added the HOOC-CH_2 group. This is accomplished by an application of the Mannich reaction. This involves the reaction with formaldehyde and dimethylamine in an acidic medium; this is a well-known example of electrophilic substitution on the ring by the iminium ion $CH_2{=}NMe_2^+$ to introduce the aminomethyl group (**9.4**). The amino group can be replaced with cyano by converting **9.4** to a quaternary salt (**9.5**), followed by attack of the nucleophilic cyanide ion. Hydrolysis of the cyano group in **9.6** leads to the acid **9.7**, which is clopirac.

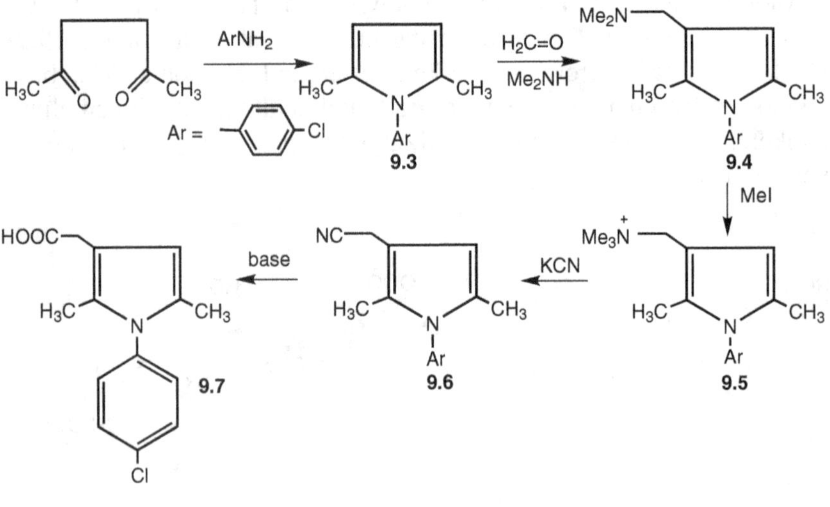

Scheme 9.5

The synthesis[3,4] of the NSAID prinomide represents another approach to pyrrolic drugs. Here, the starting material is parent pyrrole, and advantage is taken of some of its characteristic reactions to start the synthesis (Scheme 9.6). Thus, the NH group is converted to NMe by forming the anion of pyrrole and then the methylation of the anion. The carboxy group is introduced at the alpha position by forming the lithio derivative (**9.9**) and then its carbonation to give acid **9.10**. After esterification, the cyanomethyl group is introduced by displacing the methoxy group of ester **9.11** with the anion of acetonitrile. In the product **9.12**, the CH_2 group is activated by both the CN and the C=O groups; the base Et_3N can abstract a proton and the resulting carbanion adds to the C=N bond of phenyl isocyanate, PhN=C=O, to give prinomide.

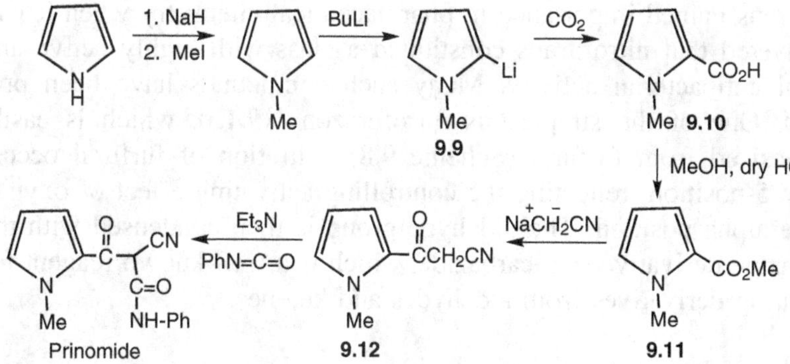

Scheme 9.6

9.3. FURANS

Two classic furan syntheses were described in Chapter 4: an adaptation of the Paal-Knorr method (section 4.2.1) and the Feist-Benary synthesis (section 4.2.3), which starts with an alpha-haloketone and a beta-ketoester. Other ring-closing processes have been devised and are outlined elsewhere.[5] It is of practical current importance to start syntheses with the readily available furfural (furan-2-carbaldehyde) or furan itself. The reason for this is that large-scale commercial syntheses of these compounds have been developed, and thus they are readily available and inexpensive.

Residues from the processing of oats, corn, and other cereal materials are a cheap source of carbohydrate polymers that can be hydrolyzed to pentoses, and from them furfural can be obtained by dehydration with strong sulfuric acid. Arabinose is shown as a typical starting pentose for furan synthesis in Scheme 9.7. In essence, the ring is formed by loss of water between C-2 and C-5, which leaves the aldehyde group in an alpha position of the 5-membered ring. Two other molecules of water must be eliminated to give the double bonds of the furan ring. Furfural is stable and has the properties of an aromatic aldehyde. The CHO group can be removed by heating with steam over a catalyst, and this is an excellent source of furan.

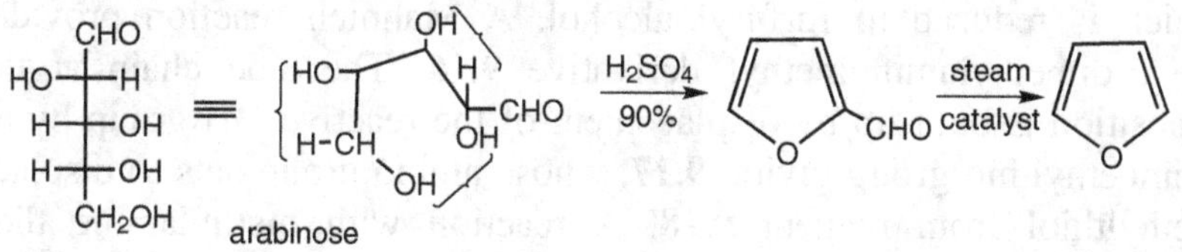

Scheme 9.7

Furans gained importance in pharmaceutical chemistry when it was discovered that nitrofurans constituted a class with highly active and useful antibacterial activity. Many such compounds have been prepared. One of the simplest is nitrofurazone (**9.13**), which is easily synthesized from furfural (Scheme 9.8). Nitration of furfural occurs at the 5-position, reflecting the controlling activating effect of oxygen at the alpha position. The aldehyde group is then condensed with the hydrazine derivative semicarbazide, which is a well-known reagent for preparing derivatives from aldehydes and ketones.

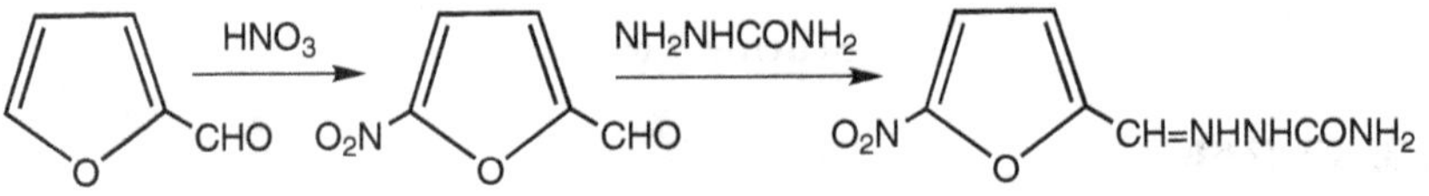

HNO_3 ; $NH_2NHCONH_2$; CHO ; O_2N ; CHO ; O_2N ; $CH{=}NHNHCONH_2$; 9.13

Scheme 9.8

Nidroxyzone (**9.14**) is formed similarly from 5-nitrofurfural and another hydrazine derivative shown in Scheme 9.9.[1a]

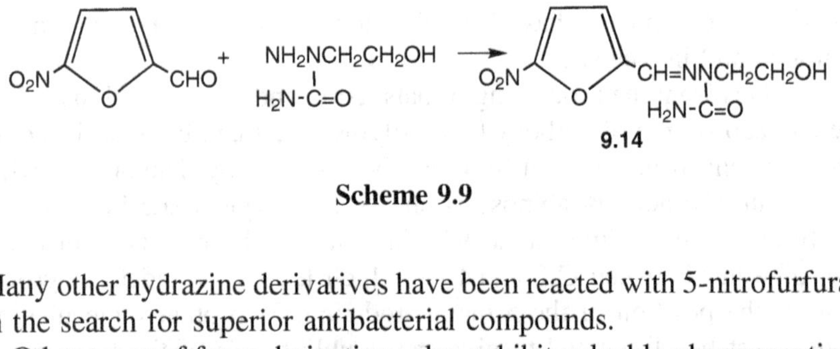

O_2N ; CHO ; + $NH_2NCH_2CH_2OH$; $H_2N\text{-}C{=}O$; O_2N ; $CH{=}NNCH_2CH_2OH$; $H_2N\text{-}C{=}O$; 9.14

Scheme 9.9

Many other hydrazine derivatives have been reacted with 5-nitrofurfural in the search for superior antibacterial compounds.

Other types of furan derivatives also exhibit valuable pharmaceutical properties. ZANTAC (Boehringer Ingelheim Pharmaceuticals, Inc.) (**9.15**, ranitidine) is a well-known H_2-receptor antagonist used in ulcer treatment. Its synthesis (Scheme 9.10) also starts with furfural, which is reduced to furfuryl alcohol. A Mannich reaction provides the 5-dimethylaminomethyl derivative **9.16**. The side chain at the 2-position is built up by displacement of the reactive OH group by an aminoethylthio group giving **9.17**, whose amino group acts to displace methylthiol from reagent **9.18**. A reaction with methylamine then displaces the second methylthio group to give ZANTAC.[6]

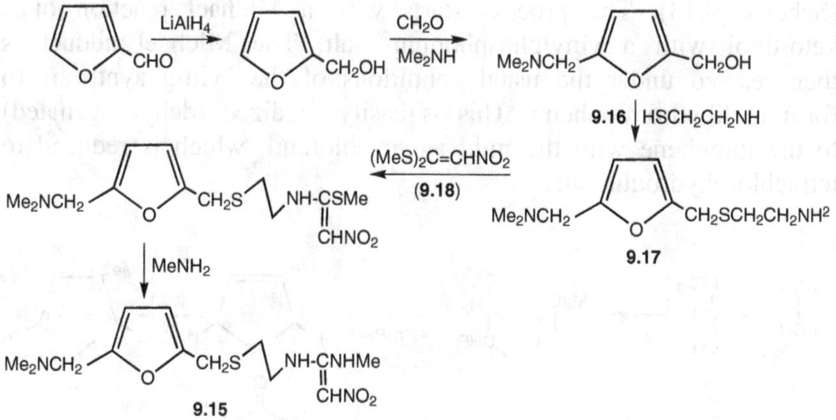

Scheme 9.10

9.4. THIOPHENES

The Paal-Knorr method was presented in Chapter 4 (section 4.2.1) as being useful for the preparation of thiophenes (Scheme 9.11).

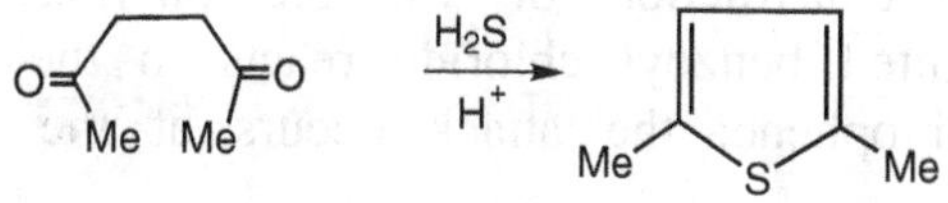

Scheme 9.11

Another classic approach to thiophenes is that of Hinsberg, where an alpha-diketone and a sulfide with two methylene groups activated by carbalkoxy groups are reacted in the presence of sodium ethoxide in a double aldol condensation (Scheme 9.12). If the carbalkoxy groups are not desired in the final product, they can be hydrolyzed with NaOH; acidification then gives the free di-acid, which can be decarboxylated thermally.

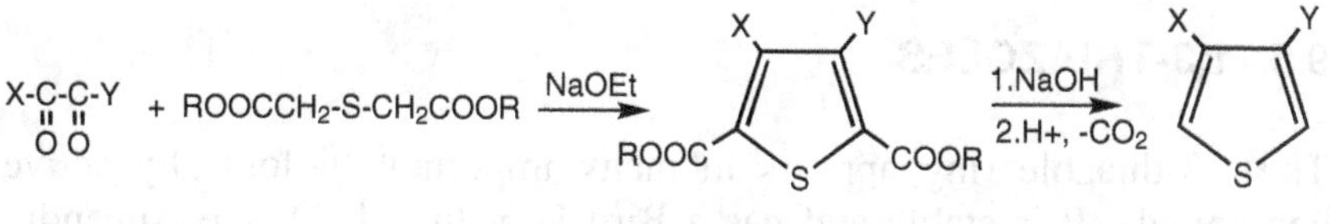

Scheme 9.12

A thiophene synthesis from more recent times, the McIntosh method,[7] employs the Wittig reaction in the ring-closing step

(Scheme 9.13). The process starts with a Michael reaction of a keto-thiol with a vinylphosphonium salt. The Michael adduct is then reacted under the usual conditions of the Wittig synthesis to form a dihydrothiophene. This is easily oxidized (dehydrogenated) to the thiophene with the mild agent chloranil, which is reduced to tetrachlorohydroquinone.

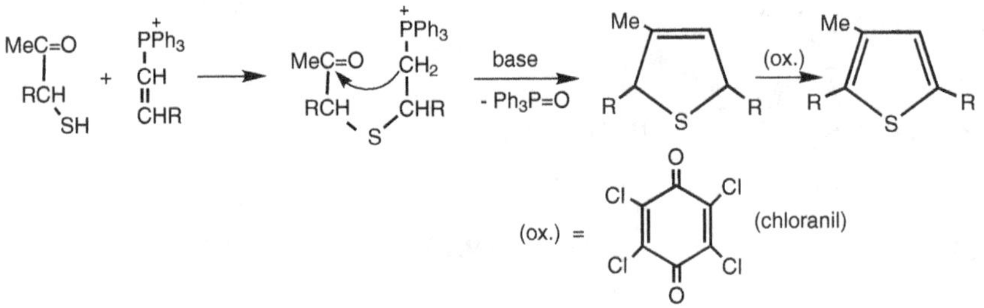

MeC=O RCH SH + $\overset{+}{P}Ph_3$ CH CHR → MeC=O RCH S CHR CH₂ $\overset{+}{P}Ph_3$ base, - $Ph_3P{=}O$ → (ox.) →

(ox.) = (chloranil)

Scheme 9.13

The synthesis of the anti-inflammatory agent Prifelone[1b] is an example of the application of the basic chemistry of the thiophene system in the construction of valuable pharmaceutical agents. Here, a substituted benzoyl chloride reacts in the Friedel–Crafts manner with thiophene; the attack occurs at the alpha position (Scheme 9.14).

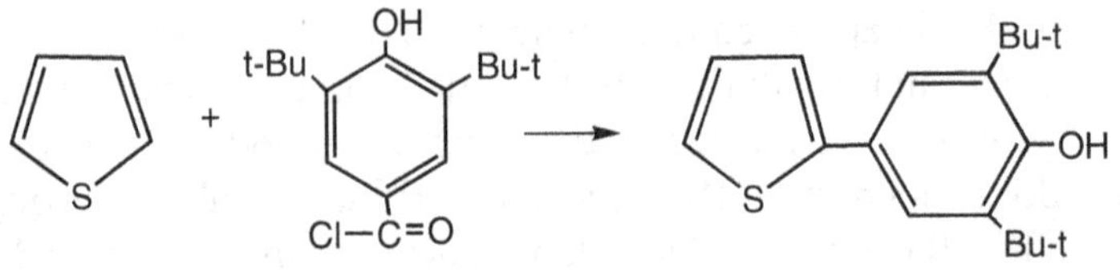

Scheme 9.14

9.5. 1,3-THIAZOLES

The 1,3-thiazole ring appears in many important biologically active compounds. It is stable and has a Bird I_A value of 79 (see Appendix for a table of values); it has both an electron-donating group (-S-) and an electron-accepting group (C=N), and the interaction of these groups is shown in the resonance hybrid **9.16**.

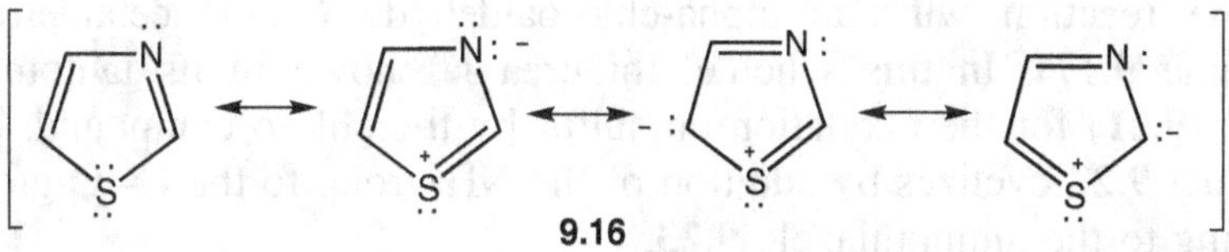

9.16

The Hantzsch synthesis of thiazoles was presented in Chapter 4 as an example of a modified Paal–Knorr method. The reaction is reproduced here as Scheme 9.15.

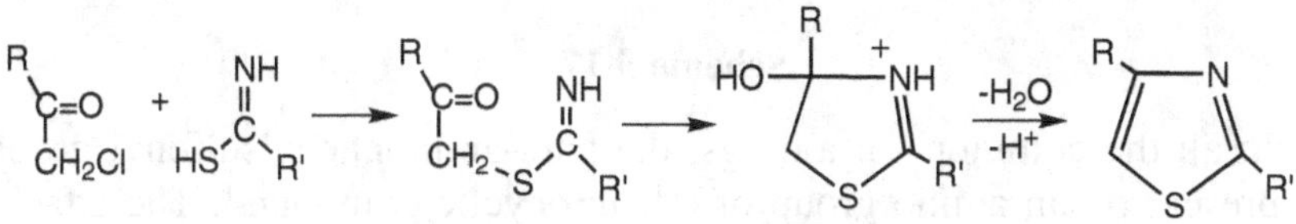

Scheme 9.15

Thiazoles can also be made by the reaction of alpha-aminonitriles and dithioacids. In the latter, the—SH group is easily displaced by nucleophiles, and it is this process that ties the two molecules together, giving intermediate **9.17** (Scheme 9.16). The proton on nitrogen can shift (tautomerize) to sulfur, and the resulting structure **9.18** cyclizes by addition of the—SH group to the nitrile group, giving structure **9.19**. This too undergoes a proton shift, from CH to NH, and this is the final product (**9.20**).

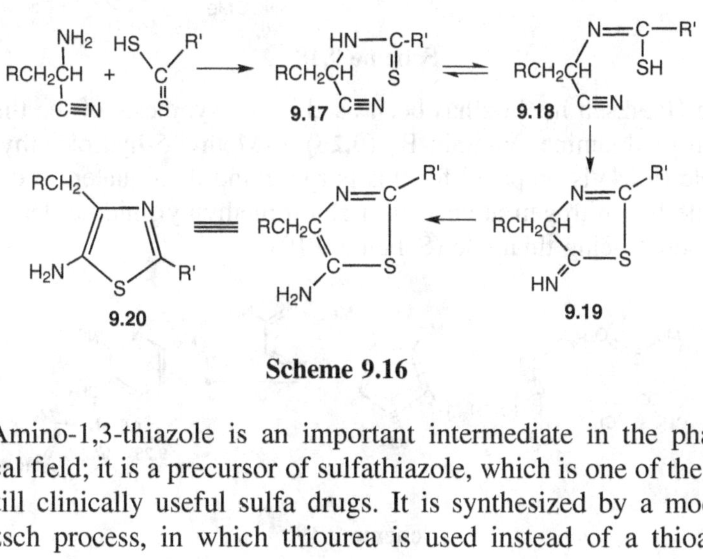

Scheme 9.16

2-Amino-1,3-thiazole is an important intermediate in the pharmaceutical field; it is a precursor of sulfathiazole, which is one of the early but still clinically useful sulfa drugs. It is synthesized by a modified Hantzsch process, in which thiourea is used instead of a thioamide in the reaction with an alpha-chloroaldehyde (chloroacetaldehyde; Scheme 9.17). In this scheme, thiourea is shown in its tautomeric form (**9.21**) for the alkylation on sulfur by the chloro compound. The product **9.22** cyclizes by addition of the NH group to the C=O group, leading to the aminothiazole **9.23**.

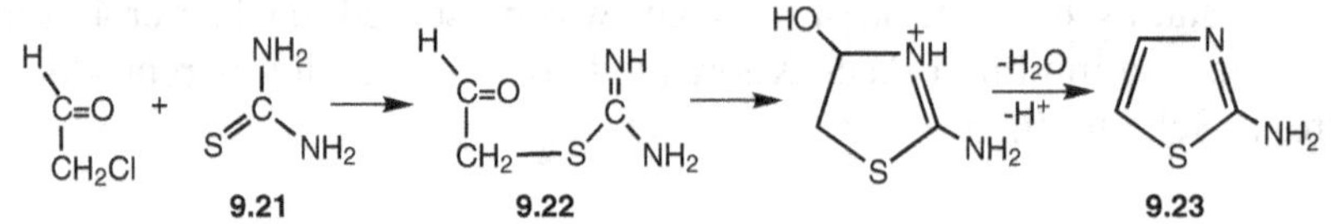

Scheme 9.17

In all the common sulfa drugs, the para-aminophenylsulfonyl group is present on an amino group of a heterocyclic compound. The attachment of this group to aminothiazole is shown in Scheme 9.18. The active sulfa drug is then prepared by hydrolysis of the amide group. Many sulfa drugs have been prepared by this method, as noted in Chapter 8.

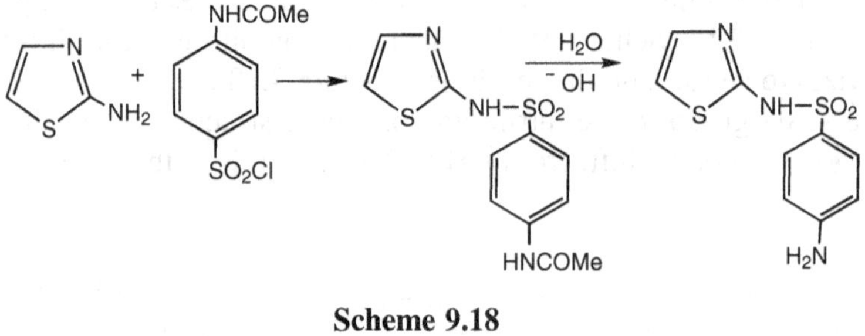

Scheme 9.18

The Hantzsch method has been used in the synthesis of the thiazole portion of thiamine, Vitamin B_1 (**9.25**). 4-Methyl-5-hydroxyethyl-1,3-thiazole (**9.24**) is prepared for this purpose and then quaternized at the pyridine-like nitrogen atom with a chloromethylpyrimidine derivative, the product being thiamine (Scheme 9.19).

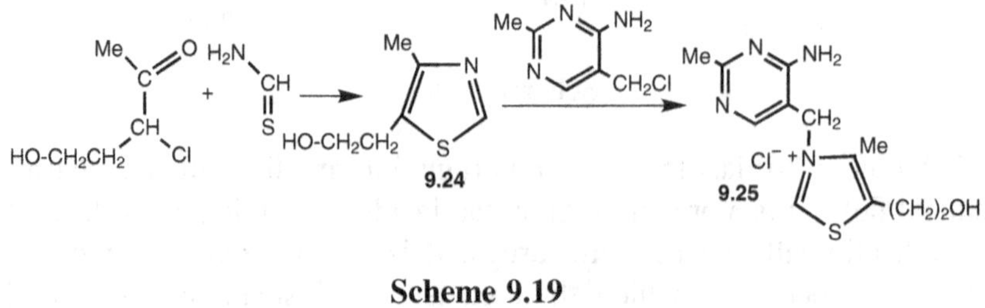

Scheme 9.19

Another application of the Hantzsch process is found in the synthesis of a monobactam antibiotic, Tigemonam.[8] The monobactams are follow-ups to the penicillins, both families being unique as derivatives of azetidinones (beta-lactams). Tigemonam also possesses a thiazole ring, which component is synthesized as in Scheme 9.20 from thiourea (shown in the -SH tautomeric form) and the chloroketone **9.26**.

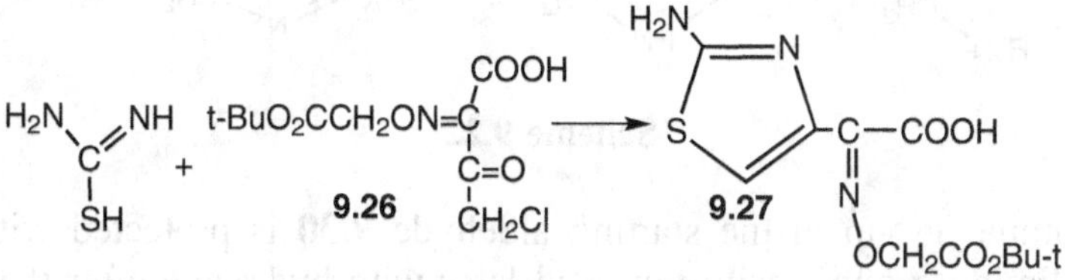

Scheme 9.20

The thiazole (**9.27**) and azetidinone **9.28** are condensed with dicyclohexylcarbodiimide (DCC), which in effect removes a mole of water from an amino group and a carboxy group to form an amide. The t-butyl protecting group is then removed by treatment with trifluoroacetic acid (TFA) to give Tigemonam (**9.29**; Scheme 9.21).

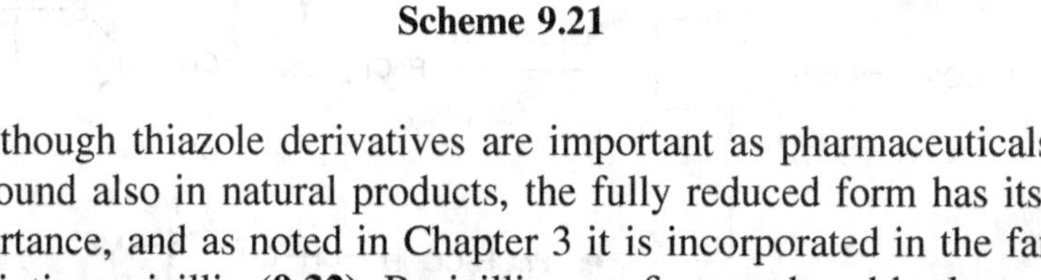

Scheme 9.21

Although thiazole derivatives are important as pharmaceuticals and are found also in natural products, the fully reduced form has its own importance, and as noted in Chapter 3 it is incorporated in the famous antibiotic penicillin (**9.32**). Penicillin was first produced by large-scale fermentation procedures, but it can also be made synthetically. The first process was that of J. Sheehan (see Chapter 3, section 3.2.5). A major hurdle to be overcome was the construction of the beta-lactam moiety fused to the thiazolidine ring. The Sheehan synthesis of benzylpenicillin is shown in Scheme 9.23, but many other derivatives can be made by modifications of the procedure. The first step involves the net elimination of water from the condensation of an aldehyde group of **9.30** with the SH and the NH_2 groups of aminothiol **9.31**. This is a well-known procedure for thiazolidine synthesis, and details can be visualized as in the general expression of Scheme 9.22.

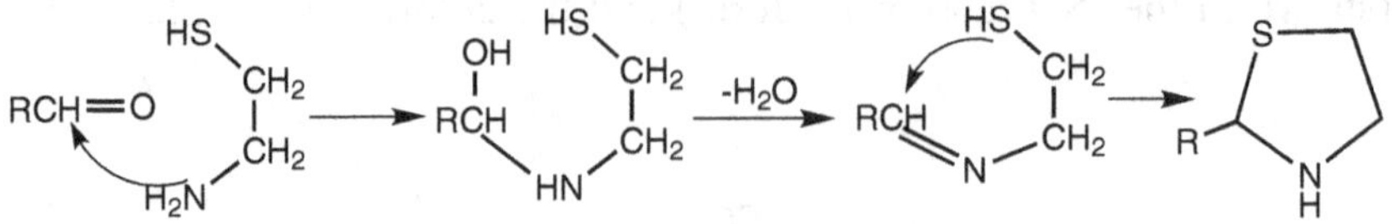

Scheme 9.22

The amino group in the starting aldehyde **9.30** is protected with the phthalimido group, easily removed later with hydrazine after the thiazolidine ring is formed. This process is known as the Gabriel method of protection. Similarly, the ester group in **9.30** is protected with the t-butyl group, and also it is easily removed using dry HCl. By these techniques, derivative **9.31** was prepared. The key step in the entire synthesis is the closing of the beta-lactam ring by dehydration with DCC between the COOH and NH groups, affording benzylpenicillin (**9.32**).

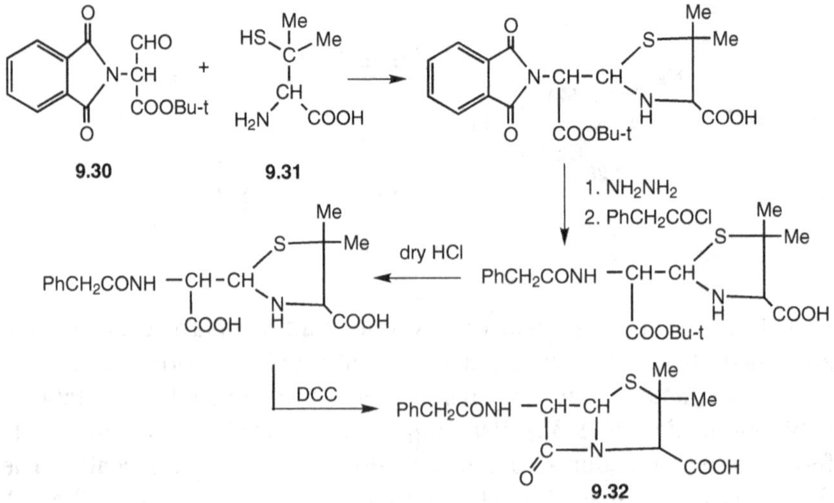

Scheme 9.23

The fermentation procedures can be modified properly to produce the penicillin molecule without the acyl group on the amino group. This compound (**9.33**) is known as 6-aminopenicillanic acid (6-APA).

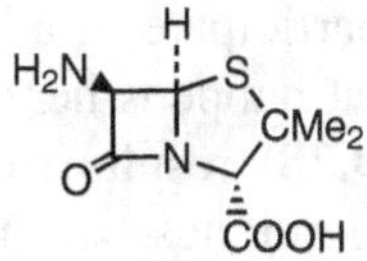

9.33

It is now produced in bulk quantities by fermentation and is a readily available starting material for the synthesis of many other antibiotics. It is used in the synthesis of oxacillin[9] (**9.36**, Scheme 9.24). First formed is the isoxazole moiety of oxacillin, using a general method for construction of this ring system involving the addition of the OH group of an oxime to a carbonyl group. Thus, from compound **9.34** is formed the isoxazole **9.35**. The COOEt group of **9.35** is then hydrolyzed and converted to the acid chloride form for acylating the amino group of 6-APA to yield oxacillin. It is obvious that many acid chlorides could be used to produce novel potentially antibiotic substances from 6-APA.

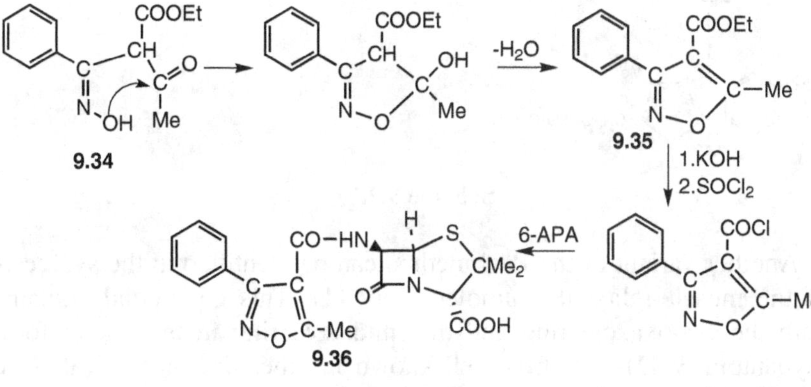

Scheme 9.24

9.6. 1,3-OXAZOLES

The 1,3-oxazole ring is also of considerable importance in pharmaceutical chemistry. Of several methods for forming the ring, we will discuss here only the use of variations of the Paal synthesis, which has been used to prepare some notable biologically active compounds. The synthesis of the anti-inflammatory agent romazarit[10] (**9.40**) is illustrative

(Scheme 9.25). The synthesis starts with the formation of compound **9.37**. Here, the unit O=C–O–C–C=O is analogous to the gamma-dicarbonyl unit so useful in the Paal synthesis. It is not common for a C=O group of an ester to participate in addition-elimination reactions like a ketone, but that is what happens here on reaction with ammonia, and the 1,3-oxazole ring (**9.38**) results from a Paal-like mechanism. The synthesis of romazarit continues with the reduction of the COOEt group to CH_2OH and its chlorination to CH_2Cl. Coupling of the chloride with an alkoxide (the Williamson reaction) leads to ether formation as shown in **9.39**. Hydrolysis of the ester group to the acid completes the synthesis.

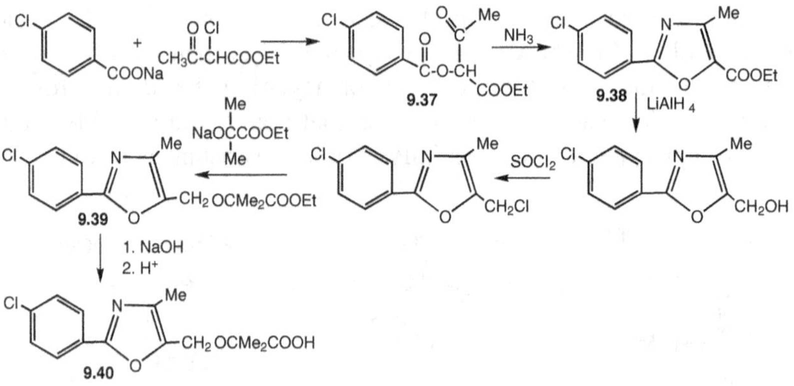

Scheme 9.25

Another variant of the Paal method can be identified in the synthesis of the muscle-relaxant azumolene[11] (**9.41**). This compound contains both the 1,3-oxazole ring and the imidazole ring in the diketo form (hydantoin, **9.42**), which is well known in other pharmaceuticals (see section 9.7.).

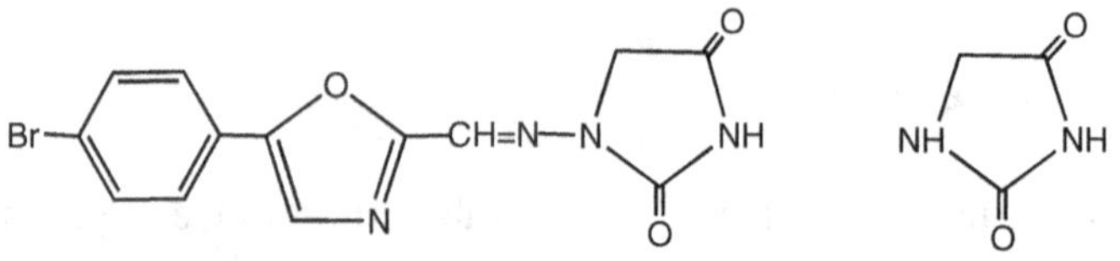

9.41 **9.42**

The synthesis of azumolene is shown in Scheme 9.26. The intermediate amide **9.43** is formed in the usual way by reaction of an acyl chloride with an amine. In **9.43** the unit O=C–C–N–C=O can be identified, which has the gamma-arrangement of two carbonyl groups. One of these is in the amide group, not normally given to ketone-like chemistry. On dehydration with $POCl_3$, the oxazole ring is formed. This may occur through the amidic carbonyl acting to accept an OH group from the enolic form of the carbonyl group in compound **9.43**, but another possibility is that the enol (**9.44**) from the amide group interacts with the ketonic group. It is not certain which mechanism applies, but that involving the enol of the amide is shown in Scheme 9.26.

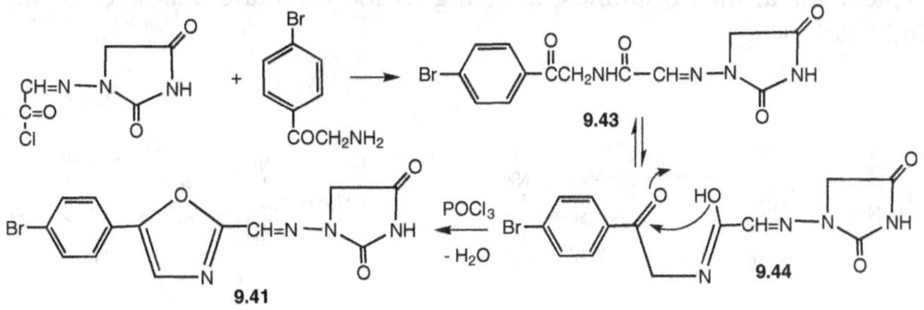

Scheme 9.26

9.7. IMIDAZOLES

Imidazole is the common name used for 1,3-diazole. This is a stable, aromatic ring system, with a Bird Unified Aromaticity Index (I_A) value of 79. Its resonance and tautomeric forms were discussed in Chapter 7 (section 7.1.8). This ring frequently appears in pharmaceuticals; fused-ring systems are also common. Thus, the all-important purines are formed from fusion of imidazole to pyrimidines and are the subject of section 9.17. There is no general method for forming the imidazole ring such as the Paal, Knorr, and Hantzsch syntheses we have discussed for other cycles, so specific syntheses must be performed.

Simple nitro derivatives of imidazole are effective as antibacterial agents. They also are useful in treating infections caused by protozoans, such as *Trichomonas*. 2-Nitroimidazole was the first to be discovered (1953); it actually was isolated from cultures of a strain of *Streptomyces* and came as a surprise because naturally occurring nitro compounds are rare. It was named azomycin. A practical synthesis was reported in 1965 and is reproduced in Scheme 9.27.[12] The imidazole ring is formed from the reaction of cyanamide (**9.46**) with the dimethylacetal (**9.45**) formed from aminoacetaldehyde. The aldehyde is introduced in this form to prevent premature condensation of the aldehyde group with the amino group of cyanamide and even of itself. Presumably, the nitrile group first accepts the amino group of **9.45** to form the guanidine derivative **9.47**, and the aldehyde group is then released from the acetal with acid to allow it to condense with the amino group of cyanamide. The product is 2-aminoimidazole (**9.48**). To form azomycin (**9.49**), the amino group is first diazotized with cold nitrous acid, followed by reaction with sodium nitrite and Cu as catalyst. Both reactions are typical for aromatic amines, attesting to the aromatic character of the imidazole ring.

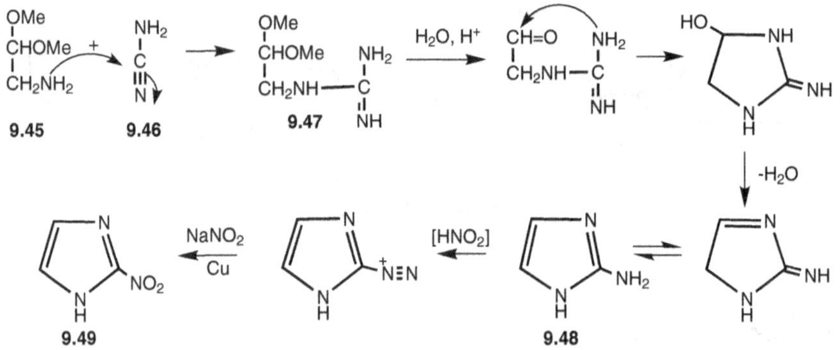

Scheme 9.27

The tautomeric character of the imidazole system is involved in the synthesis of the related drug dimetridazole[13] (**9.52**, 1,2-dimethyl-5-nitroimidazole), which is an effective agent against trichomonal infections in veterinary medicine (Scheme 9.28). The synthesis involves the nitration of 2-methylimidazole to form 2-methyl-5-nitroimidazole (**9.50**), followed by N-methylation. However, compound **9.50** is involved in tautomeric equilibrium with compound **9.51** (2-methyl-4-nitroimidazole), which would give an isomer (**9.53**) on methylation. This can be avoided by the proper choice of solvent. Using a nonpolar solvent for methylation with dimethyl sulfate, the desired **9.52** is the predominant product. Tautomerism is a complicating factor in imidazole chemistry where an NH unit is present, and N-alkyl derivatives are more commonly encountered. We will see this factor again with the purines.

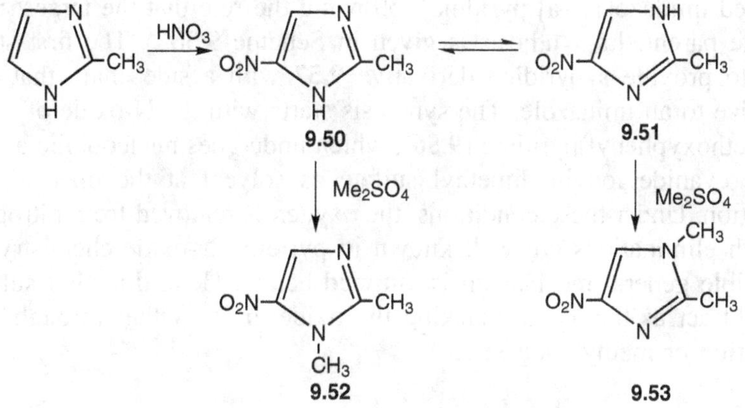

Scheme 9.28

2,4-Dihydroxyimidazoles are involved in keto-enol tautomerism, and the diketo structure, which is known as hydantoin (**9.42**), is favored. This is the framework for many valuable compounds with anticonvulsant activity. These are nonhypnotic and are used in the treatment of epilepsy. Dilantin (**9.55**) is the best known of these. Its synthesis is shown in Scheme 9.29. The Strecker reaction of benzophenone with HCN gives an alpha-aminonitrile, which adds ammonia to form compound **9.54** (an amidine). The ring is then closed with phosgene or diethyl carbonate. Hydrolysis gives phenytoin (dilantin, Pfizer, Inc.).

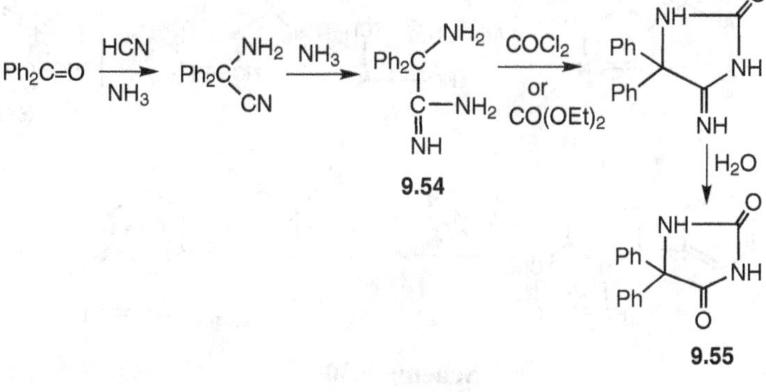

Scheme 9.29

The drug Fadrozole (**9.61**), which is an inhibitor of the enzyme aromatase and used for the treatment of estrogen-dependent neoplasms, is an example of a fused imidazole derivative. The ring system is named imidazo[1,5-a] pyridine, following the rule that the largest ring is the parent. Its synthesis is given in Scheme 9.30.[1c] The first steps are to provide a pyridine derivative **9.57** with a side chain that can cyclize to an imidazole. The synthesis starts with the N-oxide of 2-(4-carbethoxyphenyl)pyridine (**9.56**), which undergoes nucleophilic attack with cyanide ion in dimethyl sulfate as solvent at the open alpha-position. Under these conditions, the oxygen is removed from nitrogen. (Such eliminations are well known in pyridine N-oxide chemistry. A possible general mechanism is outlined below. Here, dimethyl sulfate would act as the agent attacking the oxide group, either through sulfonation or methylation.)

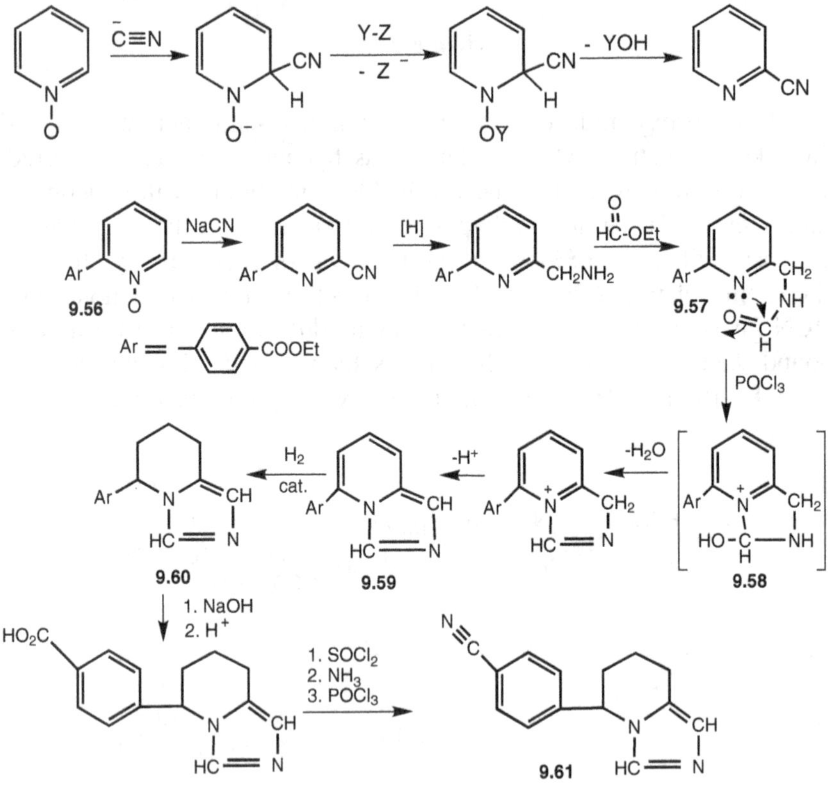

Scheme 9.30

Reduction of the nitrile group to amino, followed by acylation with ethyl formate, gives the desired formamide derivative **9.57**. Treatment with $POCl_3$ accomplishes the cyclization, wherein the pyridine nitrogen adds to the acyl carbonyl group (giving **9.58**) and dehydration by $POCl_3$ drives the reaction forward. After the loss of a proton, the imidazopyridine system (**9.59**) is formed. The pyridine ring is then reduced by catalytic hydrogenation to form **9.60**. The remaining steps are standard for accomplishing the conversion of the ester group on the phenyl substituent to the nitrile group of fadrozole (**9.61**).

9.8. PYRAZOLES

Pyrazoles have a high degree of aromaticity with a Bird I_A of 90. They also are involved in a tautomeric equilibrium, with the NH tautomer being preferred (Scheme 9.31). Generally, they are synthesized from a preformed compound with a N-N bond, usually a hydrazine. Reaction with a beta-dicarbonyl compound gives the pyrazole structure.

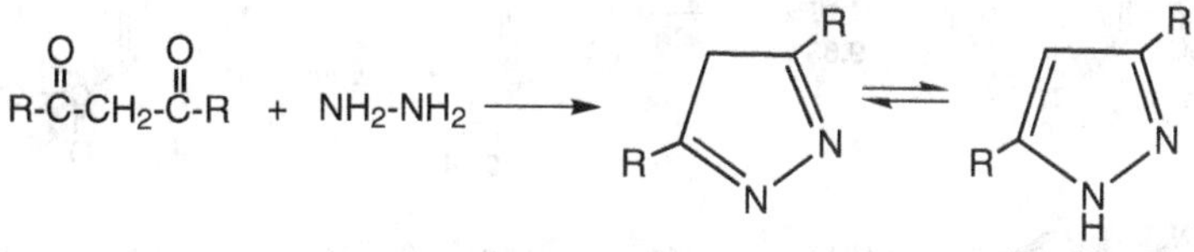

$R-C(=O)-CH_2-C(=O)-R + NH_2-NH_2 \rightarrow$ 3,5-R-pyrazole (N tautomer) $\rightleftharpoons$ NH tautomer (preferred)

Scheme 9.31

The synthesis of the anti-inflammatory drug tepoxalin is based on this general method.[14]

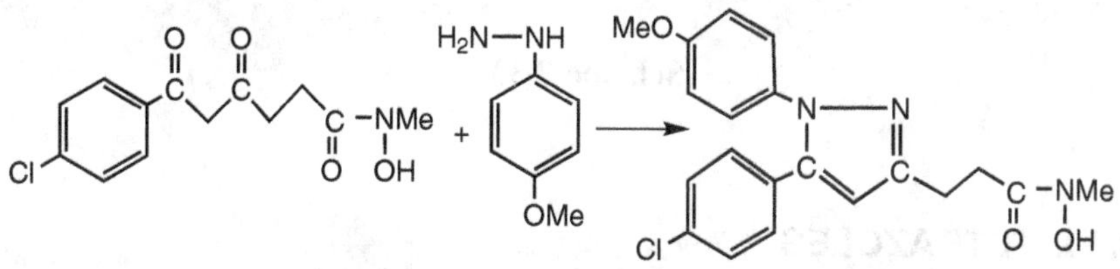

Scheme 9.32

9.9. 1,2,4-TRIAZOLES

The triazole ring system has one of the highest Bird I_A values (100). As for pyrazole synthesis, a starting material with a pre-formed N-N bond is also used in the synthesis of 1,2,4-triazoles, as illustrated with the antiulcer agent lavoltidine[1d] (GlaxoSmithKline) (**9.67**, Scheme 9.33). One starting material is a hydrazone (**9.62**), which has the terminal NH_2 of a hydrazine temporarily protected by the benzylidene group from benzaldehyde. In this compound, there is also an SMe group that can be replaced by a nucleophile, in this case the amine **9.63**. The reaction product (**9.64**) is subjected to acid hydrolysis to convert the hydrazone back to the hydrazine (**9.65**), whereupon the freed amino group adds to the carbonyl group in the usual manner with elimination of water to create the 1,2,4-triazole ring (**9.66**). The synthesis of lavoltidine is completed by the hydrolysis of the acetate group.

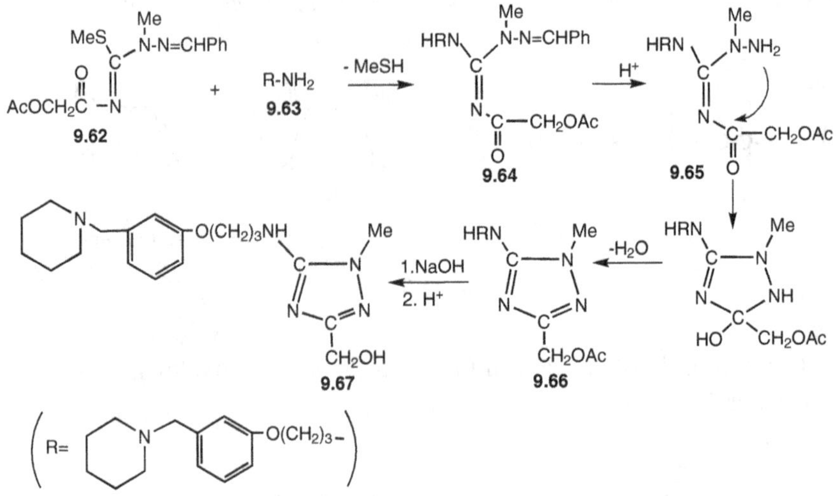

Scheme 9.33

9.10. TETRAZOLES

1,3-Dipolar cycloaddition of an azide with a nitrile constitutes the route to the tetrazole ring system, which has a high Bird I_A value of 89. The cycloaddition is used in the synthesis of losartan, which is an angiotensin II antagonist for the treatment of hypertension. This molecule (**9.71**) has also an imidazole ring. Alkylation of the NH group of the imidazole starting material **9.68** with the aromatic compound **9.69** gives structure **9.70**, which has the nitrile group needed for the addition of the dipole, in this case hydrazoic acid (Scheme 9.34). The product is losartan[1e] (**9.71**).

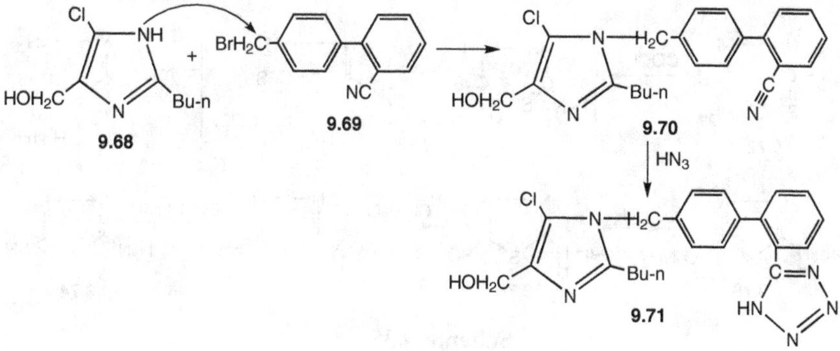

Scheme 9.34

9.11. 1,3,4-THIADIAZOLES AND OTHER 5-MEMBERED SYSTEMS

A final type of 5-membered heterocycle to be considered has the 1,3,4-thiadiazole structure. It is noted, however, that other less common medicinally useful compounds with 5-membered ring systems exist that have two or three heteroatoms and are not included in this chapter. These include 1,2-oxazoles (isoxazoles), 1,2-thiazoles, 1,3-dithioles, 1,2,5-thiadiazoles, and 1,2,4-oxadiazoles. Examples of drugs based on these systems can be found in the series *The Organic Chemistry of Drug Synthesis* by Daniel Lednicer.[1]

The 1,3,4-thiadiazole **9.76** is known as acetazolamide and is a diuretic agent. It also has been used in the treatment of glaucoma. The synthesis[15] (Scheme 9.35) starts with the hydrazine derivative **9.72**. Treatment with phosgene causes ring closure with the elimination of one amino group, forming **9.73**. A possible but unproven mechanism involves acylation of one amino group by phosgene, which converts it to a better leaving group. The tautomeric SH form at the end of the chain displaces the acylated fragment—NH–COCl or an equivalent, giving the 1,3,4-thiazole **9.74**, after a tautomeric shift. The remaining amino group is then protected by acetylation, and the thiol group is oxidized with aqueous chlorine. The product is the sulfonyl chloride **9.75**, which resists hydrolytic displacement of chloride. Treatment with ammonia then provides the sulfonamide acetazolamide (**9.76**).

Scheme 9.35

9.12. INDOLE

Derivatives of indole are common in nature. Alkaloids based on indole were discussed in Chapter 3. Indole itself is a colorless solid with a melting point (m. p.) of 52°C, and its derivatives also are solids. Unless a polar group is present, water solubility is low; indole itself has the solubility of 1 part in 540 parts of water at 20°C, but it is freely soluble in many organic solvents. Resonance properties and behavior with electrophilic agents were discussed in Chapter 7.

Because of its biological importance, numerous methods have been developed for the synthesis of indoles. The most important of these is that devised by Emil Fischer in 1883; as we shall discuss, it is still of great value in the synthesis of complex indoles. The Fischer synthesis is simple, but the mechanism is far from that. It begins by making phenylhydrazones of aldehydes or ketones by reaction with phenylhydrazine, which was also discovered by Fischer. The hydrazone is heated with a protic [alcoholic HCl or polyphosphoric acid (PPA)] or a Lewis acid ($ZnCl_2$, BF_3, etc.), and the indole ring is so formed. Countless indoles have been made by this process, but there are some restrictions. The ketone from which the phenylhydrazone is to be made must have at least one CH_3 or CH_2 group on the carbonyl group. The phenyl group may have electron-releasing substituents, but electron-withdrawing groups deactivate the ring and in general cannot be tolerated. The synthesis fails with acetaldehyde as starting material, and thus indole itself cannot be prepared by the acid-catalyzed method, although it and other indoles can be formed by high-temperature treatment (about 200°C) of hydrazones. The synthesis is illustrated in Scheme 9.36 with acetone as the starting carbonyl compound. Key features are that an alpha-carbon of the original carbonyl group is incorporated in the ring at C-3, and the terminal nitrogen of the original hydrazone is expelled.

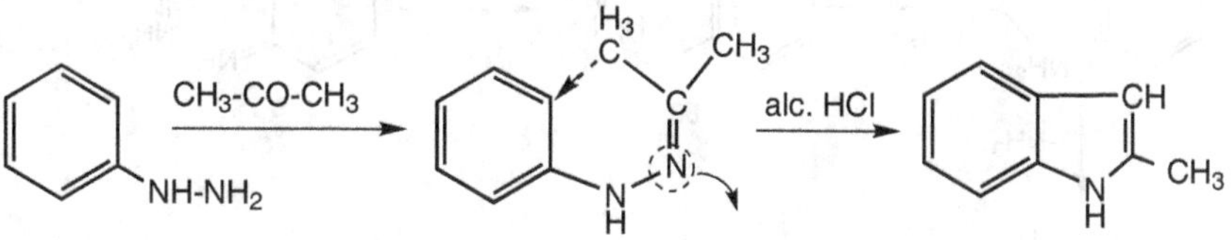

CH_3-CO-CH_3 NH-NH_2 H_3C CH_3 C N N H alc. HCl CH CH_3 N H

Scheme 9.36

The Fischer synthesis was one of the first reactions to be considered from a mechanistic standpoint. As early as 1918, before the development of electronic theory, R. Robinson proposed the basics of the atom changes involved that we now know to account for indole formation. The present view of the mechanism for protic acid catalysis is presented in Scheme 9.37. The protonated form (**9.77**) of the hydrazone undergoes a proton shift to the enamine salt structure **9.78**, which then has a bond rearrangement known as a [3,3]-sigmatropic shift. (This defines a well-known concerted rearrangement of a six-atom unit with 1,5-double bonds, wherein the terminal atoms form a new sigma bond while the sigma bond between the central atoms is being broken. The transition state has a 6-membered cyclic structure, possibly chair shaped). The Cope isomerization of 1,5-dienes and the Claisen rearrangement of allyl phenyl ethers to ortho-allyl phenols are examples. (See J. March, *Advanced Organic Chemistry*[16] for a more complete description of [3,3]-sigmatropic rearrangements.) The net effect is breakage of the N–N bond with creation of the energetically stronger C–C bond, giving the intermediate **9.79**. As a result, an imino group would appear on the benzene moiety (**9.79**), but a H-shift will restore the structure to that with an amino group on the benzene ring (**9.80**). This intermediate has been isolated and identified. In typical fashion, the amino group adds to the C=N bond and loss of NH_3, then of a proton, gives the indole structure.

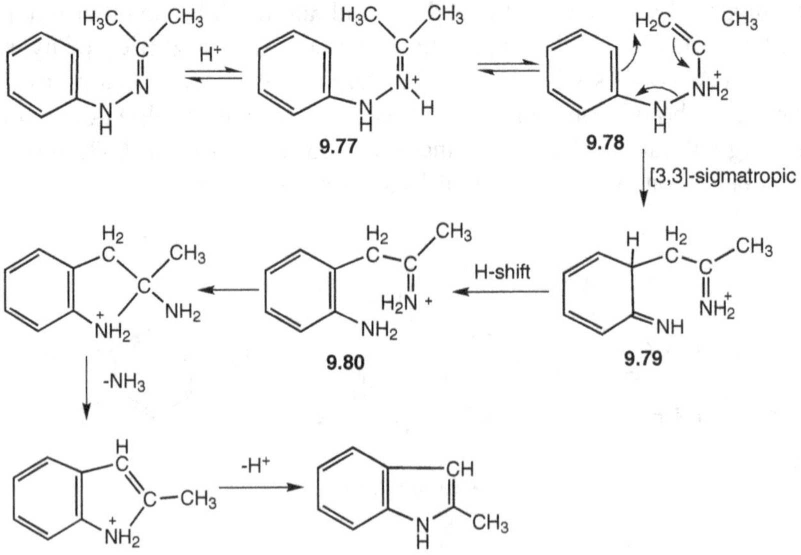

Scheme 9.37

Among other classic but useful indole syntheses are those of Bischler (1892) and Madelung (1912). In the Bischler synthesis, an aniline derivative is alkylated by an alpha-haloketone. Heating the product (**9.81**) effects the electrophilic attack of the carbonyl group on the benzene ring; the loss of water from the product gives the indole structure (Scheme 9.38).

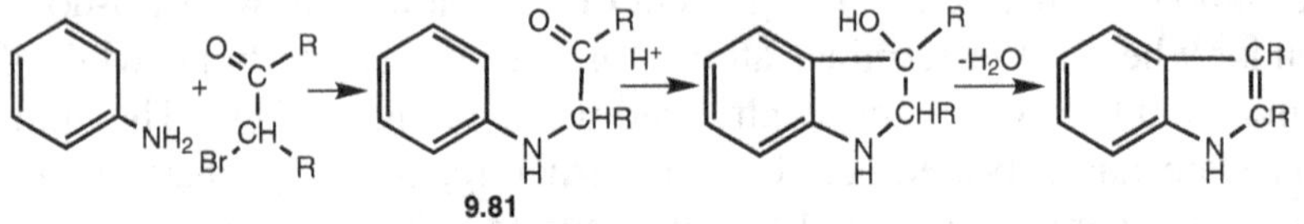

Scheme 9.38

In the Madelung process, an acetanilide with an ortho-methyl group is heated with NaOEt at 300°C. A condensation of the aldol type occurs and an indole results (Scheme 9.39).

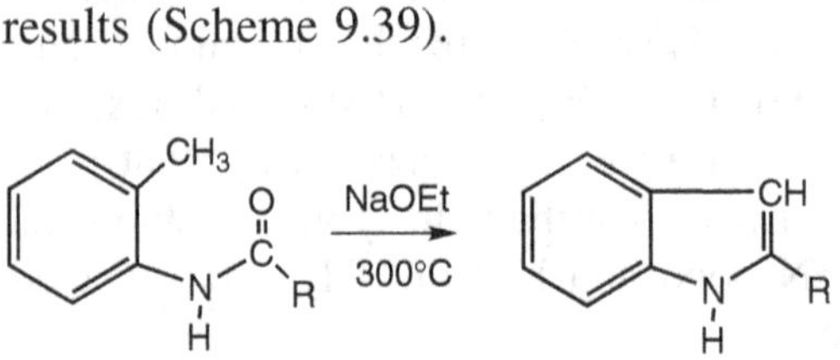

Scheme 9.39

Reflecting the importance of indole chemistry, many new methods have been devised since these major early accomplishments. Illustrative is a process employing palladium complexes, which was described in Chapter 4. The synthesis is repeated here as Scheme 9.40.

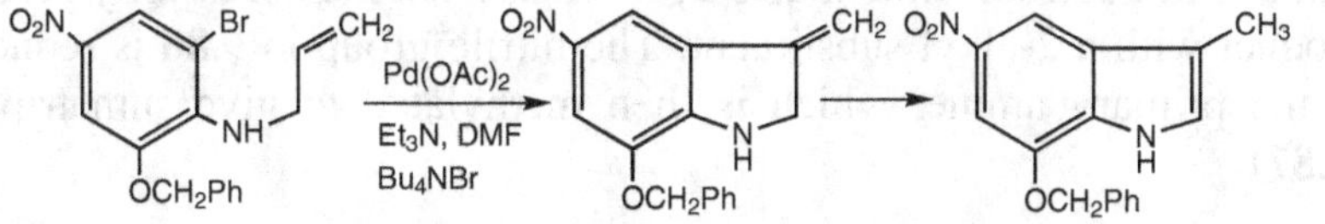

Scheme 9.40

In another advance outlined in Scheme 9.41, the Madelung synthesis was modernized by employing an intramolecular Wittig reaction (Chapter 4, section 4.3) instead of the aldol-like condensation of Scheme 9.39.[17]

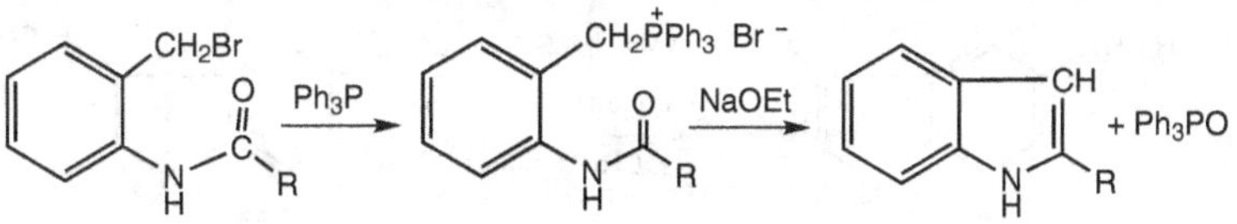

Scheme 9.41

A practical application of the Fischer method is found in the synthesis of the anti-inflammatory drug indomethacin[1f] (Scheme 9.42). Hydrazone **9.82** gives the expected indole product **9.83** on acid treatment. The COOH group is protected by conversion to the t-butyl ester **9.84** employing DCC as a dehydrating agent. Ester **9.84** is subjected to benzoylation with para-chlorobenzoyl chloride and base, and the t-butyl group is removed thermally in the final reaction to give indomethacin (**9.85**).

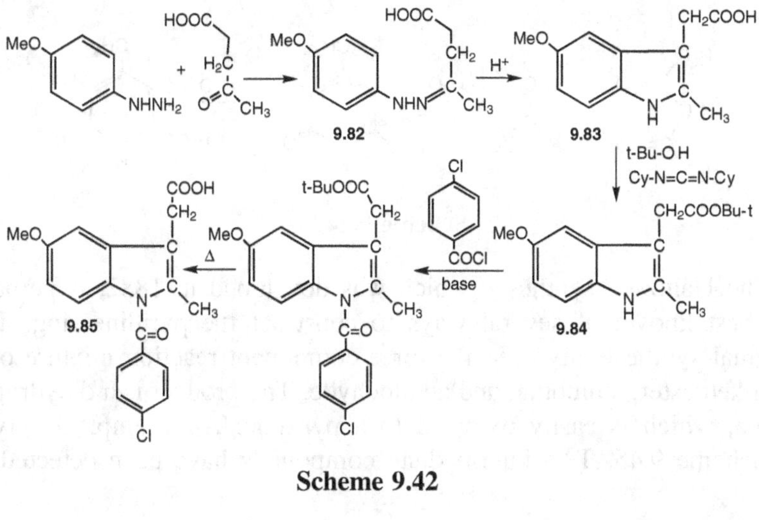

Scheme 9.42

Sumatriptan is a drug used in the treatment of migraine headaches; it too is synthesized by the Fischer method[1g] (Scheme 9.43). Here, an aldehyde is used to make the phenylhydrazone, which on acid treatment gives indole **9.86**. Note that using an aldehyde reactant gives an indole without a substituent at C-2, whereas a ketone necessarily gives a product with a 2-alkyl substituent. The nitrile group in **9.86** is reduced to the primary amine, which is then methylated to give sumatriptan (**9.87**).

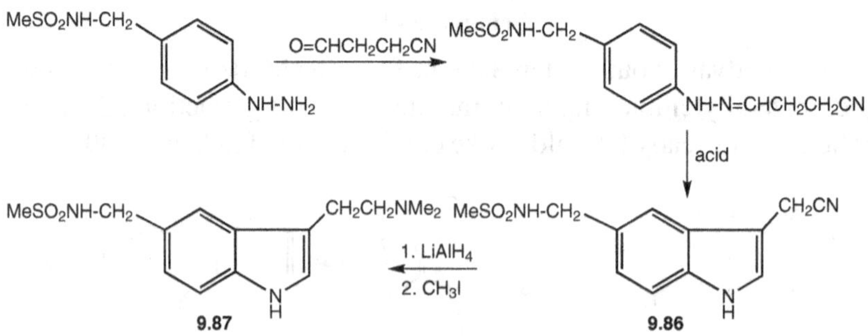

Scheme 9.43

9.13. PYRIDINES

In Chapter 4, section 4.2.3, was presented a method employing the aldol condensation for the synthesis of the pyridine ring, in the form of an alpha-pyridone. The process is summarized in Scheme 9.44.

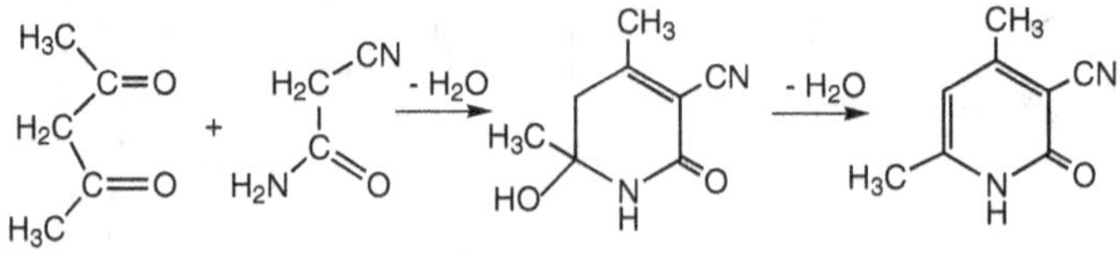

Scheme 9.44

The Hantzsch synthesis, which was developed in 1882, is perhaps the best known of several ways to construct the pyridine ring. The original synthesis involved the three-component reaction mixture of a beta-ketoester, ammonia, and an aldehyde. The product is a dihydropyridine, which is easily oxidized to a pyridine. An example is given in Scheme 9.45. Two intermediate compounds have been detected in the mixture, both formed from the ketoester. One (**9.87**) is an enamine from reaction with ammonia; the other (**9.88**) is an aldol product from reaction with the aldehyde. These intermediates, which have been detected and confirmed, then combine in a Michael reaction to give a dihydropyridine (**9.89**). In this process, the electron-rich beta-carbon of the enamine adds to the electron-deficient beta-carbon of the aldol product, and the ring is closed by addition of the amino group to the keto carbonyl. The resulting dihydropyridine is then oxidized to the pyridine.

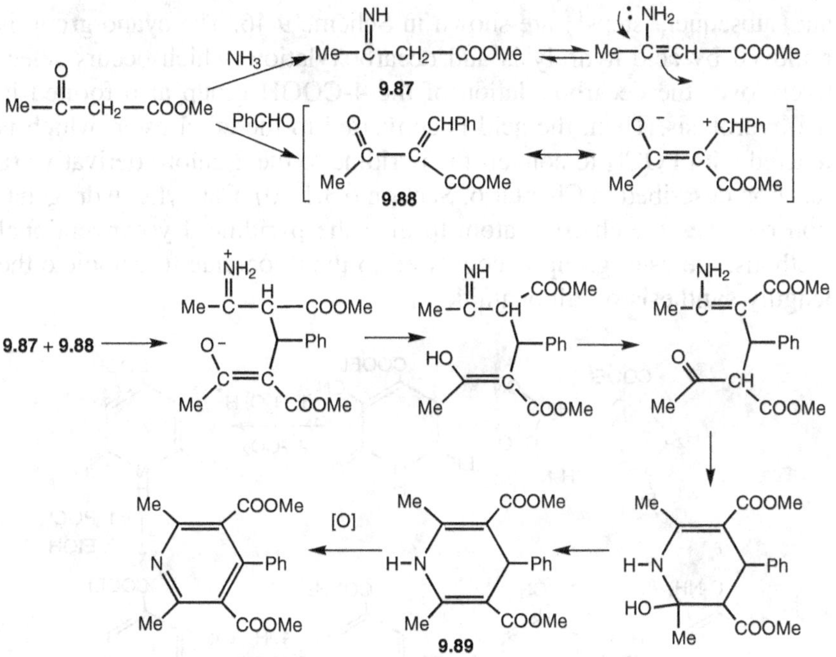

Scheme 9.45

The ester groups on the ring may of course be hydrolyzed to the acids and then removed by decarboxylation if desired.

Old as it may be, the Hantzsch process still finds use in drug synthesis. Nifedipine, which is used to relieve the chest pain of angina pectoris, is a dihydropyridine derivative prepared by the procedure outlined in Scheme 9.45 from ethyl acetoacetate and ortho-nitrobenzaldehyde, rather than benzaldehyde, and without the step of oxidation to the pyridine.[1h]

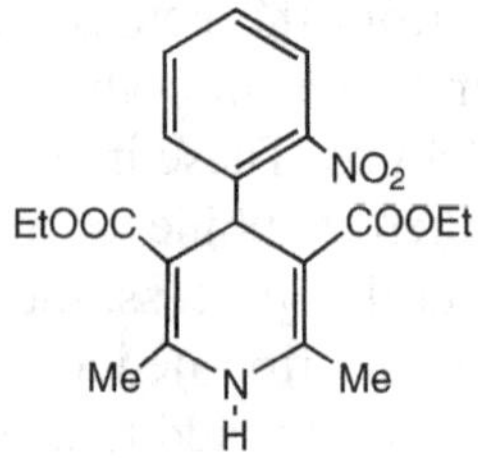

Nifedipine

The pyridine ring of the tuberculostatic drug ethionamide (**9.90**) is generated from the condensation of a beta-diketone with cyanoacetamide. This is a useful way of synthesizing alpha-pyridones. This and subsequent steps[18] are shown in Scheme 9.46. The cyano group is removed by acid hydrolysis and decarboxylation, which occurs selectively over the decarboxylation of the 4-COOH group also formed in the hydrolysis. Then, the acid is converted to the ethyl ester, which is reacted with $POCl_3$ to convert the pyridine to the 2-chloro derivative (a reaction described in Chapter 6, Section 6.3.1.10). Catalytic hydrogenation removes the chlorine atom to give the pyridine. By conventional methods, the ester group is converted to the thioamide to complete the lengthy synthesis of ethionamide.

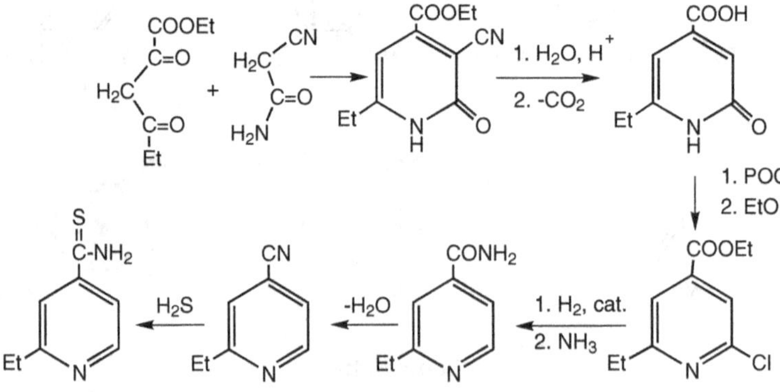

9.90

Scheme 9.46

Several pyridine-containing drugs are prepared by taking advantage of the easy displacement of halogen at the alpha- and gamma-positions by nucleophiles or metals. The halo compounds are generated from pyridones or pyridine N-oxides with phosphorus halides as presented in Chapter 6. An example is the antihypertensive ofornine,[1i] **9.91**, which is summarized in Scheme 9.47.

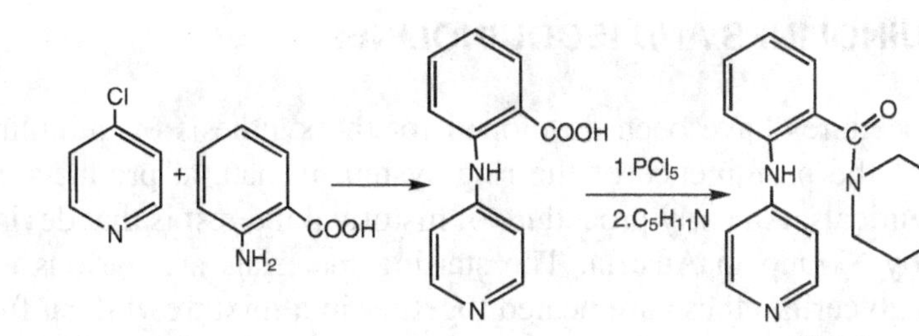

Scheme 9.47

In another example that gives the antihistamine acrivastine[1j] (Scheme 9.48), one bromine atom of 2,6-dibromopyridine can be replaced by lithium; a reaction with p-toluonitrile gives the ketone for a Wittig reaction to form **9.92**. The remaining bromine is replaced by lithium using an alkyl lithium reagent, and the product then reacted with dimethylformamide to install the aldehyde group on the ring. Compound **9.93** can be reacted with the anion of triethyl phosphonoacetate (known as the Wadsworth-Emmons reagent), which acts like a Wittig reagent in creating the carbon-carbon double bond from a carbonyl group. The other product is diethyl phosphate anion. The final step is hydrolysis of the ester group, giving acrivastine (**9.94**).

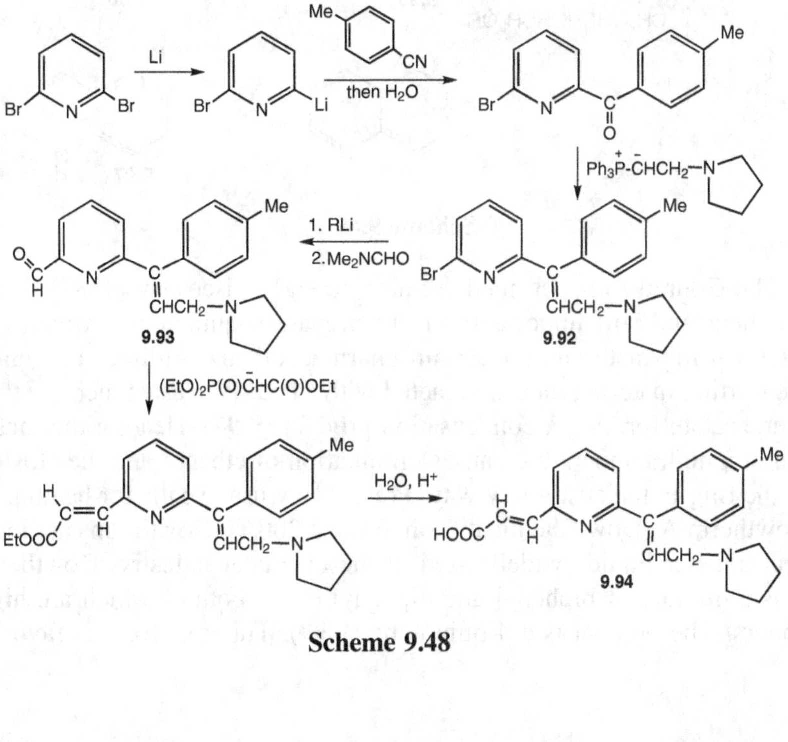

Scheme 9.48

9.14. QUINOLINES AND ISOQUINOLINES

Many procedures have been developed for the synthesis of quinolines because of the prominence of the ring system in natural products and pharmaceuticals. An early procedure of historical interest is that devised in 1881 by Skraup in Austria. The starting materials are various anilines and glycerine; these are heated together in a mixture with sulfuric or phosphoric acid, as well as an oxidizing agent of a wide variety of types (nitrobenzene, arsenic acid, lead dioxide, ferric salts, etc.). The overall process is shown as Scheme 9.49. It is believed the acid causes dehydration of the glycerine to acrolein ($CH_2{=}CH{-}CHO$); Michael addition of the aniline to the double bond then takes place, forming **9.95**. The carbonyl group attacks the benzene ring to form the usual hydroxyl intermediate (**9.96**), which after dehydration gives a dihydroquinoline (**9.97**), the substrate for the action of the oxidizing agents. The final product will necessarily be a quinoline with no substituent in the pyridine ring; this restriction is avoided by the use of other methods.

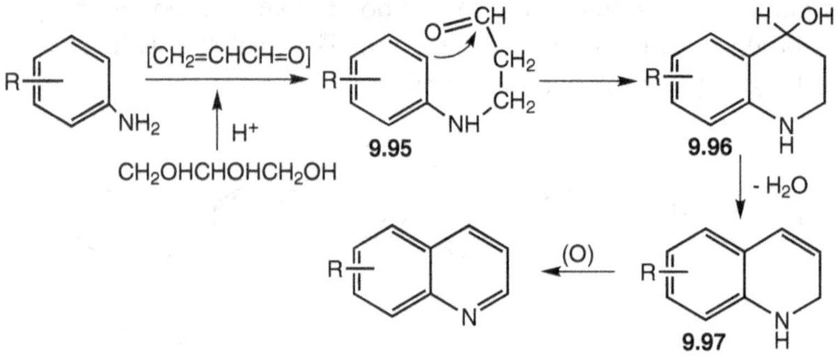

[$CH_2{=}CHCH{=}O$] H^+ $CH_2OHCHOHCH_2OH$ **9.95** **9.96** $-H_2O$ (O) **9.97**

Scheme 9.49

The Conrad-Limpach method, also an early discovery (1887), is an excellent and still important way to prepare 4-quinolones, which are vital starting materials for certain pharmaceuticals. Anilines are again the starting material; they are reacted with a beta-keto ester such as ethyl acetoacetate forming a condensation product (**9.98**). Heating the material at a high temperature causes elimination of ethanol and the closing of the ring in the customary way. Present-day usage calls for heating in Dowtherm A (Dow Chemical Company) at 200°C. Dowtherms are inert heat-transfer liquids widely used in the chemical industry. Dowtherm A is a mixture of biphenyl and diphenyl ether, both of which are high boilers. The product is a 4-quinolone (**9.99**). This oxo form is favored in the tautomeric equilibrium with the enol form (**9.100**), as is expected for a pi-deficient heterocycle. The subsequent reactions are also typical of pi-deficient heterocycles and were discussed in the study of pyridine (Chapter 6). A reaction with PCl_5 causes the displacement of hydroxyl by chlorine to give **9.101**; the chlorine is easily displaced by amines. Many 4-aminoquinolines have been prepared by these reactions.

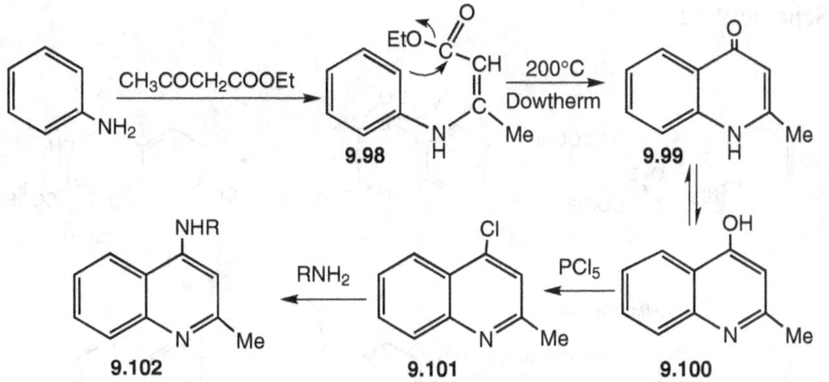

Scheme 9.50

Friedländer in 1883 devised a synthesis that has very broad scope for creating quinolines with various substituents in both rings. It is still in use today, with some modifications. It consists basically of reacting an aniline with an ortho-carbonyl group with a carbonyl compound that has a methyl or methylene group at the alpha-position. The amine group condenses with this carbonyl group while the aromatic carbonyl enters an aldol condensation with the methyl or methylene group. The overall reaction is shown as Scheme 9.51. As noted, many substituents can be tolerated in the reacting components, and there are no restrictions on the type of substituent on the benzene ring. The limiting factor in the Friedländer synthesis is the availability of the required starting materials.

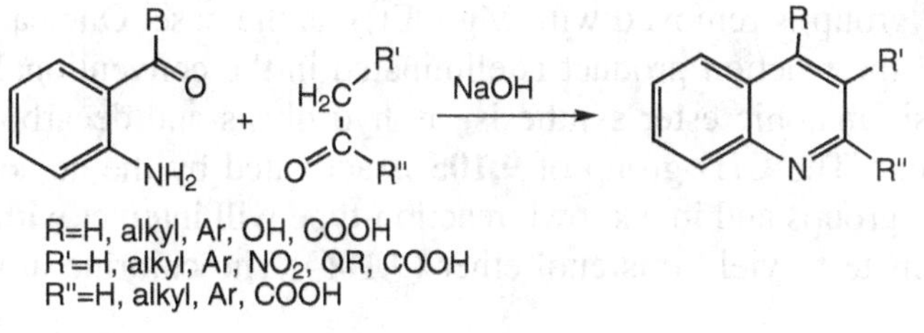

R=H, alkyl, Ar, OH, COOH
R'=H, alkyl, Ar, NO_2, OR, COOH
R"=H, alkyl, Ar, COOH

Scheme 9.51

Synthetic pharmaceuticals based on the quinoline ring are numerous. During World War II when supplies of the highly important antimalarial quinine became scarce, research was conducted in many laboratories to find a synthetic substitute. From this research came the drug chloroquine (**9.103**) and many related highly active compounds. Chloroquine is still in use today in the treatment of malarial infections. It is synthesized by a modification of the Conrad-Limpach synthesis as shown in Scheme 9.52.

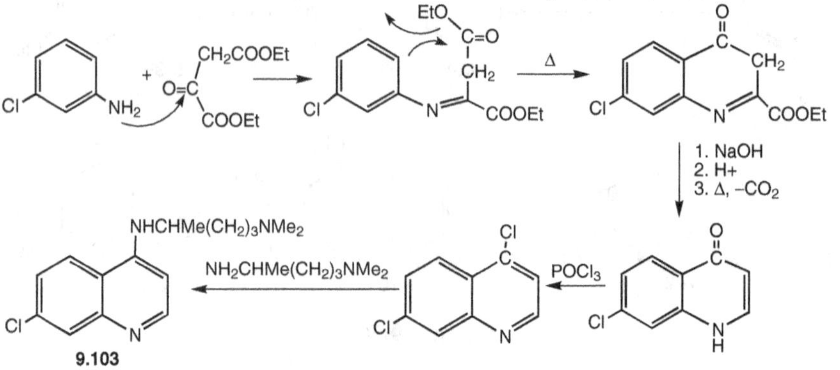

Scheme 9.52

Ciprofloxacin (**9.109**) is an antibiotic of considerable importance. Among other uses, it is prescribed for the treatment of anthrax infections, and it was so employed in the anthrax-letter scare in the United States in 2001. It is classed as a member of the 4-quinolone family. Two other heterocyclic rings are present, piperazine and aziridine. Its synthesis[19] (Scheme 9.53) contains a novel feature for the closure of the pyridone component; the displacement of an alkoxy group of an enol ether by an amino group. The synthesis can be considered to start with the trihalobenzoyl chloride **9.104**, which is used to aroylate the central carbon of diethyl malonate. To activate the latter, a proton of the CH_2 group is removed with $Mg(OEt)_2$ as the base. One carbethoxy group of the reaction product is eliminated in the conventional way of the classic malonic ester synthesis, by hydrolysis and decarboxylation of the acid. The CH_2 group of **9.105** is activated by the adjacent keto and ester groups and in a known reaction-type will interact with triethyl orthoformate to yield this enol ether (**9.106**). The chlorine atom ortho to the carbonyl group and para to the electron-attracting fluorine atom is sufficiently activated to nucleophilic attack that it can be displaced by an amine, in this case cyclopropylamine to form **9.107**. The NH group displaces the ethoxy group from the enol ether function, thereby closing the ring to form **9.108**. In the final steps, the ester group is hydrolyzed to the acid, and the remaining chlorine atom, which is activated by the para carbonyl group and ortho fluorine atom, is displaced by attack of piperazine to yield ciprofloxacin (**9.109**).

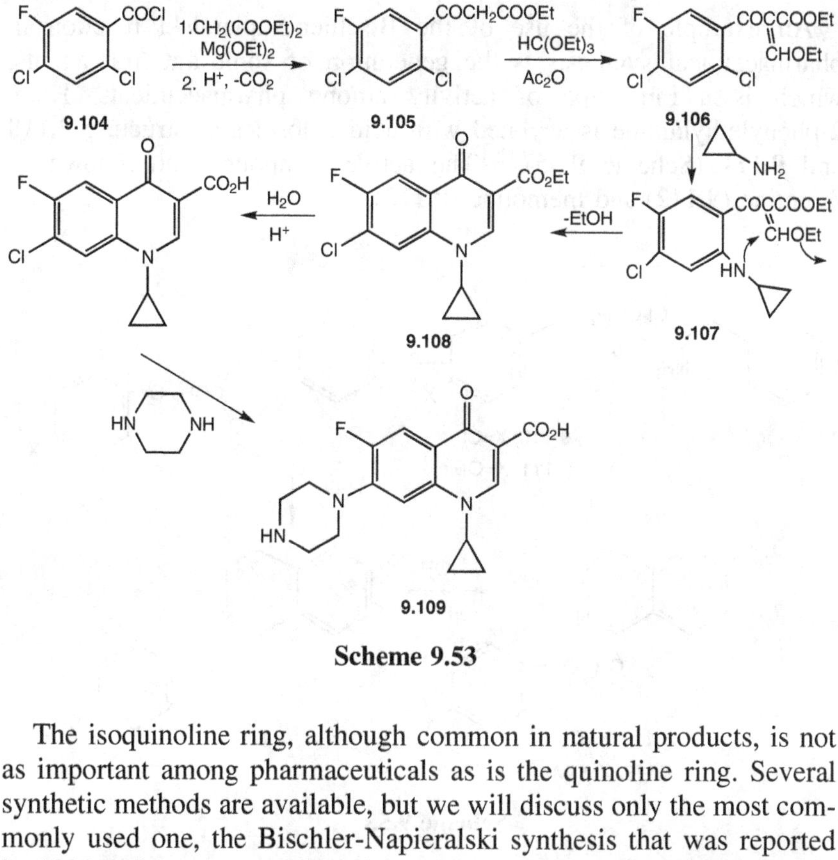

Scheme 9.53

The isoquinoline ring, although common in natural products, is not as important among pharmaceuticals as is the quinoline ring. Several synthetic methods are available, but we will discuss only the most commonly used one, the Bischler-Napieralski synthesis that was reported in 1893. With some modifications, it is still of great value today for the synthesis of isoquinolines. The process is simple and involves the reaction of an acyl derivative (an amide) of a 2-arylethylamine with phosphorus oxychloride or phosphorus pentoxide (Scheme 9.54). The product is a dihydroisoquinoline, which is easily oxidized to the isoquinoline. Dehydrogenation with palladium is also effective.

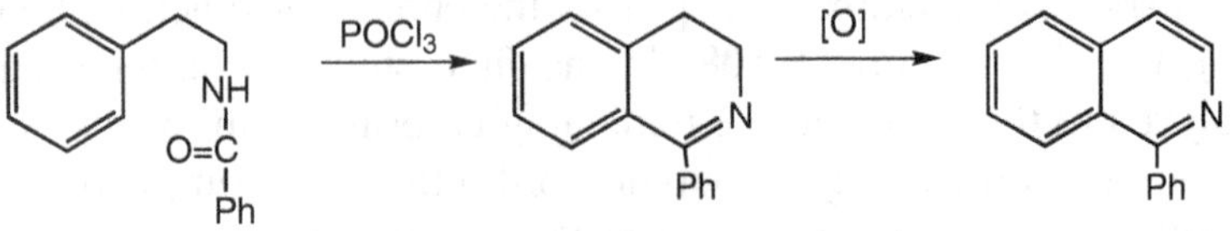

Scheme 9.54

An example of the use of the Bischler-Napieralski reaction in pharmaceutical synthesis is the generation of some antiviral agents, which is a rare type of activity among pharmaceuticals. Here, 2-phenylethylamine is acylated with acid chlorides of structure **9.110** and **9.111** (Scheme 9.55).[1k] The active compounds are known as famotine (**9.112**) and memotine (**9.113**).

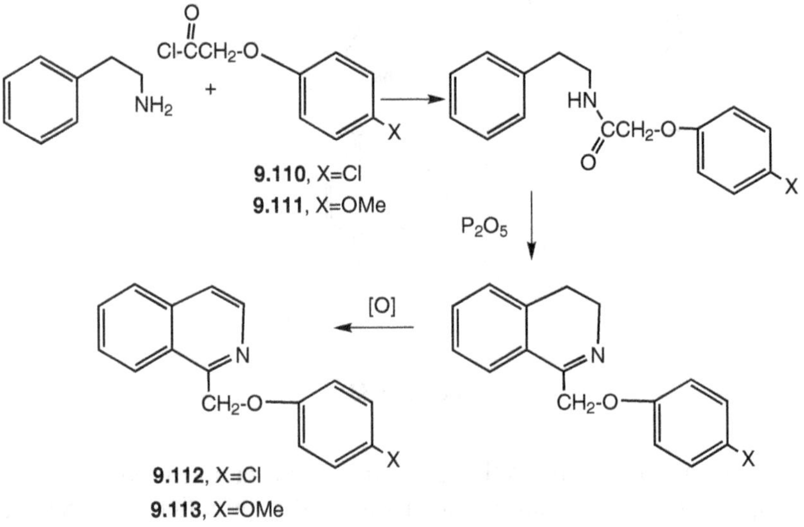

9.110, X=Cl
9.111, X=OMe

9.112, X=Cl
9.113, X=OMe

Scheme 9.55

9.15. BENZODIAZEPINES

The diazepine ring gained prominence in the pharmaceutical field when the tranquilizing effect of certain benzo derivatives was discovered by Leo Sternbach. The name Librium (Hoffmann-La Roche, Inc.) became almost a household word. Many valuable derivatives of benzodiazepine have since been prepared.

Diazepines cannot have the cyclic electron delocalization found in all other heterocycles that have been discussed, although they are in the state of maximum unsaturation for a 7-membered ring (cycloheptatriene).

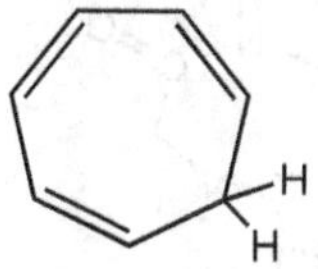

cycloheptatriene

Many biologically active derivatives are based on the dihydrobenzodiazepine structure. The tranquilizer medazepam is of this type; its synthesis[20] (Scheme 9.56) is of interest because of its close resemblance to the Bischler-Napieralski isoquinoline synthesis, in that an amide is subjected to cyclodehydration with PPA. Here, a second nitrogen atom must be present in the side chain, as in the starting amide **9.114**. This product is the tranquilizer medazepam (**9.115**).

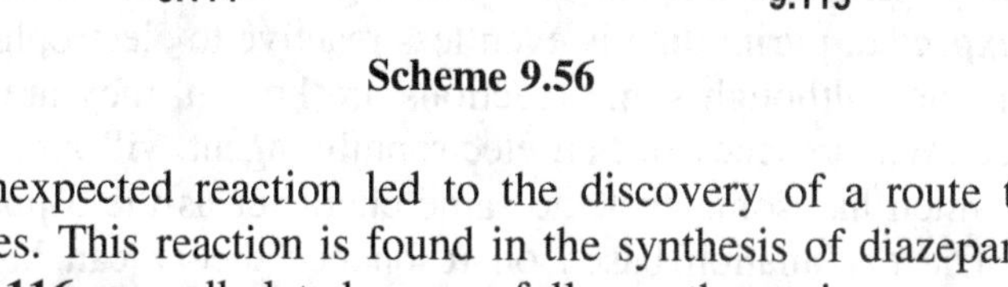

9.114 —PPA→ **9.115**

Scheme 9.56

An unexpected reaction led to the discovery of a route to benzo diazepines. This reaction is found in the synthesis of diazepam.[21] The oxime **9.116** was alkylated successfully on the amino group to form **9.117**. It was expected that the OH group of the oxime would displace the chlorine atom of the CH_2Cl group, but instead the nitrogen of the oxime performed this reaction. This provides a 7-membered ring rather than an 8-membered ring as would form from attack on oxygen. The proton of the OH group was removed with the chlorine atom, leaving the nitrogen in the form of an N-oxide (**9.118**). The amine oxide group was then reduced to form a derivative of the diazepine ring (**9.119**). This product is the valuable tranquilizer diazepam.

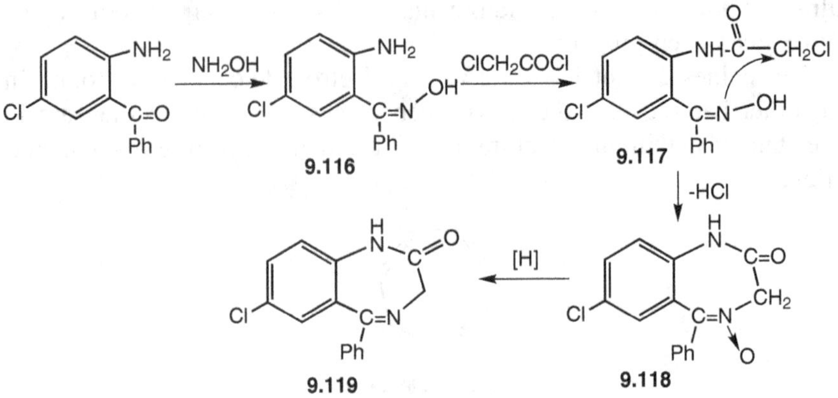

Scheme 9.57

9.16. PYRIMIDINES

The family of compounds based on the pyrimidine ring is one of the most important in the field of heterocyclic chemistry. Pyrimidines are found among the "bases" of nucleic acids, as was discussed in Chapter 3, and many pharmaceutically useful compounds have been discovered. The pyrimidine ring is stable, with resonance energy of 26 kcal/mol. Its numerous resonance forms were shown in Chapter 6, section 6.3.2.1. The pyrimidine ring has properties of a typical strongly pi-deficient system. Thus, it is but weakly basic, with $K_b 10^{-13}$. This basicity is less than that of pyridine (K_b 2×10^{-9}), which can be explained by the added electron-withdrawing effect of the second C=N unit. As expected, pyrimidine is even less reactive to electrophiles than is pyridine, and although some reactions are known, they are of little importance. When successful, an electrophilic agent will attack the 5-position, which has somewhat the same character as the 3-position of pyridine. The explanation based on resonance theory can be applied to pyrimidine substitution. An example of a successful substitution by bromine is shown in Scheme 9.58.

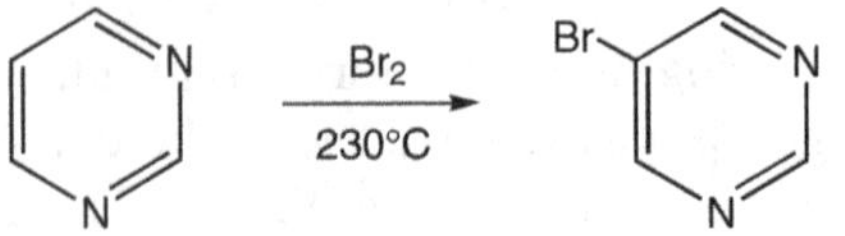

Scheme 9.58

Much more is known about substitutions on pyrimidines bearing the common electron-releasing groups, for just as with pyridine these substituents greatly activate the ring. Thus, 2-aminopyrimidine can be brominated in water at 80°C and can react with diazonium ions in cold media (Scheme 9.59). This order of reactivity can be likened to that of a phenol. Because amino and hydroxyl derivatives are easily prepared directly by ring synthesis and indeed are found in natural products, much more work has been done with these derivatives than with pyrimidine itself.

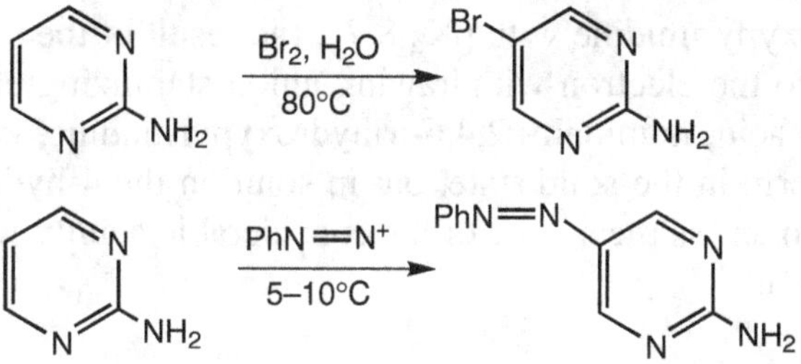

Br_2, H_2O, 80°C

$PhN{\equiv}N^+$, 5–10°C

Scheme 9.59

As a typical pi-deficient system, hydroxyl groups on the pyrimidine are involved in tautomeric equilibrium with the oxo form, and it is this form that greatly dominates in the equilibrium. Nevertheless, electrophilic substitution is easily effected on the oxopyrimidine structure. As an example, the diketo pyrimidine uracil (**9.120**) reacts readily with nitric acid to give the 5-nitro product **9.121**. This suggests that the reactive species may be the hydroxyl form, albeit in a low concentration (Scheme 9.60).

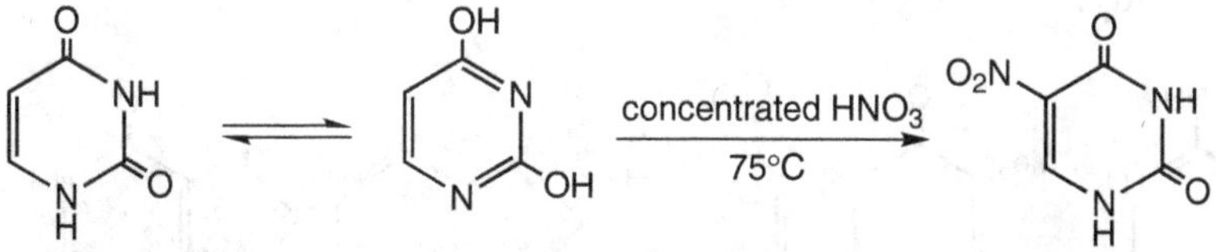

9.120 concentrated HNO_3, 75°C **9.121**

Scheme 9.60

The stabilization effect of pyrimidine resonance is not lost in the oxo form, because it can be written with a cyclic resonance form (Scheme 9.61).

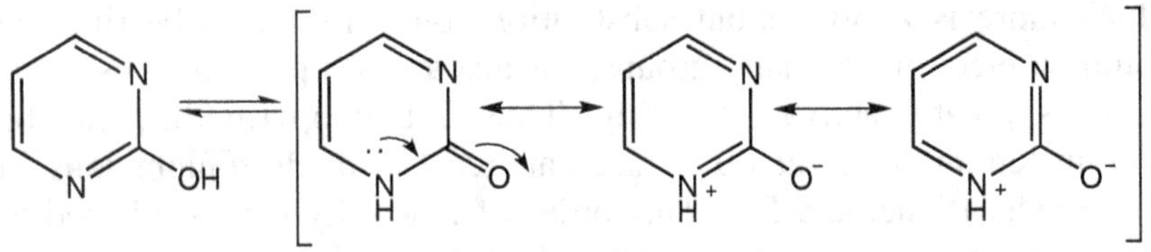

Scheme 9.61

When OH is placed at the 5-position, tautomerism is not important, and the normal phenol-like acidity is present. Thus, the pK_a of 5-hydroxypyridine is 6.78. This compound is therefore a stronger acid than 3-hydroxypyrimidine with pK_a 8.72, the result of the second C=N unit adding to the electron-withdrawing anion stabilizing effect.

Barbituric acid, nominally 2,4,6-trihydroxypyrimidine, is entirely in the tri-oxo form in the solid state, but in solution the 4-hydroxy makes a contribution and is the source of the considerable acidity with pK_a 3.9 (Scheme 9.62).

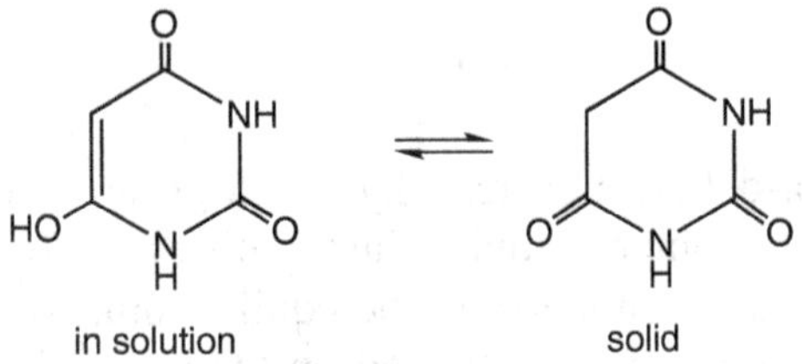

in solution solid

Scheme 9.62

The highly important pyrimidine "bases" of nucleic acids and the general structural features of the acids were described in Chapter 3, section 3.2.3. The structures of the bases are reproduced here to aid in their study.

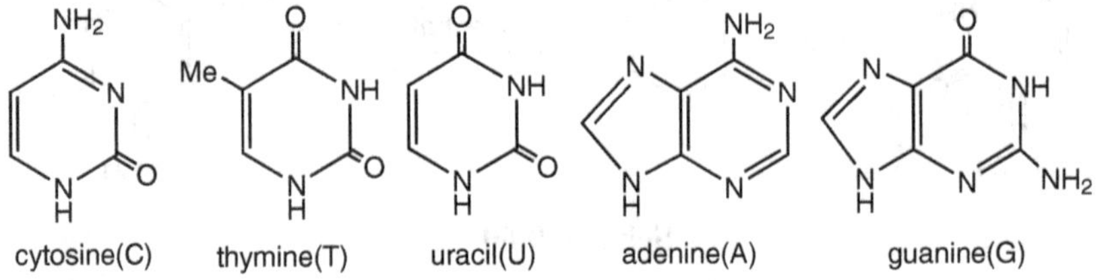

cytosine(C) thymine(T) uracil(U) adenine(A) guanine(G)

All the bases except adenine have carbonyl groups from the tautomeric shift of an OH group, and another chemical property has to be considered: the capability for carbonyl oxygen to accept a proton in the

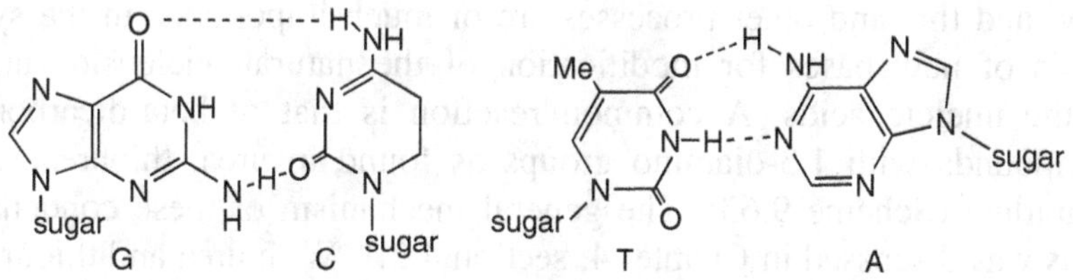

Figure 9.1. Example of hydrogen bonding.

phenomenon of hydrogen bonding. Hydrogen bonds are weak (typically 3–5 kcal/mol) but are of great importance in biological structures, and nowhere is this more true than in the nucleic acids. Nitrogen in a C=N unit can also serve as an acceptor. The bases all have NH or NH_2 groups that serve as hydrogen donors. The H-bonding takes place specifically between certain pairs of bases, and this is the source of binding between the two strands of the nucleic acid DNA in the well-known double helix structure. Two of the bases involved are pyrimidines (C and T) and two are purines where an imidazole ring is fused to the pyrimidine (G and A). The H-bonding in the double helix of DNA occurs between G of one chain and C of the other, and between A and T, as diagrammed in Figure 9.1. The sequence of the bases on the DNA strands is of course the basis for the genetic code.

RNA is single stranded but involved in base-pairing with one of the strands of DNA after the double helix is opened. The RNA-DNA pairing involves G-C, but adenine is paired with U rather than T. All the bases are planar, and the H-bonded pairs together lie in a plane. A study of the fascinating structure and chemistry of the nucleic acids can be found in any basic organic chemistry textbook.

Many useful transformations are possible with pyrimidine derivatives, and these play a role in the synthesis of particular compounds. For example, as with pyridones, reaction of the oxo derivatives with PCl_5 or $POCl_3$ gives the chloro derivatives, which can be used to attach various nucleophiles to the ring. Chlorine can be removed by catalytic hydrogenation. Amino groups can be created by the reduction of the diazo group from coupling with diazonium ions. Pyrimidines also form N-oxides, which again have reactivity of pyridine N-oxides. We will discuss examples of the use of these reactions in connection with the synthetic methods for pyrimidines.

An extensively used synthesis of pyrimidines involves the combination of two 3-atom units, although other combinations are also in use. All the pyrimidine bases of the nucleic acids can be made this way, and this and other processes are of much importance in the synthesis of new bases for modification of the natural nucleoside units in the nucleic acids. A common reaction is that of beta-dicarbonyl compounds with 1,3-diamino groups as found in urea, thiourea, and guanidine (Scheme 9.63). The general mechanism of these condensations was discussed in Chapter 4, section 4.2.1. With urea and thiourea, the C=O and C=S groups, respectively, are retained at the 2-position of the pyrimidine product, but tautomerism occurs with the guanidine product to yield a 2-aminopyrimidine.

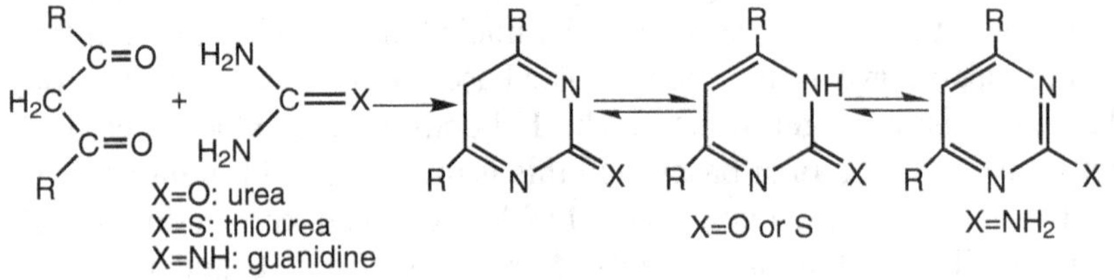

X=O: urea
X=S: thiourea
X=NH: guanidine
X=O or S
$X=NH_2$

Scheme 9.63

Diesters also react smoothly with the urea derivatives, which undergo displacement of the alkoxy groups to create amide bonds. Barbituric acid can be made this way by reaction of diethyl malonate (**9.122**, R=R'=H) with urea (Scheme 9.64). An important family of pharmaceutical agents results from the use of substituted malonic esters. Thus, with a phenyl and an ethyl substituent the important hypnotic agent, phenobarbital (**9.123**) is produced. Veronal, which is another important hypnotic, is prepared from diethyl-substituted malonates.

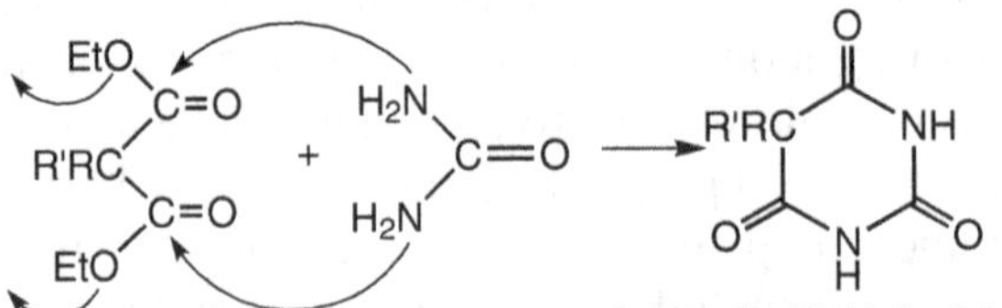

9.122 R=R'=H, barbituric acid
9.123 R=Ph, R'=Et, phenobarbital

Scheme 9.64

A valuable synthesis of 4-alkylpyrimidines consists of the use of beta-ketoesters in reactions with the urea derivatives (Scheme 9.65).

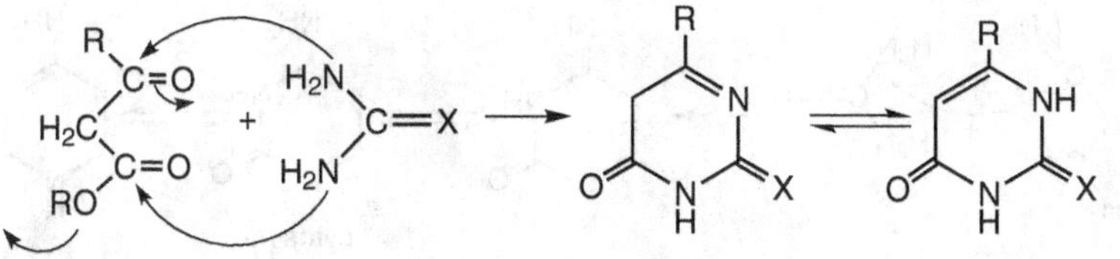

Scheme 9.65

The bases uracil and thymine can be made by this route (Scheme 9.66).

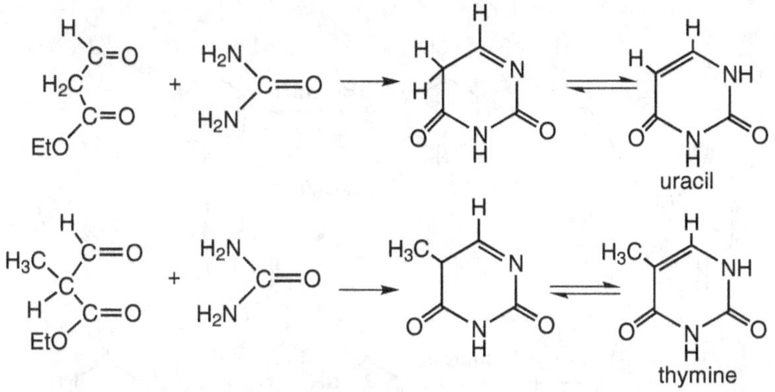

Scheme 9.66

2-Alkylpyrimidones can be made by using amidines (**9.124**) instead of urea derivatives in condensations with carbonyl compounds and diesters (Scheme 9.67).

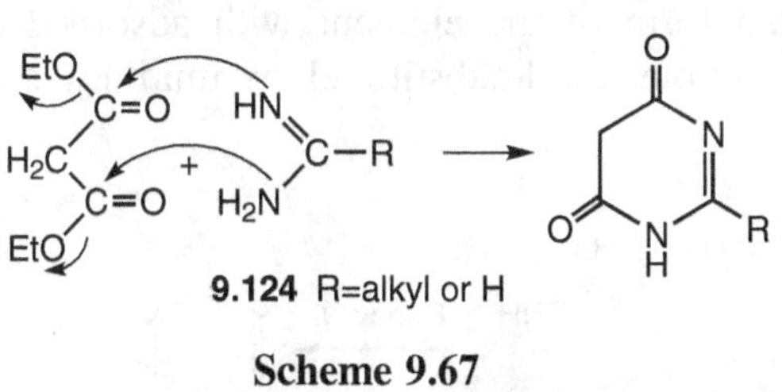

9.124 R=alkyl or H

Scheme 9.67

Another combination is that of nitrile group or groups with the urea or amidine derivatives. From the nitrile group will originate an amino group. In Scheme 9.68, the synthesis of cytosine is shown by this approach.

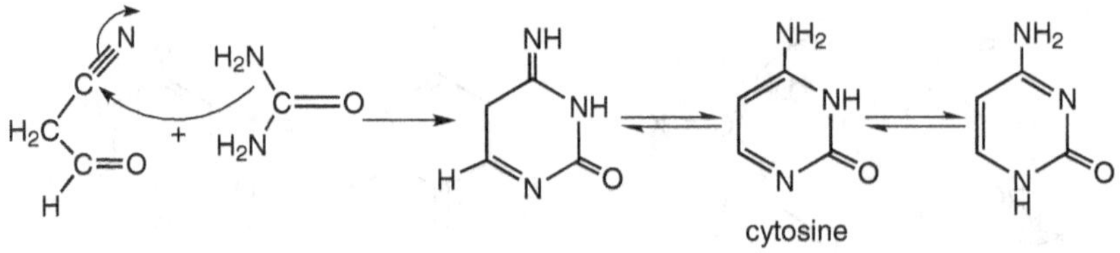

cytosine

Scheme 9.68

Some of the useful transformations that can be performed with the pyrimidones are shown in Scheme 9.69, using uracil as the starting material.

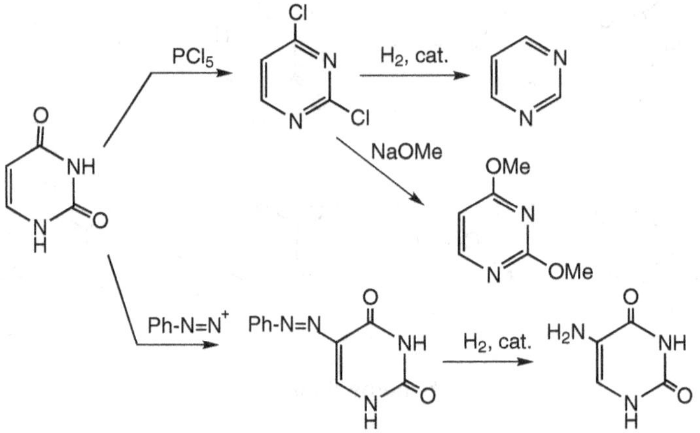

Scheme 9.69

The products from condensations with thiourea offer another opportunity: the replacement of S by H by treatment with Raney nickel (a finely divided form of the element with adsorbed hydrogen). By this approach, simple alkyl-substituted pyrimidines can be obtained (Scheme 9.70).

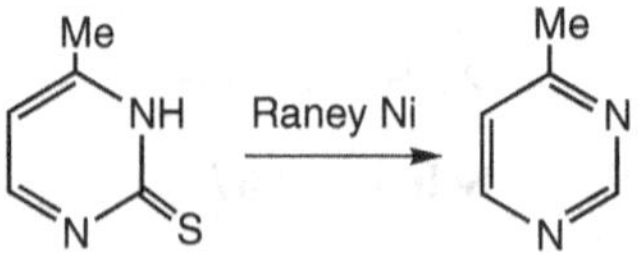

Scheme 9.70

Other types of ring closures are possible that are useful for pyrimidine synthesis, such as a 5 + 1 condensation. Thus, the condensation of a 1,3-diamino derivative with a one-carbon acylating reagent can be used effectively. An example is shown as Scheme 9.71 where the product is barbituric acid.

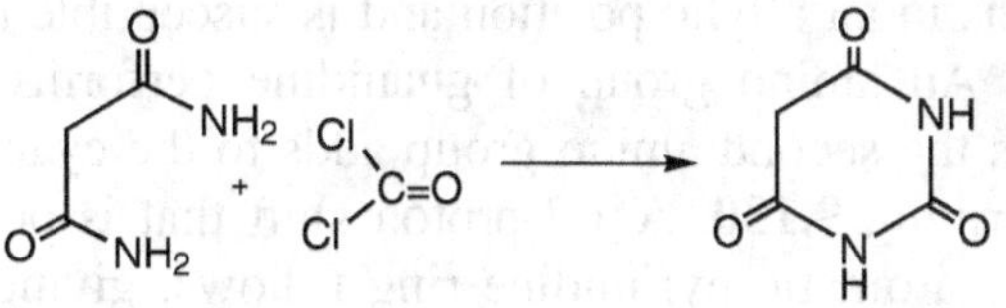

Scheme 9.71

An example of the practical use of the 3 + 3 condensation approach in medicinal chemistry is provided by the synthesis of the vasodilator and antihypertensive drug minoxidil[11] (later discovered to increase hair growth and sold as ROGAINE). As shown in Scheme 9.72, the starting materials are guanidine and ethyl cyanoacetate. This combination will give a 2,4-diaminopyrimidine (**9.125**). The oxo group reacts with $POCl_3$ to give the chloro derivative **9.126**. Peroxide oxidation occurs selectively at N-3 to give the N-oxide **9.127**. As in pyridines, ring halogen is made reactive by being alpha to nitrogen and gamma to the N-oxide function. In **9.127**, the chlorine is easily displaced by amines, here with piperidine, to give the drug minoxidil (**9.128**).

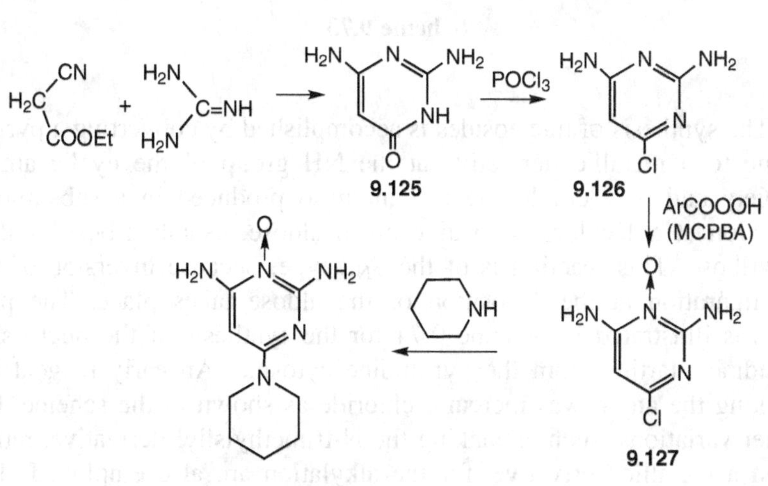

Scheme 9.72

The synthesis[11] of the antibacterial drug trimethoprim is accomplished by a different form of 3 + 3 condensation (Scheme 9.73). The synthesis starts with the aldol condensation of a benzaldehyde derivative with 3-ethoxypropionitrile. In the product (**9.129**), the ethoxy group is in an allylic position and is susceptible to nucleophilic displacement. An amino group of guanidine performs this displacement, whereas the second amino group adds to the cyano group in the usual way, forming **9.130**. A 1,3-proton shift that is promoted by the creation of the aromatic pyrimidine ring follows, giving trimethoprim (**9.131**).

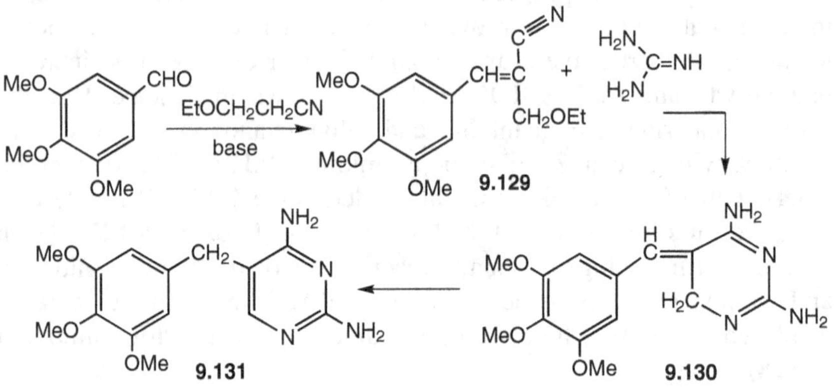

MeO, MeO, OMe, CHO, $EtOCH_2CH_2CN$, base, MeO, MeO, OMe, CH=C, C≡N, CH_2OEt, **9.129**, +, H_2N, C=NH, H_2N, NH_2, H, C, N, H_2C, N, NH_2, MeO, MeO, OMe, **9.130**, MeO, MeO, OMe, CH_2, NH_2, N, N, NH_2, **9.131**

Scheme 9.73

The synthesis of nucleosides is accomplished by converting a pyrimidine to a metallic derivative at the NH group of the cyclic amide moiety and then employing the anion so produced in a substitution reaction at a 1-chloro derivative of an aldose, usually ribose or desoxyribose. This reaction is of the S_N2 type, because inversion of the configuration at the 1-position of the aldose takes place. The process is illustrated in Scheme 9.74 for the synthesis of the nucleoside cytidine, starting from the pyrimidine cytosine. An early reagent for making the anion was mercuric chloride as shown in the scheme, but other variations, such as making the N-trimethylsilyl derivative, rather than a metallic derivative, for the alkylation are also employed. The acetyl protecting groups are removed from N and O to obtain cytidine.

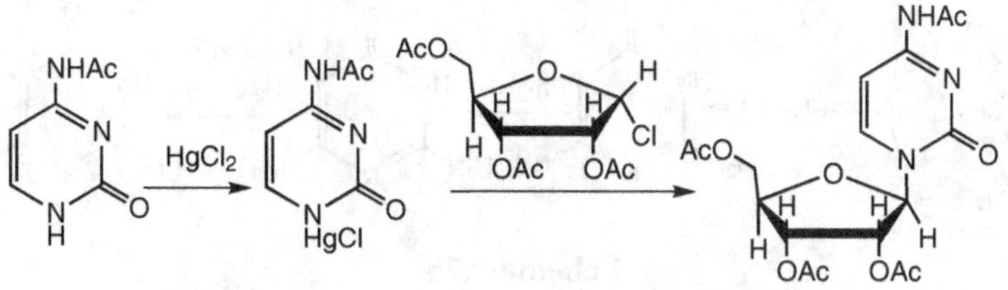

Scheme 9.74

9.17. FUSED PYRIMIDINES: PURINES AND PTERIDINES

Purine and pteridine compounds are of great importance as medicinals and in other biochemical studies. The compounds of natural and synthetic interest have amino groups, or carbonyl groups from OH tautomerism. Hydrogen bonding is especially strong in the carbonyl-containing structures and causes them to have high melting points and low water solubility.

As was discussed for imidazole, tautomerism makes it difficult to predict the position of the proton in this component of the purine structure. In adenine, the proton is largely found at the 9-position, whereas in guanine a mixture of the two forms is present in solution. In nucleosides, it is the 9-position that is attached to the sugar moiety as noted in section 3.2.3. The numbering in purines is exceptional to the rules of nomenclature but is accepted by the International Union of Pure and Applied Chemistry (IUPAC). It is illustrated in Figure 9.2.

4,5-Diaminopyrimidines are frequently the starting framework for the construction of the second ring in both cyclic systems, as will be shown in the examples that follow, but other methods have also been developed.

To obtain the requisite diaminopyrimidines, a monoaminopyrimidine is first prepared by the conventional routes and the second amino group added by the sequence of coupling with benzenediazonium ion followed by reduction, or by nitrosation with HNO_2 and reduction. The former method is illustrated in Scheme 9.75.

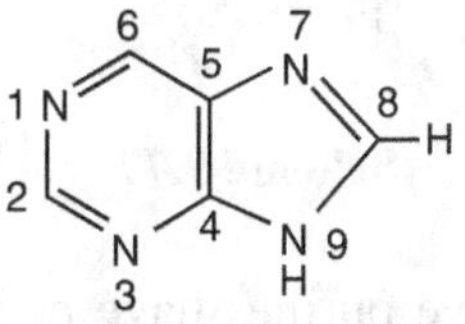

Figure 9.2. Numbering in purine.

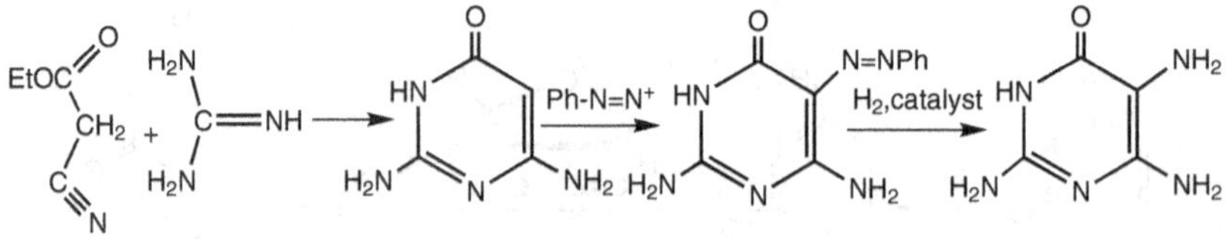

Scheme 9.75

To add the fused ring (a pyrazine) of a pteridine, the diamino compound is condensed with an alpha-dicarbonyl compound. Thus, using compound **9.132** would give pteridine **9.133** (Scheme 9.76).

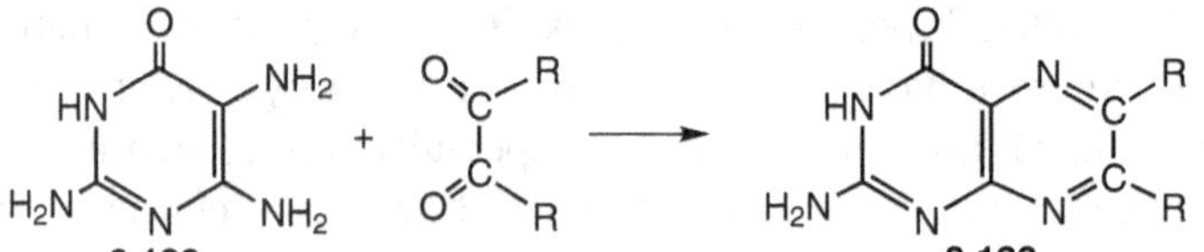

Scheme 9.76

A well-known purine is the central nervous stimulant (CNS) caffeine (**9.134**), which is found in coffee and chocolate. Its synthesis (Scheme 9.77) illustrates the technique of forming the fused imidazole ring of purines. It was noted in Chapter 4 that diaminobenzenes reacted with carboxylic acids or esters to form benzimidazoles; this process is known as the Phillips synthesis. It is this reaction that is used to fuse imidazoles to pyrimidines. Here, the process is known as the Traube purine synthesis. The initially formed purine is tri-methylated with methyl chloride and base to form caffeine (**9.134**). Note that under these conditions it is the 7-nitrogen that is methylated.

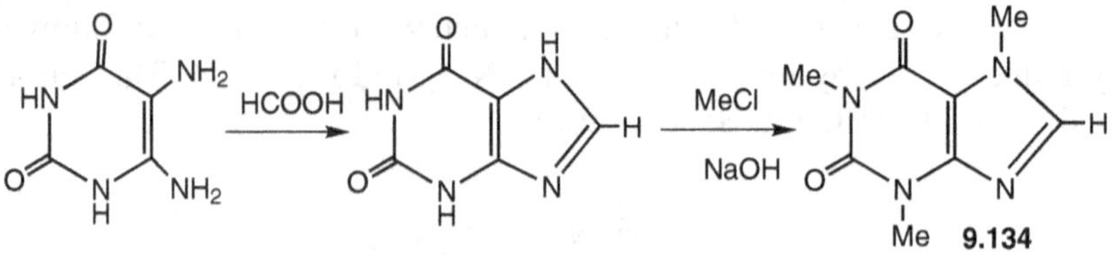

Scheme 9.77

Many biologically active purines have been prepared by the Traube synthesis, where various carboxylic acids provide novelty at the imidazole carbon. Alkylation of the imidazole NH also leads to novel structures. An illustration of the practical use of these methods is found in the synthesis of the effective CNS stimulant bamiphylline (Scheme 9.78).[1m] The alkylation takes place preferentially on the imidazole tautomer shown in Scheme 9.78.

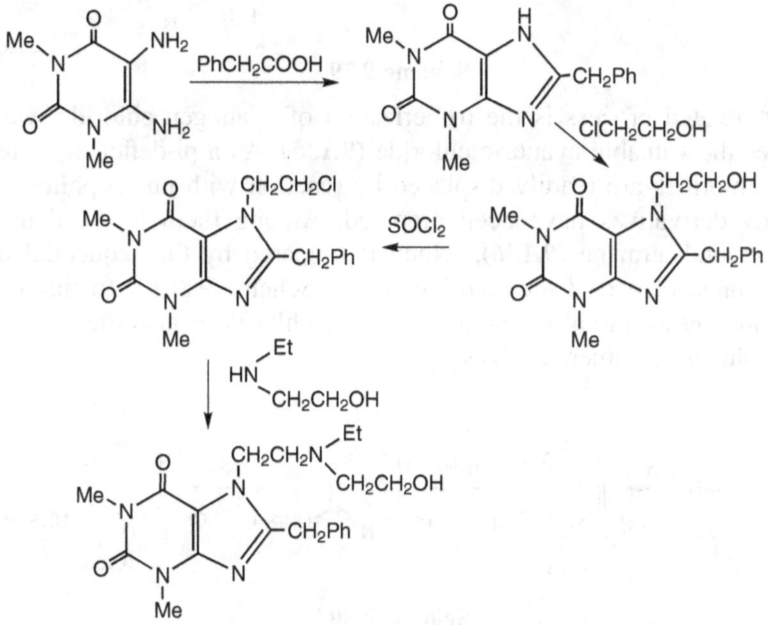

Scheme 9.78

9.18. 1,3,5-TRIAZINES

Triazines are of limited value in the pharmaceutical field, but they deserve attention because of their great importance in the chemical industry. The best known triazine is the triamino derivative melamine, which gives the widely used resin Melmac (American Cyanamid) on reaction with formaldehyde. Melamine has achieved notoriety for other reasons. It has been used improperly in pet foods to increase the nitrogen level, and it has caused the death of many animals. For the same reason, it has been put in milk with danger to humans. Melamine is a trimer of cyanamide. This compound can undergo dimerization to form a valuable intermediate called dicyandiamide (more properly cyanoguanidine), which is a valuable starting material for heterocyclic synthesis. The reaction can also be continued to give the trimeric product (Scheme 9.79). Melamine has been produced in huge quantities by this process.

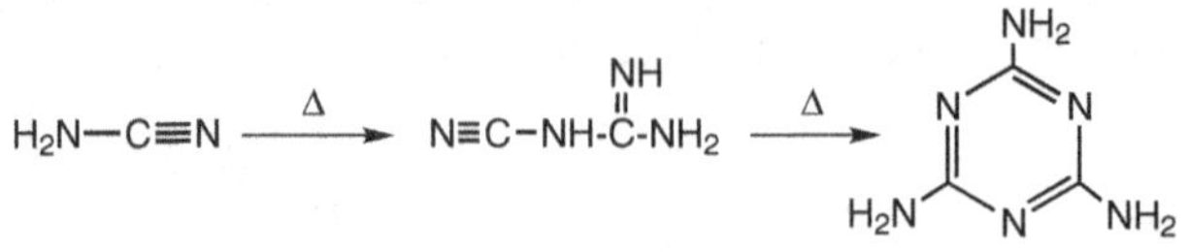

$H_2N{-}C{\equiv}N \xrightarrow{\Delta} N{\equiv}C{-}NH{-}C(=NH){-}NH_2 \xrightarrow{\Delta}$ (melamine: triazine ring with three NH_2 groups)

Scheme 9.79

A related process is the trimerization of cyanogen chloride, which gives the valuable cyanuric chloride (**9.135**). As a pi-deficient system, the chlorines are readily displaced by reaction with nucleophiles, and many derivatives have been prepared. Among them is the diamino compound atrazine (**9.136**), which is prepared by the sequential displacement of two of the chlorine atoms (Scheme 9.80). Atrazine is on the market as a total herbicide. Cyanuric chloride is also the precursor of valuable commercial dyes.

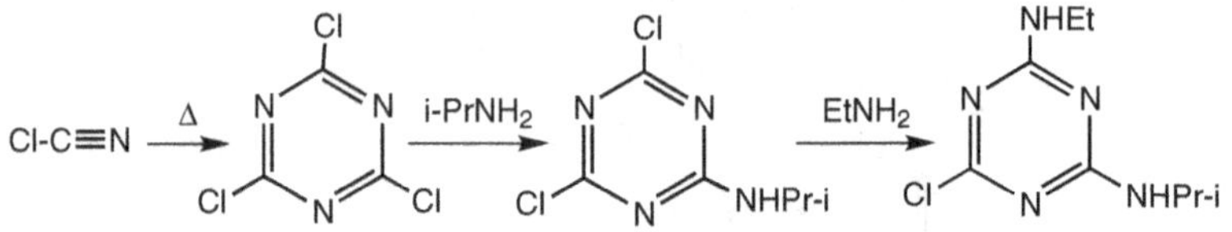

$Cl{-}C{\equiv}N \xrightarrow{\Delta}$ **9.135** (Cl, Cl, Cl) $\xrightarrow{i\text{-}PrNH_2}$ (Cl, Cl, NHPr-i) $\xrightarrow{EtNH_2}$ **9.136** (NHEt, Cl, NHPr-i)

Scheme 9.80

The trihydroxy compound rearranges to the tautomeric tri-oxo form, again a predictable property as the compound is derived from a pi-deficient ring system. The tri-oxo compound has measurable acidity and is known as cyanuric acid. Although it can be prepared by hydrolysis of cyanuric chloride, a better process involves urea as a starting material. Thermolysis causes loss of ammonia and gives isocyanic acid, which trimerizes to cyanuric acid (Scheme 9.81).

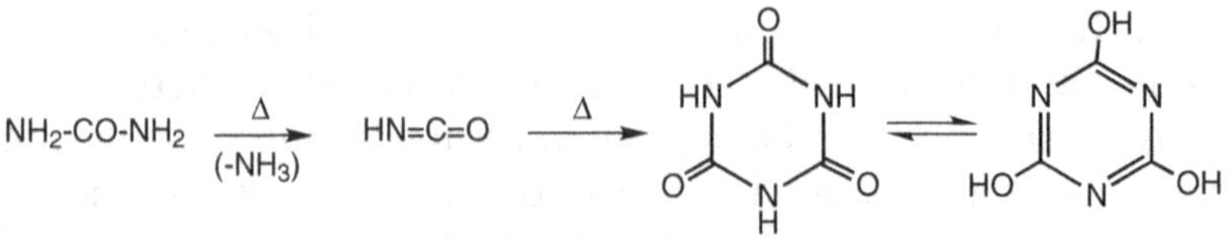

$NH_2{-}CO{-}NH_2 \xrightarrow[(-NH_3)]{\Delta} HN{=}C{=}O \xrightarrow{\Delta}$ (tri-oxo form: HN, NH, N-H, three C=O) $\rightleftharpoons$ (trihydroxy form: OH, HO, OH on triazine ring)

Scheme 9.81

Cyanuric acid forms a framework for the construction of a class of biological alkylating agents, which were among the earliest of the anticancer agents. It was desirable to have more than one alkylating group in the agent for cross-linking of DNA, and this structure becomes a possibility with cyanuric acid through replacement of the three amidic protons.[22] Epoxides are useful alkylating groups for action on DNA; these were therefore attached to cyanuric acid by reaction with epichlorohydrin (1-chloro-2,3-epoxypropane) (Scheme 9.82). This gave the drug teroxirone (**9.136**). Carbon-2 of epichlorohydrin is chiral, but this reactant was used as a racemic mixture. There were isolated mixtures (diastereoisomers), which were separated to give one form with considerable anticancer activity.[23]

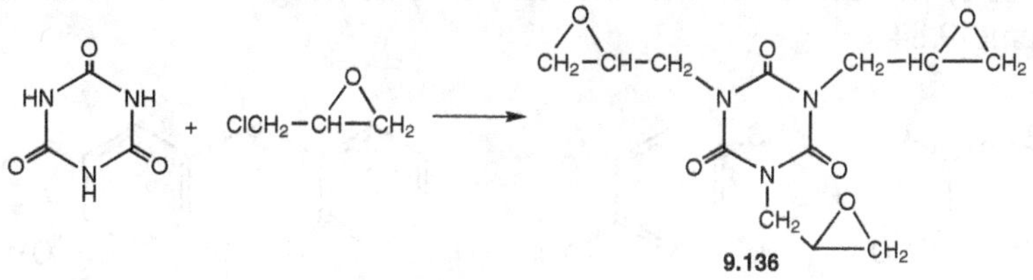

Scheme 9.82

Triazines can be obtained by a different procedure that has provided an antineoplastic agent called decitabine[24] (**9.141**). The sugar isocyanate **9.137** condenses with the O-methyl ether of urea (**9.138**) by the addition of the HN=C group of the latter to the N=C group of the isocyanate. The intermediate **9.139** is then reacted with trimethyl orthoformate to supply the carbon needed for completion of the triazine ring (**9.140**). The pi-deficient ring activates the methoxy group for displacement, and this is accomplished with ammonia. At the same time the protecting benzoyl groups (Bz) are removed, giving the product decitabine (Scheme 9.83).

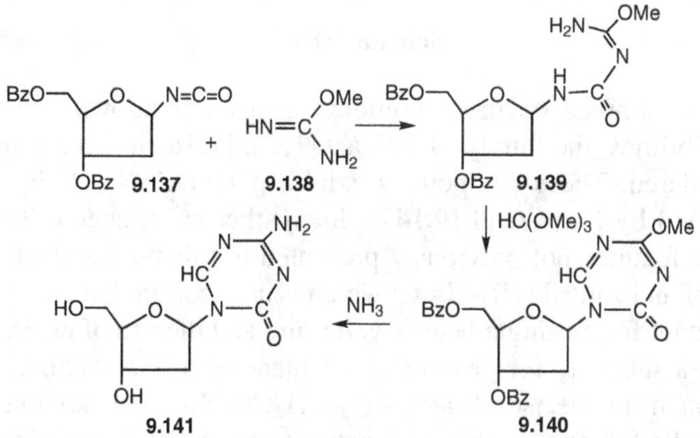

Scheme 9.83

9.19. MULTICYCLIC COMPOUNDS

Multicyclic ring systems are also found to be present in synthetic heterocyclic pharmaceuticals. These drugs may be benzo derivatives of monocyclic heterocycles or they may contain other fused-on (said to be annelated or annulated) heterocyclic rings. An important example of a dibenzo derivative, namely, a derivative of the tricyclic phenothiazine ring system, is the valuable antihistamine methopromazine (**9.142**). Its synthesis[1n] employs conventional reactions of aromatic chemistry to construct the interior thiazine ring, as is outlined in Scheme 9.84.

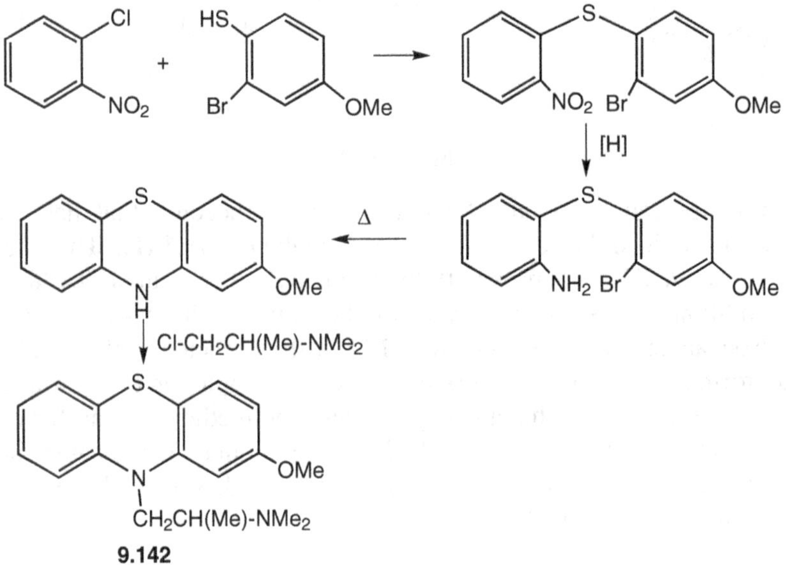

Cl NO2 + HS Br OMe → S NO2 Br OMe
[H]
S NH2 Br OMe
Δ
S N H OMe
Cl-CH2CH(Me)-NMe2
S N OMe CH2CH(Me)-NMe2
9.142

Scheme 9.84

Many examples of more complex drugs are known. To suggest the possibilities, the family of imidazo[1,5-a][1,4]benzodiazepines may be considered. These compounds exhibit potent CNS activity and are exemplified by flumazenil (**9.147**). Its synthesis[25] (Scheme 9.85) contains two features not previously presented in this book, which include the use of an isonitrile (R−N=C: which is an electron deficient carbene) as a reactant for closing a heterocyclic ring and the use of an enol phosphate as a substrate for group displacement by a nucleophile, here the carbanion from the isonitrile C=N−CH_2COOEt. The laboratory procedure calls for conducting all reactions sequentially in a single flask (called a "one-pot" reaction). The exact sequence of steps is uncertain, but Scheme 9.85 represents an attempt to describe possible events leading to bond formations. The synthesis is shown as starting with a preformed benzodiazepinedione (**9.143**). The amide group is phosphorylated with diethyl chlorophosphate in the presence of potassium t-butoxide (KOBu-t) to give an enol phosphate (**9.144**). The isonitrile C=N–CH_2COOEt is converted with additional KOBu-t to a carbanion, which then displaces the phosphate group to give structure **9.145**. Isonitriles have carbon in the electron-deficient carbene state and add both a nucleophile and an electrophile to achieve tetracovalencey.[15] This behavior is expressed in the attack of the carbene on the imine group of **9.145**, with pickup of a proton as the electrophile to form **9.146**. Loss of a proton from the imidazole ring completes the process of forming flumazenil.

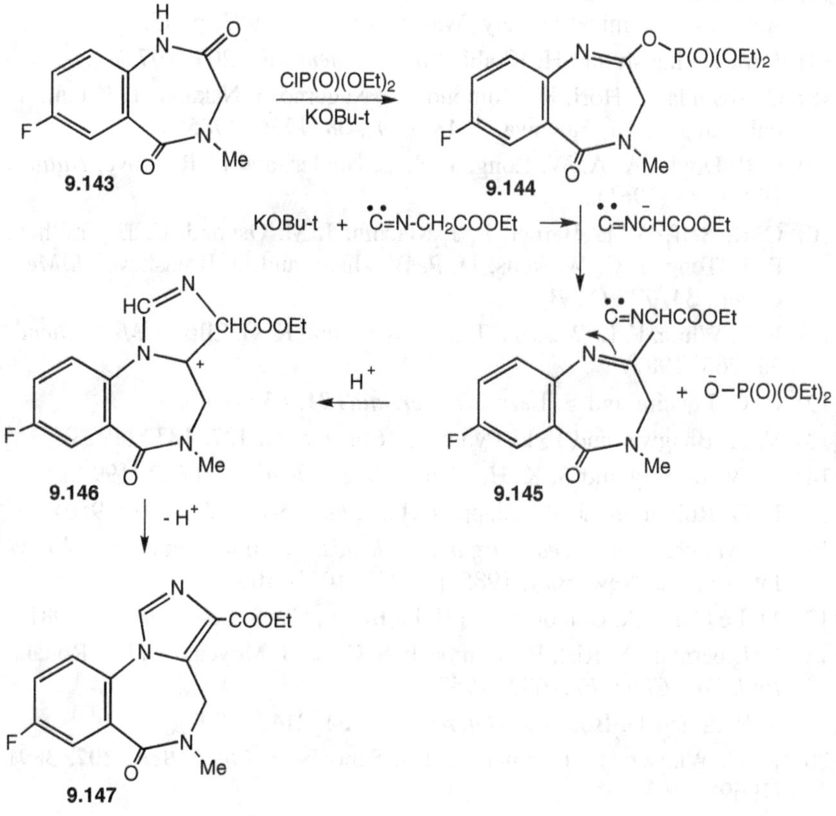

Scheme 9.85

REFERENCES

(1) D. Lednicer, *The Organic Chemistry of Drug Synthesis*, Vols. 1-7, Wiley-Interscience, New York, (a) Vol. 1, 1977, p. 228; (b) Vol. 5, 1995, p. 68; (c) Vol. 5, 1995, p. 141; (d) Vol. 5, 1995, p. 77; (e) Vol. 5, 1995, p. 73; (f) Vol.1, 1977, p. 318; (g) Vol. 5, 1995, p. 108; (h) Vol. 2, p. 283; (1980), (i) Vol. 4, p. 102; (1990), (j) Vol. 4, 1990, p. 105; (k) Vol. 2, 1980, p. 378; (l) Vol. 1, 1977, p. 262; (m) Vol. 1, p. 426; (1977), (n) Vol. 1, 1977, p. 374.

(2) D. Lednicer, *Strategies for Organic Drug Synthesis and Design*, Wiley-Interscience, New York, 1998, p. 190.

(3) W. G. Walker, *U. S. Pat. No*. 4,256,759 (1981).

(4) *Chem. Abst*., **95**, 7046 (1981).

(5) A. R. Katritzky, C. W. Rees, and E. F. V. Scriven, *Comprehensive Heterocyclic Chemistry* II, Vol. 2, Pergamon, Oxford, 1996.

(6) J. Bradshaw, in *Chronicles of Drug Discovery*, D. Lednicer, Editor, American Chemical Society, Washington, DC, 1993, p. 45.

(7) J. M. McIntosh and H. Khalil, *Can. J. Chem*., **53**, 209 (1975).

(8) C. Yoshida, T. Hori, K. Momonoi, K. Nagumo, J. Nakano, T. Kitani, Y. Fukuoaka, and I. Saikawa, *J. Antibiot*., **38**, 1536 (1985).

(9) F. P. Doyle, A. A. W. Long, J. H. C. Nayler, and E. R. Stove, *Nature*, **192**, 1183 (1961).

(10) C. R. Self, W. E. Barber, P. J. Machin, J. M. Osbond, C. E. Smithen, B. P. Tong, J. C. Wickens, D. P. Bloxham, and D. Bradshaw., *J. Med. Chem*., **34**, 772 (1991).

(11) R. L. White, F. L. Wessels, T. J. Schwan, and K. O. Ellis, *J. Med. Chem*., **30**, 263 (1987).

(12) G. C. Lancini and E. Lazari, *Experentia*, **21**, 83 (1965).

(13) V. K. Bhagwat and F. L. Pyman, *J. Chem. Soc*., **127**, 1832 (1928).

(14) W. V. Murray and S. K. Hadden, *J. Org. Chem*., **57**, 6662 (1992).

(15) R. O. Roblin and J. W. Clapp, *J. Am. Chem. Soc*., **72**, 4890 (1950).

(16) J. March, *Advanced Organic Chemistry*, third edition, Wiley-Interscience, New York, 1985, pp. 870, 1021–1034.

(17) M. Le Corre, A. Hercouet, and H. Le Baron, *Chem. Commun*., 14 (1981).

(18) D. Liberman, N. Rist, F. Grumbach, S. Cals, M. Moyeux, and A. Rouaix, *Bull. Soc. Chim. Fr*., 687 (1958).

(19) S. Radl and D. Bouzard, *Heterocycles*, **34**, 2143 (1992).

(20) K. H. Wünsch, H. Dettman, and S. Schönberg, *Chem. Ber*., **102**, 3891 (1969).

(21) L. H. Sternbach and E. Reeder, *J. Org. Chem*., **26**, 4936 (1961).

(22) M. Budowski, *Angew. Chem., Internat. Ed*., **7**, 827 (1968).

(23) M. M. Ames, J. S. Kovach, and J. Rubin, *Cancer Res.*, **44**, 4151 (1984).

(24) J. Piml and F. Sorm, *Coll. Czech. Chem. Commun.*, **29**, 2576 (1964).

(25) J. Yang, Y. Teng, S. Ara, S. Rallipalli and J. M. Cook, *Synthesis*, 1036 (2009).

REVIEW EXERCISES

9.1. V. S. Matiychuk, et al., *Chem. Heterocyc. Comp.*, **40**, 1218, (2004). Predict the product:

O H Cl H_3C + O O H_3C O + NH_3 →

9.2. T. Hickey, et al., *Invest. New Drugs*, **25**, 425–433 (2007). Predict the product (authors did not isolate this product but rather treated the lithium anion with titanium tetrachloride to give a titanocene).

N H_3C N H_3C —t-Bu Li→ NMe_2 →

9.3. A. Traversone and W. K. -D. Brill, *Tetrahedron Lett.*, **48**, 3535–3538 (2007). Draw the structure of the dicarboxylic acid formed after condensation and ester hydrolysis.

R'O S OR O O + O O Ph Ph →

9.4. M. P. Susnik, et al., *Monatsh. Chem.*, **140**, 423–430 (2009). Draw the product.

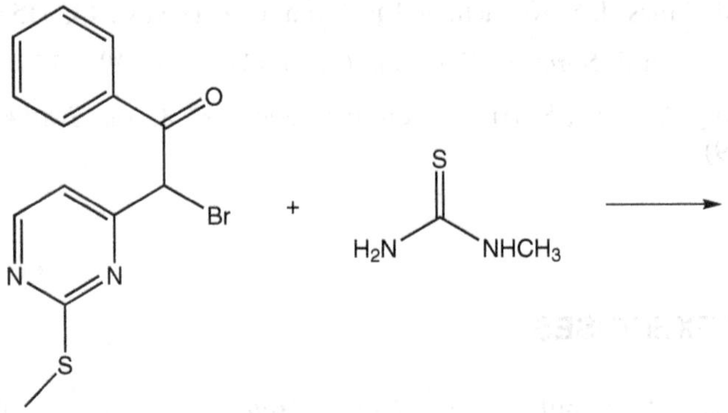

Br + H_2N NHCH$_3$ S O N N S

9.5. C. Glover, et al., *Tetrahedron Lett.*, **48**, 7027–7030 (2007). Draw the product.

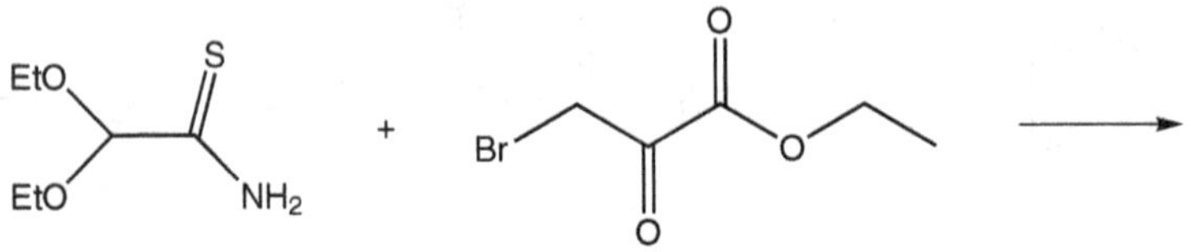

EtO EtO S NH$_2$ + Br O O O

9.6. T. J. Miller, et al., *Org. Lett.*, **4**, no. 6 (2002). Draw the product.

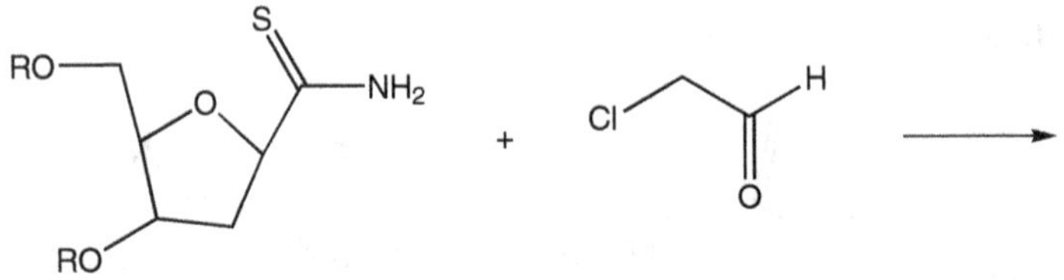

RO O S NH$_2$ RO + Cl H O

9.7. E. Biron, et al., *Org. Lett.*, **8**, no. 11 (2006). Draw the product for this part of the general solid-phase synthesis of peptides and peptidomimetics.

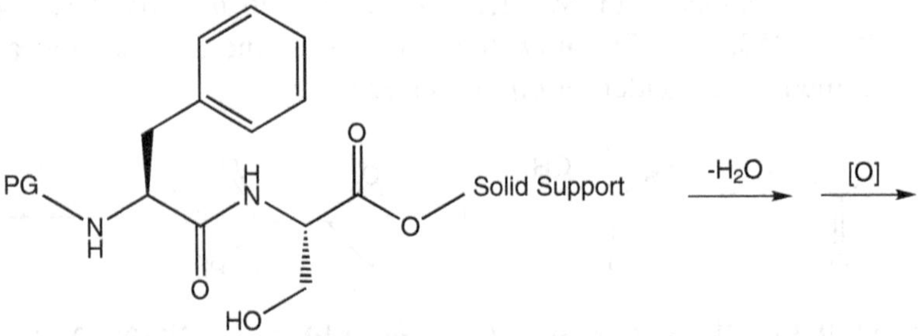

PG N H O H N O O Solid Support HO -H$_2$O [O]

PG = protecting group = Fmoc, Boc, Cbz, or Alloc

9.8. R. Di Santo, et al., *J. Med. Chem.*, **48**, 5140–5153 (2005). This product is one of a series of compounds prepared for evaluation as an antifungal agent. It is prepared by alkylation of an in situ generated imidazole with an alcohol. The reaction is done with N,N'-carbonyldiimidazole (CDI). Draw the product.

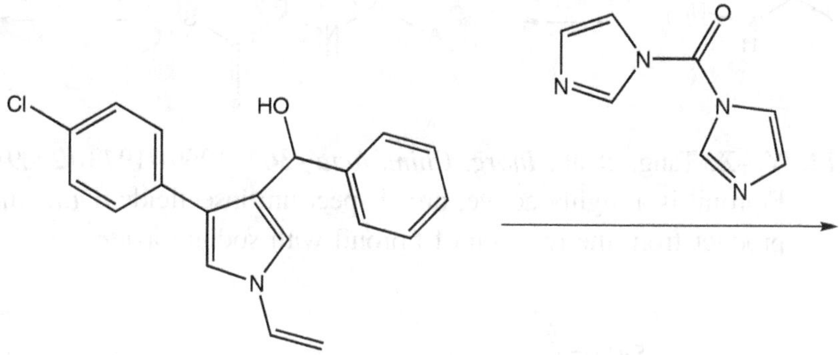

9.9. Y. -L. Zhong, et al., *Tetrahedron Lett.*, **50,** 2293–2297 (2009). These Merck researchers were evaluating a series of compounds for antiviral activity. They performed nucleophilic aromatic substitution with a hindered amine preferentially at the less hindered aryl fluoride followed by a second nucleophilic aromatic substitution at the hindered aryl fluoride. After condensation and loss of water, the product is formed. Draw the product.

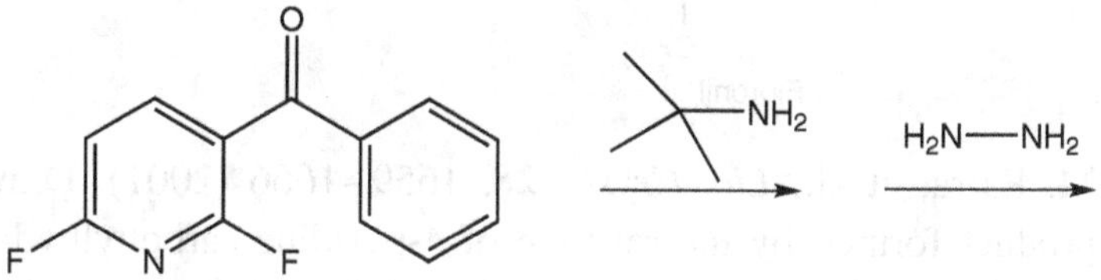

9.10. N. Sarikavakli, *Fascicula Chim.*, **15**, 62–68 (2008). This work condenses hydrazine with acetylacetone and transforms the product to form a series of ligands. Draw the original condensation product.

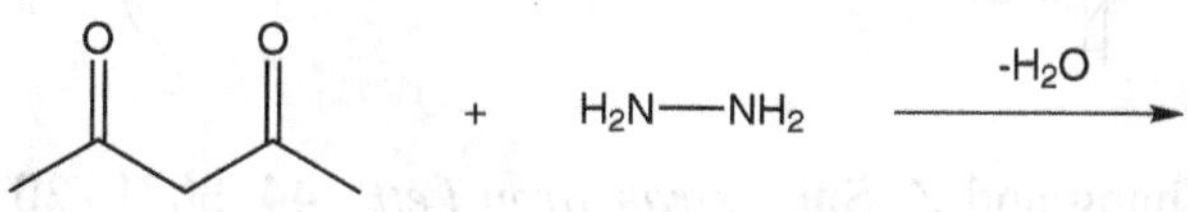

9.11. B. F. Abdel-Wahab, et al., *Monatsh. Chem.*, **140**, 601–605 (2009). These products were synthesized because of their antimicrobial potential. The acyl hydrazine was reacted with carbon disulfide followed by reaction with hydrazine. Draw the product.

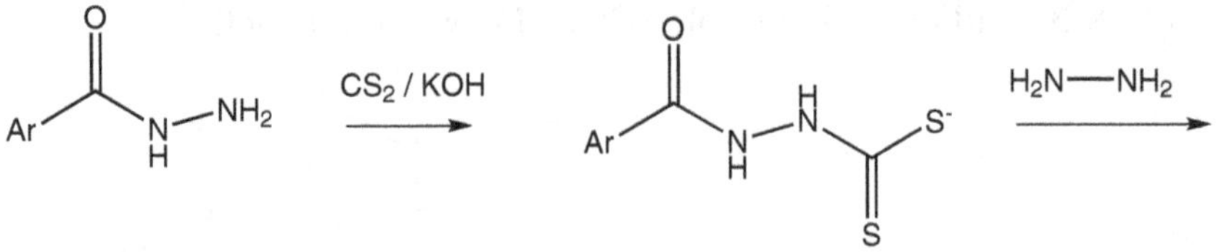

CS_2 / KOH

$H_2N—NH_2$

9.12. Y. -Z. Tang, et al., *Inorg. Chim. Acta*, **362**, 1969–1973 (2009). Fipronil is a highly active, broad-spectrum insecticide. Draw the product from the reaction of fipronil with sodium azide.

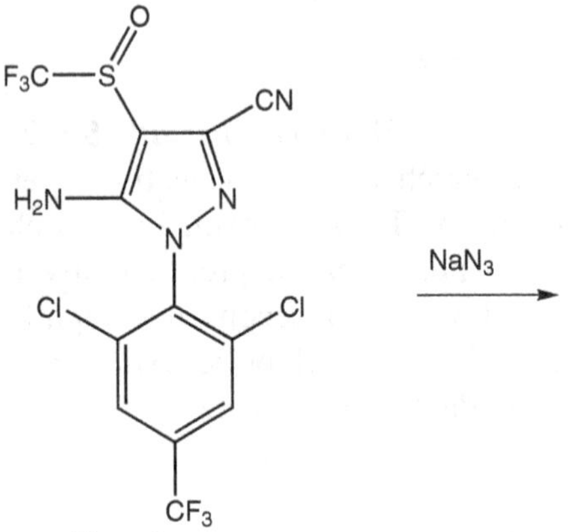

NaN_3

Fipronil

9.13. M. Parra, et al., *Liq. Cryst*., **28**, 1659–1666 (2001). Draw the product formed by the reaction of 4-pyridine-carbonylhydrazine with ammonium thiocyanate followed by dehydration with sulfuric acid.

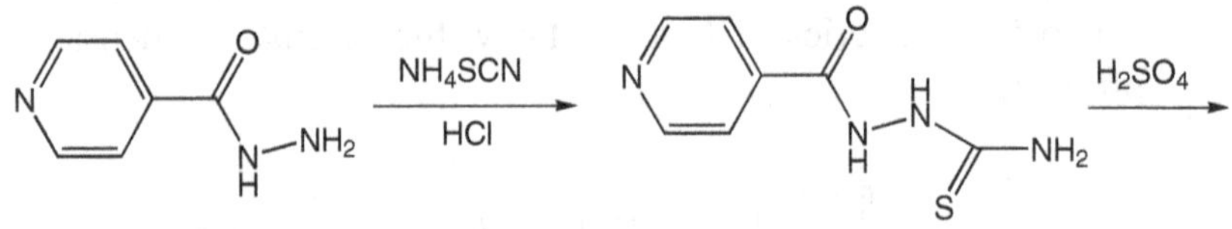

NH_4SCN

HCl

H_2SO_4

9.14. X. Zhang and Z. Sui, *Tetrahedron Lett*., **44**, 3071–3073 (2003). Steroids bearing heterocycles fused to the A-ring are of pharmaceutical interest. Researchers from Johnson & Johnson have developed a synthesis based on the amination of estrone followed by amine protection as the mesylate, cyclization, and deprotection. Write the structure for the final product.

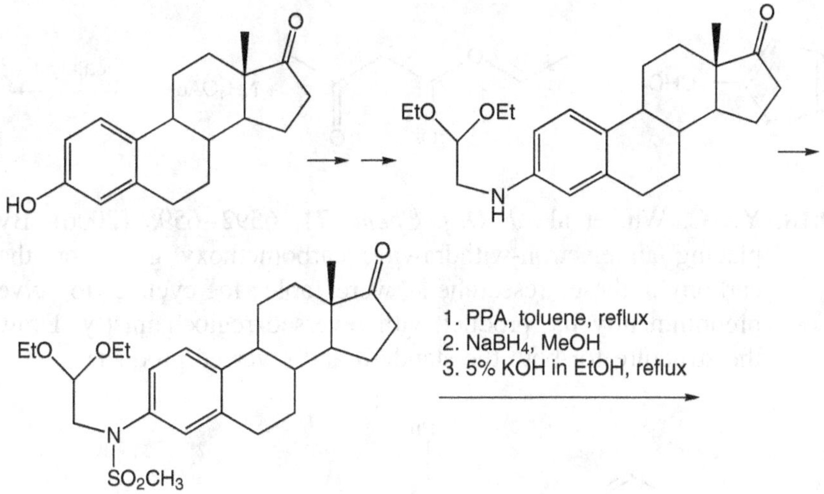
HO
EtO
OEt
N
H
EtO
OEt
N
SO_2CH_3
1. PPA, toluene, reflux
2. $NaBH_4$, MeOH
3. 5% KOH in EtOH, reflux

9.15. S. Nagy, B.P. Etherton, R. Krishnamurti, and J. Tyrell, *U.S. Pat. No. 6,232,260* (2001). These researchers reacted p-tolylhydrazine with 1-indanone to form a precursor, which was converted in subsequent reactions to an olefin polymerization catalyst. Draw the product from the first reaction.

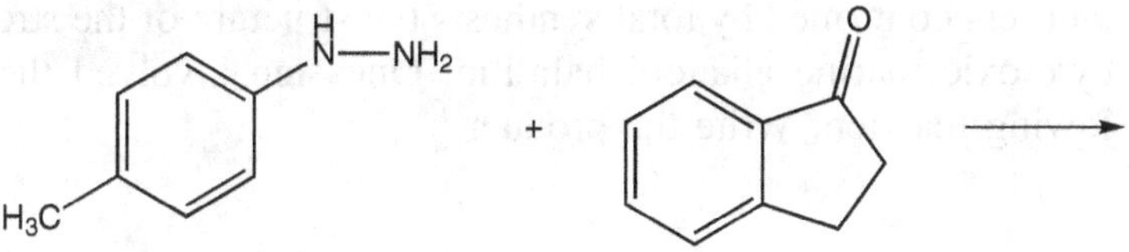
H
N—NH_2
H_3C
+
O

9.16. B. M. Tsuie, et al., *U.S. Pat No. 6,908,972* (2005). These researchers reacted p-tolylhydrazine with 2-indanone to form a precursor, which was converted in subsequent reactions to an olefin polymerization catalyst. Draw the product from the first reaction.

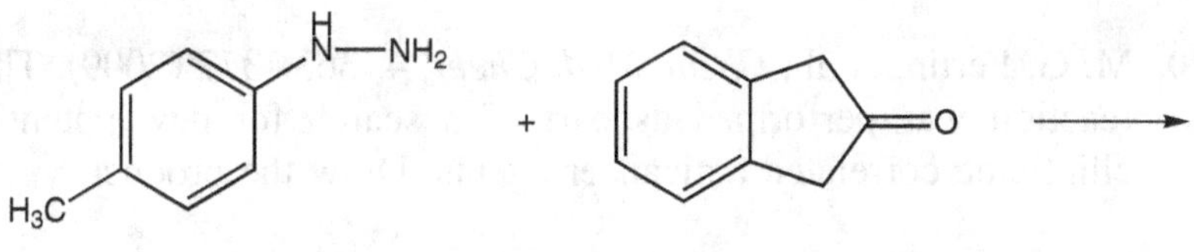
H
N—NH_2
H_3C
+
O

9.17. C. A. Antonyraj, and S. Kannan, *Appl. Catal. A Gen.*, **338**, 121–129 (2008). These researchers studied the use of pharmaceutically accepted hydrotalcites as solid base catalysts for the following reaction. Draw the product.

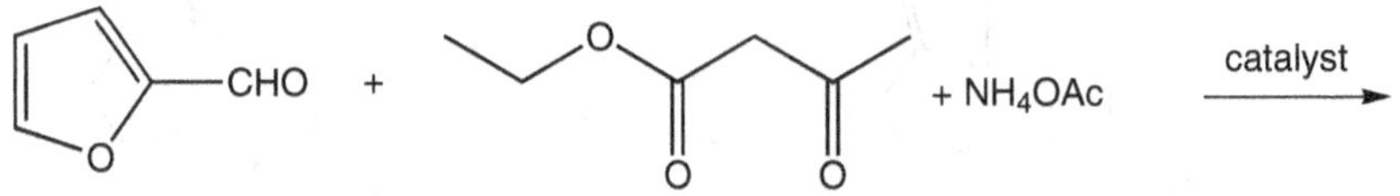

CHO + + NH_4OAc catalyst

9.18. Y. -C. Wu, et al., *J. Org. Chem*, **71**, 6592–6595 (2006). By placing an electron-withdrawing carbomethoxy group on the carbonyl, these researchers were able to cyclize to give predominantly the product with reversed regiochemistry. Draw the structure for both the standard and reversed products.

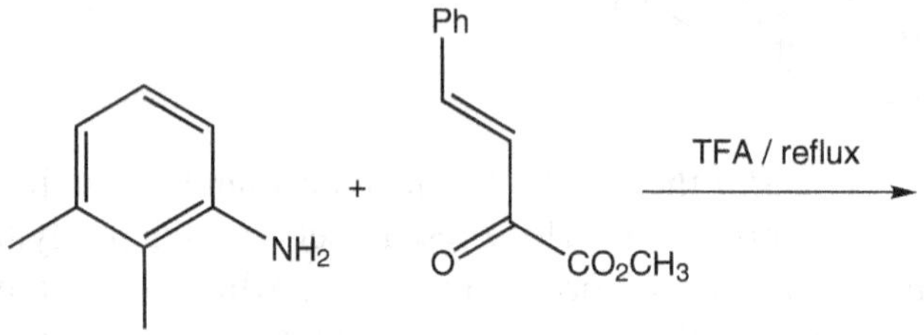

Ph

NH_2 + TFA / reflux

O CO_2CH_3

9.19. M. R. Heinrich, et al., *Tetrahedron*, **59**, 9239–9247 (2003). These workers confirmed by total synthesis the structure of the strongly cytotoxic marine alkaloid halitulin. One step involved the following reaction. Write the product.

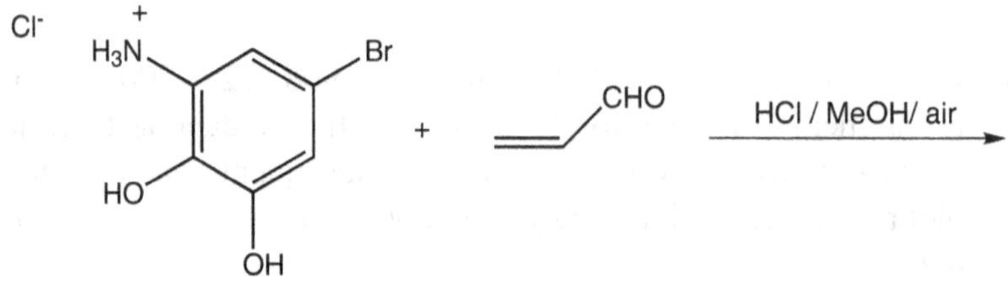

Cl^- H_3N^+ Br CHO + HCl / MeOH/ air

HO OH

9.20. M. G. Ferlin, et al., *Chem. Med. Chem*, **4**, 363–377 (2009). This reaction was performed as part of a search for new potential ellipticine-correlated anticancer agents. Draw the product.

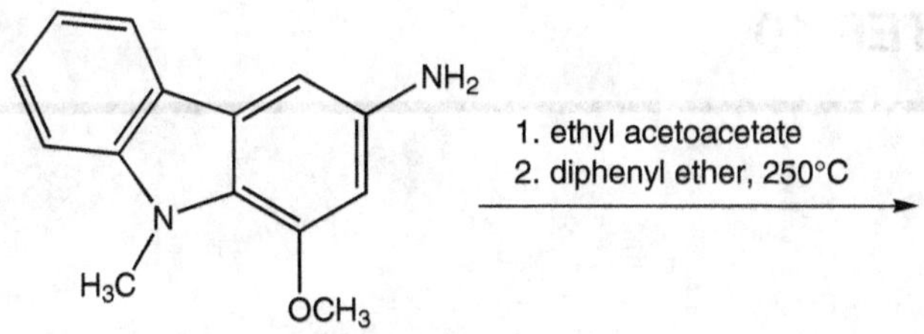
NH2
1. ethyl acetoacetate
2. diphenyl ether, 250°C
N
H3C
OCH3

9.21. J. Jacobs, et al., *Tetrahedron*, **65**, 1188–1192 (2009). Draw the cyclocondensation product.

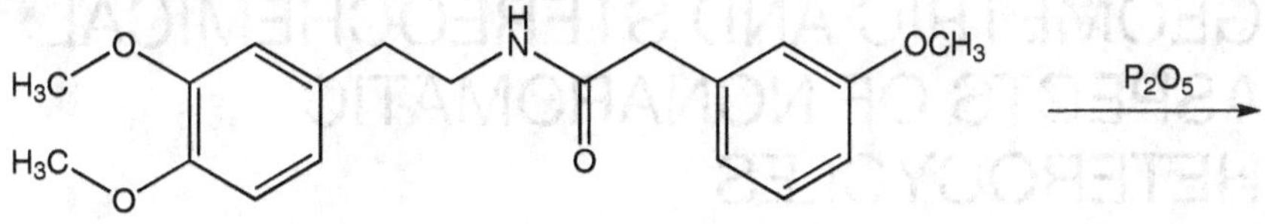
H3C
O
H3C
O
H
N
O
OCH3
P2O5

CHAPTER 10

GEOMETRIC AND STEREOCHEMICAL ASPECTS OF NONAROMATIC HETEROCYCLES

10.1. GENERAL

With the exception of 3-membered rings (and to some extent, 4-membered rings), the chemical properties of fully saturated rings do not differ greatly from those of noncyclic counterparts. It is in the area of conformational and stereochemical properties that we find unique character for the heterocyclic ring systems. The emphasis in this chapter, therefore, will be on these properties. The synthesis of the ring systems can either employ cyclization of saturated noncyclic compounds or can be accomplished by the reduction (usually catalytic hydrogenation) of partially saturated cyclic intermediates. We have observed many such intermediates on the way to the aromatic heterocycles, and indeed the latter can also be reduced completely to the saturated counterparts but with varying degrees of difficulty.

In the cyclizations to form either fully saturated or partially saturated intermediate compounds, some important structural limitations can be encountered because of geometric constraints in the formation of the required transition state. A group of rules has been developed to help understand these effects, and these rules have predictive value. The original rules and their testing by model reactions are initially from

Fundamentals of Heterocyclic Chemistry: Importance in Nature and in the Synthesis of Pharmaceuticals,
By Louis D. Quin and John A. Tyrell Copyright © 2010 John Wiley & Sons, Inc.

J. E. Baldwin,[1] but they have been used and expanded by many others over the years since his first publication in 1976. We will examine these rules in section 10.3.

In Chapters 3 and 8, numerous examples were given of valuable biological properties of heterocyclic compounds and their prominence among pharmaceuticals. A special feature is encountered when there are saturated carbon centers in the molecule: If the four groups on carbon are all different (whereupon the center is said to be asymmetric or chiral), then there can be two configurations for the carbon tetrahedron, meaning two isomeric forms (designated R and S) that are mirror images of each other. This leads us to the field of stereochemistry as involved with heterocyclic compounds; a brief review of important stereochemical ideas is presented here, but a basic organic chemistry textbook should be consulted to gain a better understanding of the ideas and terms used. The individual stereoisomers are optically active and described as chiral, but together in equal amounts they give an optically inactive racemic mixture or racemate. The two stereoisomers (called enantiomorphs or enantiomers) have identical chemical properties but differ in their reaction rate with another optically active molecule. Their physical properties are also identical, except that they rotate the plane of polarized light in opposite directions and are adsorbed on chiral surfaces to different extents. In complex compounds, there can be several chiral centers and numerous stereoisomers (calculated from 2^n, where n = number of asymmetric carbons). In this case, mixtures of diastereoisomers are possible, but these are separable by conventional means. For many years, the separation of racemic mixtures was a difficult task, and synthetic compounds were generally used as the racemic mixture in pharmaceutical testing and applications. Now there are both gas and liquid chromatographic methods, employing optically active stationary phases, which are effective for racemate separation (resolution). Also, reactions have been developed, mostly using optically active coordination complexes as catalysts or optically active coreactants (in an asymmetric synthesis), that direct a reaction toward one of the stereomeric forms. As a result of such advances, it is common practice to use optically active forms for biological testing, and it is found frequently that one isomer is more active than the other. Indeed, cases are known where the stereoisomers differ in one being useful and the other having different properties and even being unacceptably toxic. In section 10.5, we will discuss some examples of synthetic optically active drugs.

10.2. SPECIAL PROPERTIES OF THREE-MEMBERED RINGS

Saturated 3-membered rings possess a property that distinguishes them from other sizes of rings, namely, their ready ring opening by attack of nucleophiles and their polymerization on thermal treatment. Nevertheless, they can be prepared by several different techniques and are valuable as starting materials in synthetic work. Their uniqueness arises from the highly contracted angles that must be formed to connect three atoms in a ring (60° for the internuclear connections in cyclopropane, defined by the dotted lines in Figure 10.1). Here, it is necessary to distinguish the internuclear angles from the bond angles. In forming the sigma bonds, the hybridization of the atoms changes from the usual sp^3 value. The internal ring bonds acquire more p-character, making the exocyclic bonds heavier in s-character. The p-rich orbitals that lie in the plane of the ring then overlap in a nonlinear manner; a 90° angle is formed between the p-orbitals that overlap to form the sigma bonds. This is illustrated in Figure 10.1.

Such bonds are sometimes called "bent bonds" or "banana bonds" (because the sigma bond is somewhat semicircular). The internuclear angle at oxygen in ethylene oxide is known to be 61° and is similar for rings based on other heteroatoms. This special hybridization is easily recognized from ^{13}C nuclear magnetic resonance (NMR) spectroscopy. It is known that the magnitude of the one-bond coupling of ^{1}H and ^{13}C (designated $^{1}J_{CH}$) depends on the hybridization of the carbon atom; the coupling is larger when the sigma bond is formed from orbitals richer in s-character. The coupling is observed readily when the spectrum is obtained with proton-carbon coupling allowed (described as undecoupled spectra) rather than with the more customary technique of operating with C-H decoupling. Thus, in an open-chain ether, $^{1}J_{CH}$ is 140 Hz for the alpha carbons, but in ethylene oxide with its high s-character in the carbon-hydrogen bond it is 176 Hz. This suggests

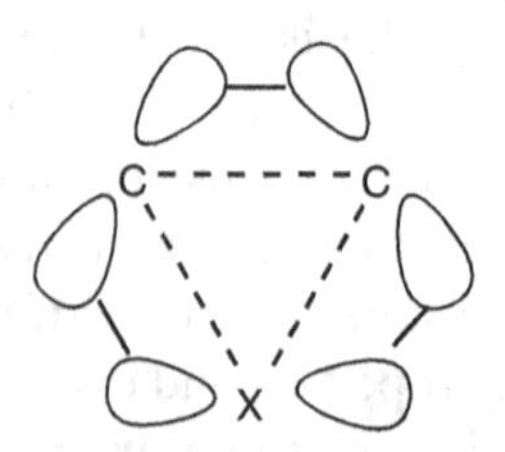

Figure 10.1. The orbital interaction in 3-membered rings.

a hybrid with 37% s-character rather than the usual 25% s-character of sp^3 hybridization. The opening of such rings is easy because of the release of strain. Using ethylene oxide (oxirane) as an example, we demonstrate this effect in Scheme 10.1, which shows some ring-opening reactions; unlike acyclic ethers, simple acid hydrolysis will open the ring, as will ammonolysis, reaction with Grignard reagents, etc. Ethylene glycol is made commercially by a hydrolysis procedure applied to ethylene oxide.

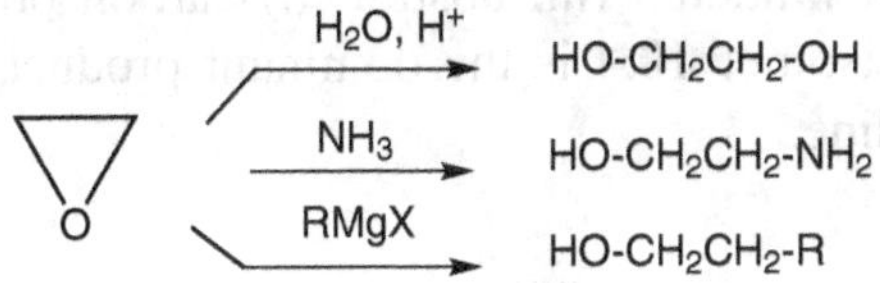

H_2O, H^+ → $HO\text{-}CH_2CH_2\text{-}OH$
NH_3 → $HO\text{-}CH_2CH_2\text{-}NH_2$
RMgX → $HO\text{-}CH_2CH_2\text{-}R$

Scheme 10.1

Ethylene oxide is also polymerizable to the commercial polyethylene glycol (**10.1**, PEG) of various chain lengths.

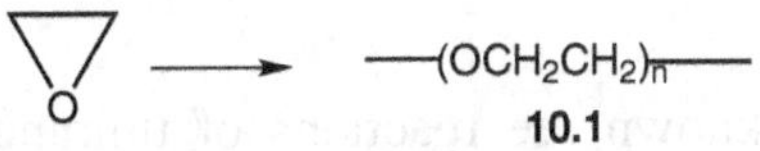

→ $-(OCH_2CH_2)_n-$
10.1

Similar ring-opening reactions are known for aziridines. Hydrolysis constitutes an important process for the synthesis of aminoalcohols (Scheme 10.2). This is usually performed with aqueous acid; an aziridinium cation is first formed and then is opened by reaction with water.

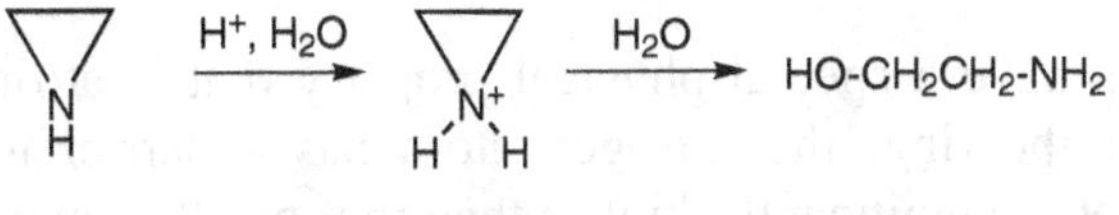

H^+, H_2O → N^+ H H, H_2O → $HO\text{-}CH_2CH_2\text{-}NH_2$

Scheme 10.2

In these ring-opening processes, the nucleophile attacks from the backside of the carbon-heteroatom bond in typical S_n2 manner. This has important stereochemical consequences, which are best illustrated with cyclic derivatives. Thus, from the bicyclic compound **10.2**, where an aziridine ring is fused to cyclohexane, the amino and nucleophile groups will have the trans orientation in the cyclohexane product, as diagrammed in Scheme 10.3.

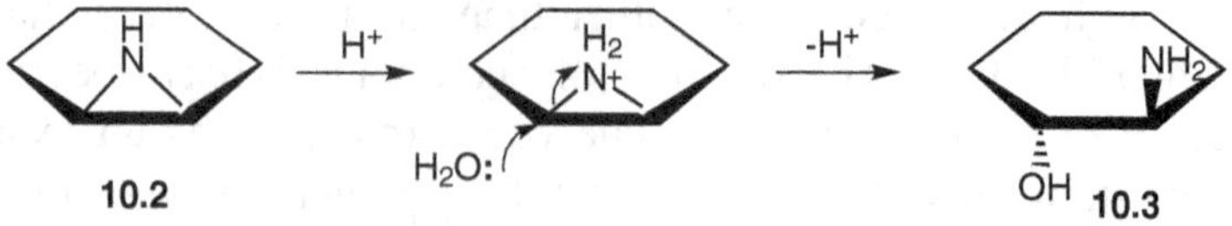

Scheme 10.3

If the 3-membered ring has a substituent on carbon, then two different ring-opened products are possible. Generally, the isomer from attack at the least hindered (unsubstituted) carbon predominates; thus in Scheme 10.4, isomer **10.4** is the dominant product from hydrolysis of 2-methylaziridine.

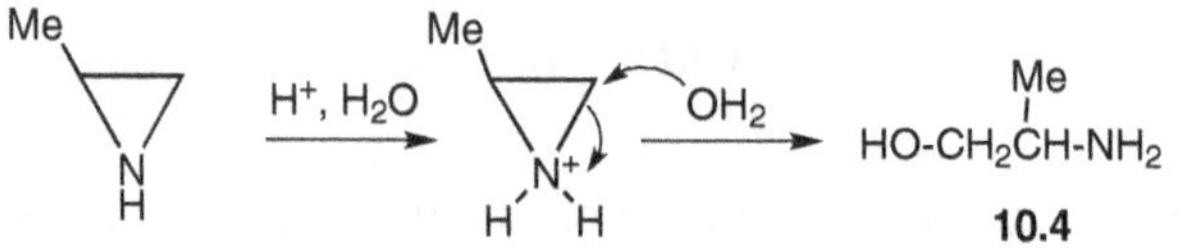

Scheme 10.4

Much less well-known are reactions of thiiranes (episulfides) and phosphiranes, but the general considerations discussed previously apply; ring opening is common from attack of nucleophiles. For all heterocyclic derivatives, reactions other than simple ring opening are known; these are discussed in detail in *Comprehensive Heterocyclic Chemistry*,[2] which includes treatment of rings containing two heteroatoms (e.g., oxaziridines) and rings with one double bond (e.g., azirines).

Aziridines have a special physical property that is attributed to the geometry of the ring; the nitrogen atom has a barrier to pyramidal inversion that is significantly higher than that of other monocyclic and acyclic amines. It will be recalled that amines have modified tetrahedral geometry where the lone pair occupies one site of the tetrahedron. If the lone pair orbital is ignored, the shape can be described as a pyramid. If the pyramid were stable, there would be two stereoisomers just as in carbon compounds. These would be nonsuperimposable mirror image (**10.7**) and have optical activity. However, the amine molecule is in a state of equilibrium wherein the nitrogen atom flattens to a trigonal planar structure (sp^2, **10.6**), which then resumes its pyramidal shape by the nitrogen moving above and below the plane. The process, which is very fast, is referred to as pyramidal inversion. It can be pictured as in Scheme 10.5.

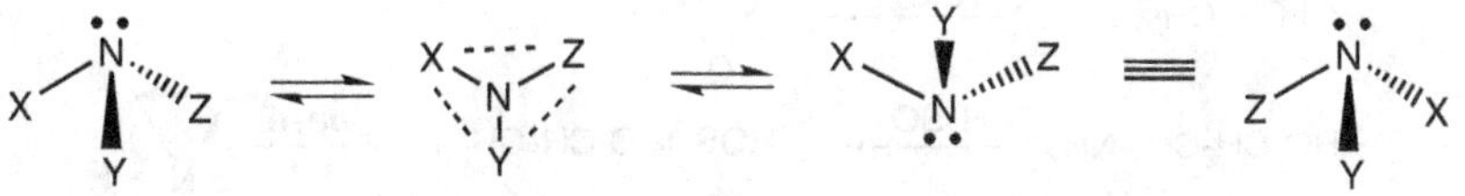

10.5 10.6 10.7

Scheme 10.5

The consequence is that although the two possible pyramids are nonidentical mirror images, their lifetime at room temperature is so short that they cannot be detected by conventional means, and amines are never optically active. However, if structural features can retard the rate of pyramidal inversion, then the two forms could be detectable, which is exactly what happens in aziridines. Stability of the pyramid in a cyclic amine means that cis,trans isomers can exist if another substituent is present on a ring carbon. In 1,2-dimethylaziridine, the pyramidal inversion rate is slow enough at −40°C that both the cis and trans forms can be detected by NMR spectroscopy.

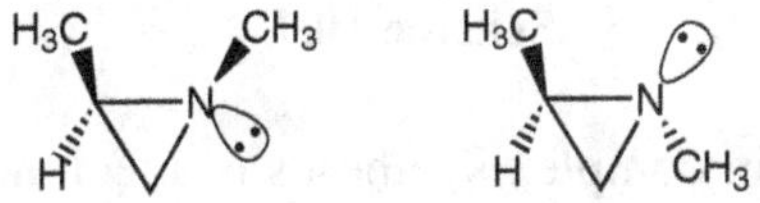

10.8, cis 10.9, trans

The effect is more dramatic when hydrogen on nitrogen is replaced by halogen. In 1968, Brois[3] separated the cis and trans isomers of 1-chloro-2-methylaziridine by gas chromatography and found that they were stable at 135°C. N-Bromo and N-fluoro derivatives possessed similar high barriers to pyramidal inversion.

Unlike their nitrogen counterparts, noncyclic phosphines have such a high energy barrier to pyramidal inversion that stable optically active forms are common. Some have half-lives at 135°C of 3–4 hr.[4] Cis and trans isomers of 1,2-disubstituted phosphiranes are therefore readily prepared.

Three-membered rings can be constructed by several methods. Of rather general occurrence is the intramolecular nucleophilic substitution process depicted for oxiranes and aziridines in Scheme 10.6. Many compounds have been prepared by these methods. Phosphiranes are made by reacting 1,2-dihalo compounds with metallic phosphides; thiiranes have other methods of synthesis.[2]

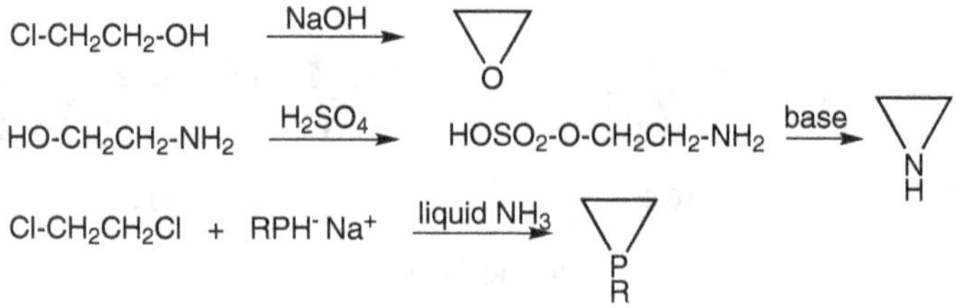

$$Cl\text{-}CH_2CH_2\text{-}OH \xrightarrow{NaOH} \text{oxirane}$$

$$HO\text{-}CH_2CH_2\text{-}NH_2 \xrightarrow{H_2SO_4} HOSO_2\text{-}O\text{-}CH_2CH_2\text{-}NH_2 \xrightarrow{base} \text{aziridine (N-H)}$$

$$Cl\text{-}CH_2CH_2Cl + RPH^- Na^+ \xrightarrow{liquid\ NH_3} \text{phosphirane (P-R)}$$

Scheme 10.6

Oxiranes and aziridines can also be prepared by special methods. A particularly useful synthesis of oxiranes involves the reaction of a C=C group with a hydrogen peroxide derivative. Commonly used are m-chloroperbenzoic acid and t-butylhydroperoxide, in what is referred to as an epoxidation reaction (Scheme 10.7).

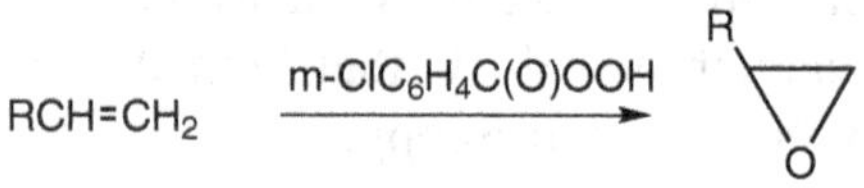

$$RCH{=}CH_2 \xrightarrow{m\text{-}ClC_6H_4C(O)OOH} \text{2-R-oxirane}$$

Scheme 10.7

Of particular value in complex syntheses is a technique for epoxidation that can be applied to allylic alcohols and that directs the approach of the oxidizing group to one or the other of the two faces of the double bond. This results in the formation of one enantiomeric form in excess of the other and, thus, stands as an asymmetric synthesis. The technique is simple and consists of the formation of a chiral catalyst, a coordination complex, from titanium tetra-isopropoxide and one of the optically active forms of a dialkyl tartrate. The allylic alcohol associates with this complex in a specific way and then is epoxidized on one face by t-butyl hydroperoxide. The epoxide is produced in high enantiomeric excess, frequently more than 95%. This process has been used widely in organic synthesis since its discovery in 1980.[5] It is now known as the Sharpless epoxidation.

A useful procedure for aziridine synthesis is the pyrolysis or photolysis of 1,2,3-triazolines; these compounds are prepared easily by reaction of azides with alkenes in a 1,3-dipolar cycloaddition reaction (Chapter 5). Nitrogen is eliminated in this process, probably to form a diradical, and the ring is closed to form the aziridine. Many applications of this process, which are outlined in Scheme 10.8, have been reported.

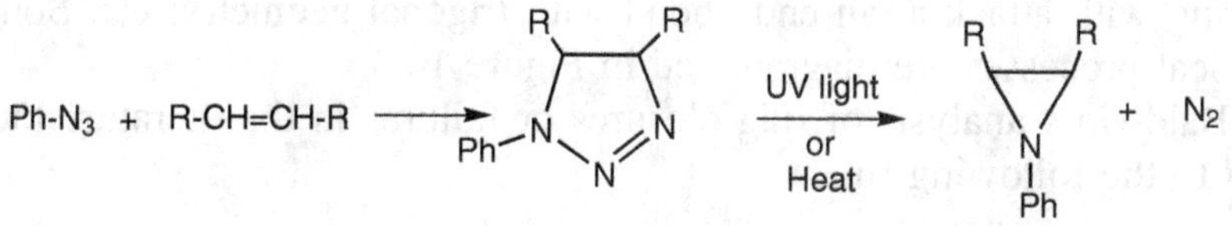

$$Ph\text{-}N_3 + R\text{-}CH\text{=}CH\text{-}R \longrightarrow \text{triazoline} \xrightarrow[\text{Heat}]{\text{UV light or}} \text{N-phenylaziridine} + N_2$$

Scheme 10.8

10.3. CLOSING HETEROCYCLIC RINGS: BALDWIN'S RULES

Many examples of intramolecular ring-closing reactions have been presented in Chapter 4 and elsewhere, which makes it clear that this is an important concept for heterocyclic synthesis. However, certain types of proposed intramolecular reactions fail. These observations led J. E. Baldwin[1] to develop a set of empirical rules that allow prediction of which structural types will be favored or disfavored for cyclization. For an intramolecular cyclization to occur, the functional group at one end of a chain must be able to reach the site where a new bond is to be formed, so that efficient orbital interaction can take place. The ease of the interaction is reflected in the activation energy for the formation of the transition state; thus, cyclization is under kinetic control. Baldwin's rules are based on the following structural features:

1. The size of the ring being formed.
2. Whether the bond being broken lies in the ring being formed (designated endo) or outside the ring (exo), as described by Figure 10.2.
3. The geometry at the atom where a bond is being broken. Designations: tet if tetrahedral; trig if trigonal; dig if digonal.
4. The attacking atom must be from the first row of the periodic table (exceptions).

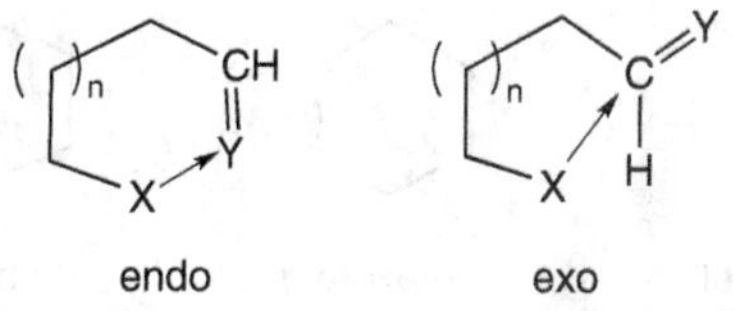

endo exo

Figure 10.2. Definition of exo and endo.

Cyclizations can be described by terminology such as the following: 6-exo-tet, meaning a 6-membered ring is to be formed from attack at an atom having an exo bond that has tetrahedral geometry; 7-endo-trig, 7 ring with attack at an endo bond with trigonal geometry; etc. Some typical processes are diagrammed in Figure 10.3.

Baldwin's analysis of ring closures or failures in the literature then led to the following rules:

Rule 1: Tetrahedral systems
- a) 3 to 7 exo-tet are favored
- b) 5 and 6 endo-tet are disfavored

Rule 2: Trigonal systems
- a) 3 to 7 exo-trig are favored
- b) 3 to 5 endo-trig are disfavored
- c) 6 and 7 endo-trig are favored

Rule 3: Digonal systems
- a) 3 and 4 exo-dig are disfavored
- b) 5 to 7 exo-dig are favored
- c) 3 to 7 endo-dig are favored

As noted, ring formation takes place when the two reacting sites have favorable geometry for orbital interaction to occur. Calculations have established the preferred angle (the approach vector) for this to occur. Two examples are as follows:

1. For exo-trig attack on a C=O group, an angle of 109° for the trajectory of group X to the plane of C=O is preferred.
2. For exo-dig attack on the terminal carbon of an alkyne, an angle of 120° between the approaching group X and the hydrogen-carbon bond at the terminus of the triple bond.

Many cases in the literature show the validity of the rules in practical ring closures. Some examples are given in Figure 10.4.

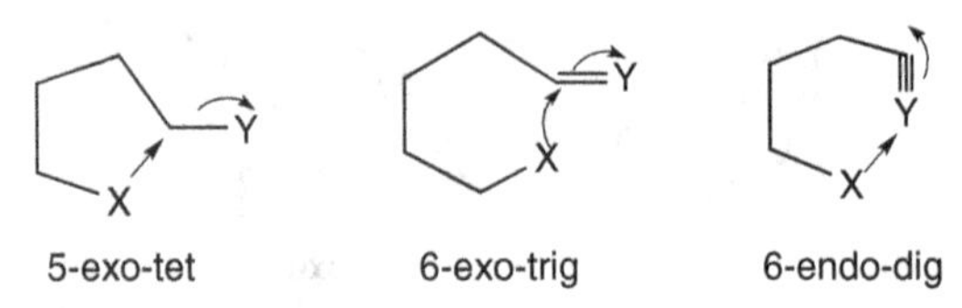

Figure 10.3. Illustrations of ring closure.

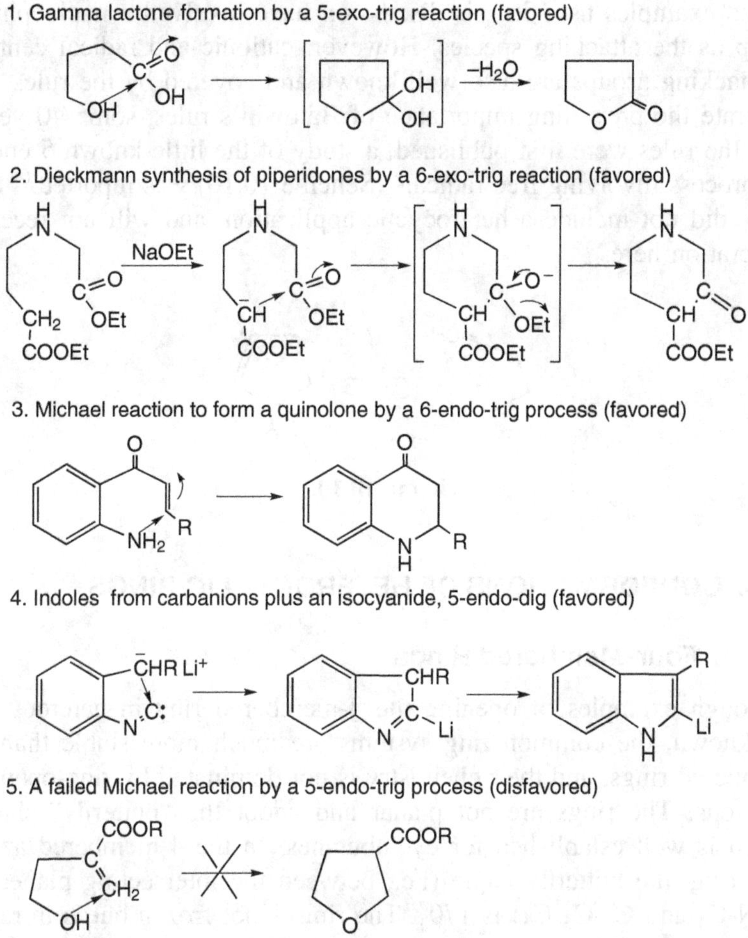

1. Gamma lactone formation by a 5-exo-trig reaction (favored)

2. Dieckmann synthesis of piperidones by a 6-exo-trig reaction (favored)

3. Michael reaction to form a quinolone by a 6-endo-trig process (favored)

4. Indoles from carbanions plus an isocyanide, 5-endo-dig (favored)

5. A failed Michael reaction by a 5-endo-trig process (disfavored)

Figure 10.4. Practical applications of Baldwin's rules.

Reaction 5 in Figure 10.4 is of special note; the Michael reaction does not occur, but there is an alternative pathway that is favored and indeed is followed. This is the 5-exo-trig attack of OH on the ester carbonyl to form a lactone (Scheme 10.9).

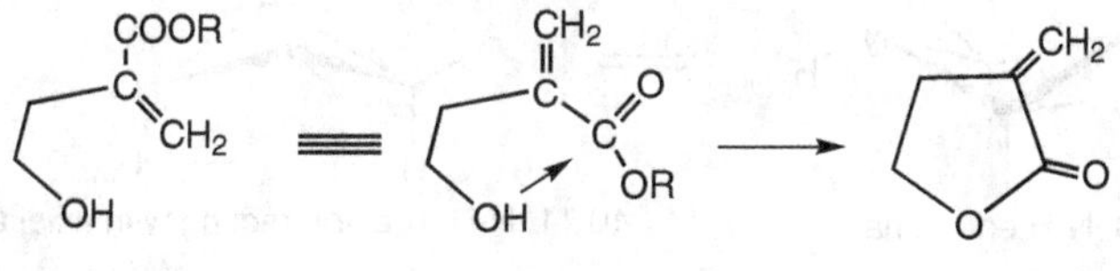

Scheme 10.9

The examples used in this discussion have involved a nucleophilic group as the attacking species. However, cationic and radical centers as attacking groups are also well known and covered by the rules. To illustrate the prevailing importance of Baldwin's rules, some 40 years after the rules were first published, a study of the little-known 5 endo-dig process involving free radicals (Scheme 10.10) was reported.[6] The study did not include a heterocyclic application, and will not receive elaboration here.

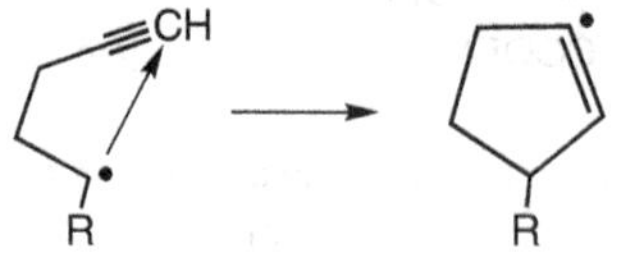

Scheme 10.10

10.4. CONFORMATIONS OF HETEROCYCLIC RINGS

10.4.1. Four-Membered Rings

Although examples of opening the 4-membered ring in heterocycles are known, the common ring systems are much more stable than 3-membered rings, and their chemistry is not dominated by ring-opening reactions. The rings are not planar and adopt the "butterfly" shape, which is well established for cyclobutanes. In the 4-membered azetidine ring, the butterfly angle (i.e., between the intersecting planes of C_2-N-C_4 and C_2-C_3-C_4) is 170°. The ring is not frozen but is in rapid equilibration with other conformers from ring inversions. The conformational equilibrium for azetidine is shown in Figure 10.5. The position of the conformational equilibrium in any substituted cycloalkane is sensitive to interactions of the substituent with ring hydrogens. This interaction is greater when the substituent is in the axial position where

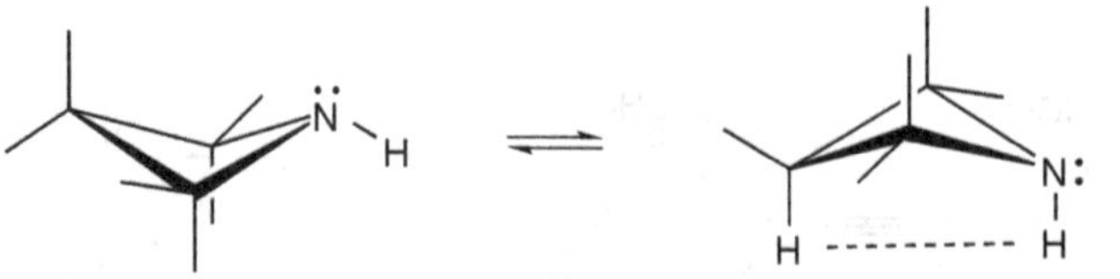

10.10, N-H equatorial

10.11, N-H axial interacting with axial C-H

Figure 10.5. Conformational equilibrium of azetidine.

it can be "crowded" by interaction with axial hydrogens on the beta carbons. This is referred to as a 1,3 interaction and causes the equilibrium to be biased toward the form with an equatorial substituent. (This is the same type of interaction that is found in noncyclic alkanes such as butane, where it will be recalled that the conformation with the staggered conformation is favored over the conformation with eclipsing hydrogens or alkyl substituents.) Heterocyclic rings are subject to the same type of conformational biasing. Therefore, with a nitrogen atom in the ring, the hydrogen or substituent on this atom will preferentially occupy the equatorial position. The ring inversion equilibrium for azetidine is shown in Figure 10.5; the form with equatorial N-H (**10.10**) is favored.

An important qualification needs to be made with these structures. The nitrogen atom is shown as if the pyramid were stable. In fact, the usual pyramidal inversion of an amine (section 10.3) is going on, so that it is not proper to view a bond from nitrogen as being fixed in the equatorial or axial positions at ordinary temperatures. The barrier to pyramidal inversion (in energy terms, the free energy of activation, $\Delta G^{\ddagger}$) is about 2 kcal/mol less than that in aziridines. Thus, conformation **10.10** (and **10.11** as well) is undergoing rapid pyramidal inversion, as expressed in Figure 10.6, which implies that there will be no such thing as stable cis,trans isomers in C-substituted azetidines. This is true for all monocyclic amines, with the exception already shown for some 3-membered rings. This matter will be developed more fully in considerations of the better known 6-membered rings.

10.4.2. Five-Membered Rings

Just as for cyclopentanes, the 5-membered heterocyclic rings are not planar but adopt a conformation in which one ring atom is above (or below) the plane formed by the other four ring members. This gives a shape resembling an opened envelope; this term is used frequently in studies of 5-membered rings. The out-of-plane position is shared by all

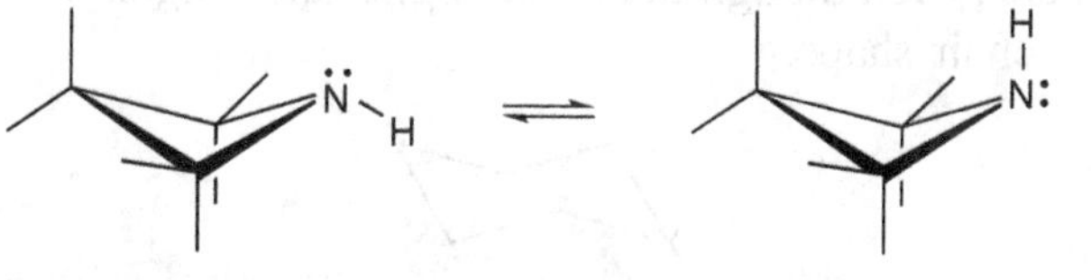

Figure 10.6. Pyramidal inversion in azetidine.

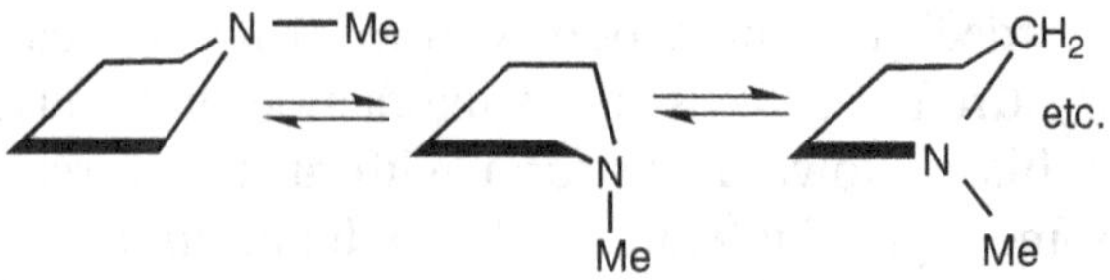

Figure 10.7. Conformational equilibrium for N-methylpyrrolidine.

ring members, so that the molecule is in a conformational equilibrium with all possible structures. The equilibration is rapid, and the individual conformers cannot be observed. Using pyrrolidine as an example, the conformational equilibrium is represented in Figure 10.7.

Again for the nitrogen heterocycles, we must remember that pyramidal inversion is also in progress and that cis,trans isomers are not observed in C-substituted pyrrolidines.

In marked contrast to tri-covalent nitrogen, tri-covalent phosphorus has a significant barrier to pyramidal inversion, as discussed in section 10.2. In C-substituted phospholanes and the 4-membered phosphetanes as well, cis,trans isomers are stable and separated easily. This possibility was not recognized until it was demonstrated in 1965 that 1,2-dimethyl derivatives **10.12** and **10.13** of dihydrophospholes, which are known as 3-phospholenes, could be separated by fractional distillation or gas chromatography.[7]

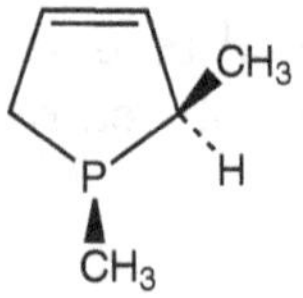

10.12, cis 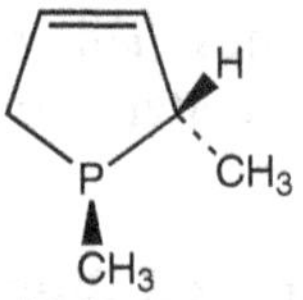 **10.13**, trans

10.4.3. Six-Membered Rings

Much more is known about 6-membered heterocyclic rings than about any other ring sizes. Just as for cyclohexanes, the saturated rings always adopt a chair conformation, which is shown in **10.13** for tetrahydropyran and thiane, even though the bond angles and lengths will differ and modify the chair shape.

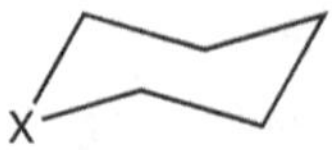

10.13, X = O or S

These rings are in conformational equilibrium with other chair forms, generally expressed as in Figure 10.8 (which ignores intermediate stages of the "ring flipping" process).

Substituents on carbon in these rings prefer the equatorial configuration just as in cyclohexanes. However, a notable and informative exception occurs with 1,3-dioxanes bearing a 5-substituent (Figure 10.9). Here, a diminished energy difference is observed between the conformers with an axial or an equatorial substituent, because the nonbonded steric 1,3-interactions with axial hydrogens on the ring, which make the axial position in cyclohexanes energetically unfavorable, are absent when oxygen is in the ring. The conformational equilibrium in 5-substituted 1,3-dioxanes, therefore, has a higher contribution of axial form **10.14** than in cyclohexanes. In fact, with certain polar groups, the axial position is actually preferred.

With polar groups alpha to an oxygen atom, the axial position is strongly preferred. This has been dubbed the anomeric effect, because it was first recognized with isomers at the anomeric carbon (the 1-position) of aldohexoses. There, the glycoside with an alpha-alkoxy substituent (**10.15**) is more stable than that with an equatorial alkoxy (**10.16**). An interaction between the ring oxygen lone pair orbital and the antibonding sigma* orbital of the oxy substituent is thought to account for this effect.[8]

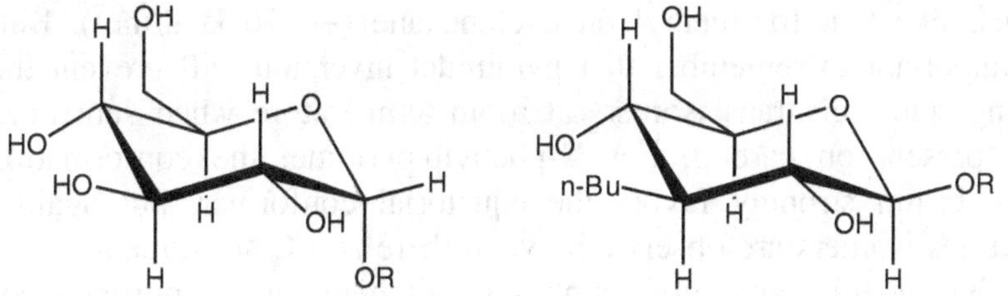

10.15, an α–D-glucoside

10.16, a β–D-glucoside

Piperidines also participate in the rapid equilibration of chair conformers. But as in 4- and 5-membered rings, the situation is complicated by the nitrogen atom additionally undergoing rapid pyramidal inversion, and once again it is not possible to observe

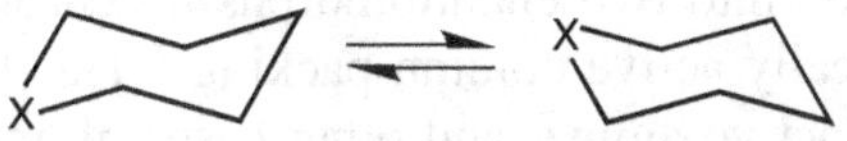

Figure 10.8. Conformational equilibrium in 6-membered rings.

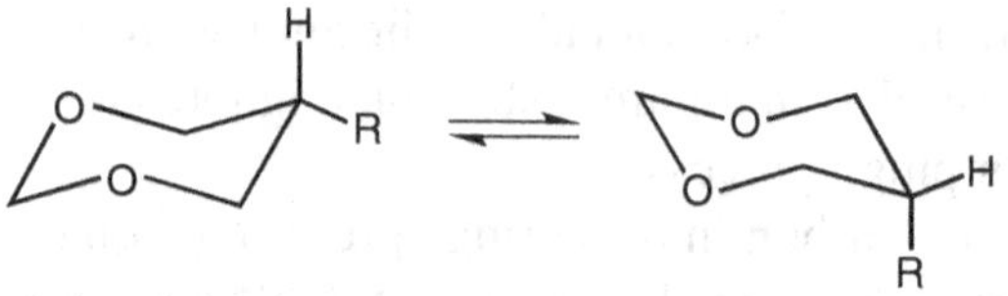

10.14

Figure 10.9. Conformational equilibrium in 1,3-dioxanes.

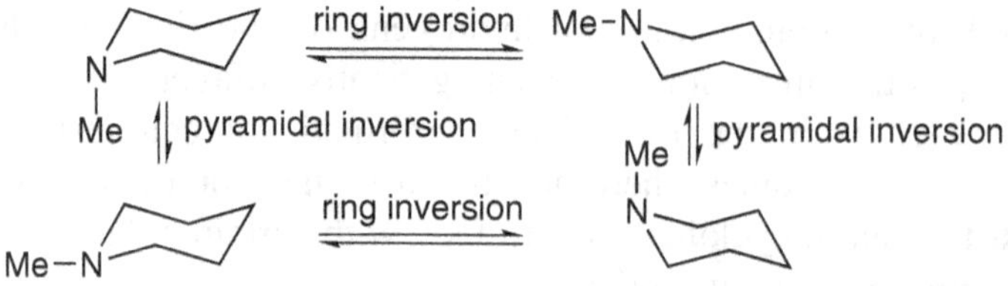

Figure 10.10. Equilibria in a piperidine.

cis,trans isomers under ordinary conditions. Both equilibria are expressed for N-methylpiperidine in Figure 10.10. The net result is that experimentally, the N-substituent is preferentially found in the equatorial position; for the methyl group, the free energy difference (ΔG^0, derived from the equilibrium constant for a reaction from the expression $\Delta G^0 = -2.3RT \log K_{eq}$) is -1.93 kcal/mol, which is much like that for methyl on cyclohexane (-1.76 kcal/mol). But it is important to remember that pyramidal inversion will prevent there being stable cis,trans isomers at room temperature when substituents are present on carbon. For N-phenylpiperidine, the conformational equilibrium strongly favors the equatorial conformer, but again no cis,trans isomers are observable when there is a C-substituent.

The "no cis,trans rule" applies to all heterocyclic nitrogen compounds with only bonds to carbon and that contain an exocyclic substituent on nitrogen. When nitrogen is at the bridgehead position of bicyclic structures such as 1-azabicyclo[2.2.2]octane (**10.17**), pyramidal inversion cannot take place. In certain complex cases, there may be such structural rigidity that molecular dissymmetry may be present and optically active forms may exist. The first example of this case was reported in 1944 for the compound known as Tröger's base (**10.18**), which was resolved into two enantiomorphs by chromatography on d-lactose as an optically active column packing.[9] This has been billed as the first optically active amine, and other cases of amines with molecular dissymmetry have since been found.[10]

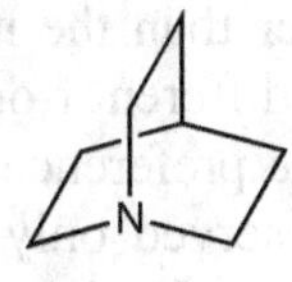

10.17 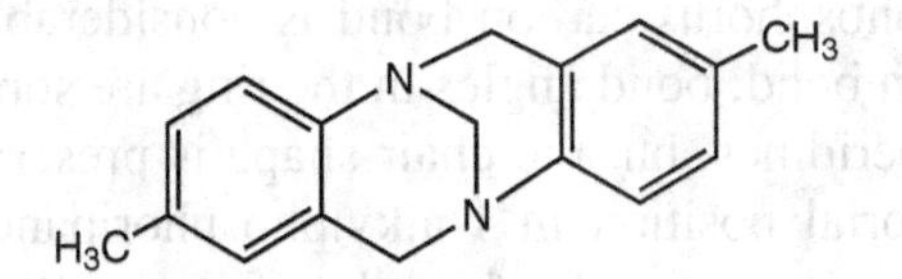 10.18

In somewhat less rigid bicyclic structures, there is little hindrance to N-inversion and the usual equilibration can take place. Thus, in the indolizidine system, found in, e.g., strychnine and erythrina alkaloids, the substituent on N is not fixed in a certain position but the equilibrium of Figure 10.11 prevails. The "trans" form is highly favored ($\Delta G^0 = -2.4$ kcal/mol).

The same situation prevails in the quinolizidine system (Figure 10.12) found in, e.g., lupinine and yohimbine alkaloids; again the "trans" form is strongly favored ($\Delta G^0 = -2.61$ kcal/mol).

Because of the considerably higher barrier to pyramidal inversion of phosphorus, stable cis,trans isomers are well known for phosphinane derivatives (also known as phosphorinanes). This is shown, for example, in the 1,4-dimethylphosphinanes **10.19** and **10.20**.[11]

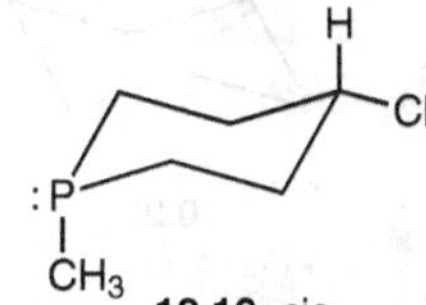

10.19, cis 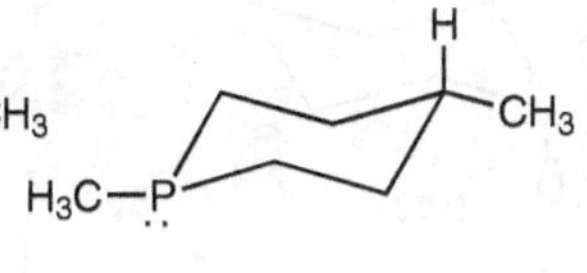 10.20, trans

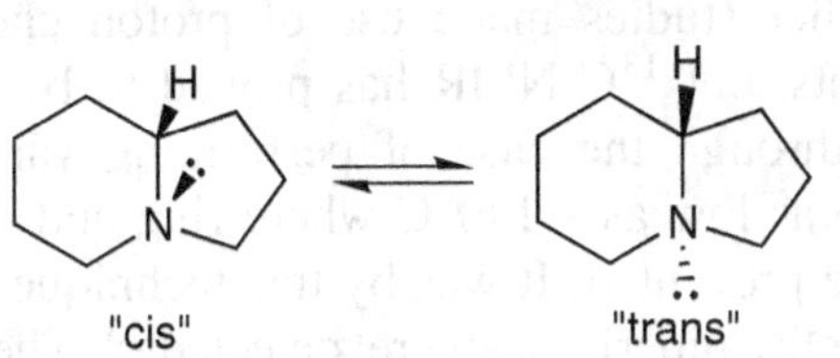

"cis" "trans"

Figure 10.11. N-Inversion equilibrium in indolizidine.

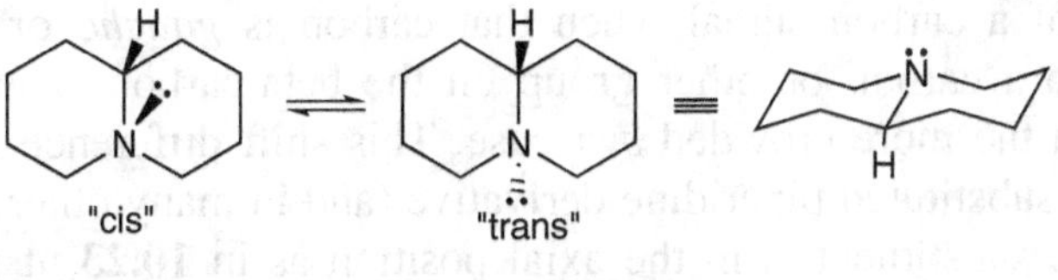

"cis" "trans"

Figure 10.12. N-Inversion equilibrium in quinolizidine.

The phosphorus-carbon bond is considerably longer than the nitrogen carbon bond; bond angles in the ring are somewhat different from those in piperidines, but the chair shape is preserved. The preference for the equatorial position in 1-alkylphosphorinanes is observed only at low temperatures (e.g., for 1-methyl, 3:1 at $-100°C$) where the ring-flipping equilibrium is stopped and both forms can be observed (e.g., by NMR). The concentration of the axial form at equilibrium increases with the temperature through an entropy effect, and actually it exceeds that of the equatorial conformer by 2:1 at 20°C. The X-ray diffraction analysis of crystalline 1-phenylphosphorinan-4-one shows only the axial conformer and reveals an important structural effect: The phosphorus pyramid is somewhat flattened. This is confirmed by X-ray analysis of a pair of isomers with an axial and an equatorial P-phenyl group; it is shown clearly that the isomer with the axial substituent is flatter at P than in the equatorial isomer. This may be the result of the 1,3-interactions with the axial H at C-3,5; the strain is relieved by the flattening. Similar results have been reported for thiane derivatives where sulfur is tri-covalent; the axial conformer is favored over the equatorial in structures **10.21** and **10.22**.[12]

NMR spectroscopy, in various experimental modes, has played a major role in elucidating the stereochemistry of saturated heterocyclic compounds. Earlier studies made use of proton chemical shifts and coupling constants, but ^{13}C NMR has proved to be particularly valuable especially through the ease of performing variable temperature measurements to as low as $-140°C$ where ring and nitrogen inversion equilibrations are prevented. It was by this technique that the two conformers of N-methylpiperidine were detected.[13] The conformer with axial N-Me (**10.23**) had δ 17.0, which is well upfield from the shift of the equatorial form with δ 23.0 (**10.24**). These relative shifts are the result of a steric effect called steric compression that causes upfield shifting of a carbon signal when that carbon is *gauche* or *syn* with respect to a carbon or other group on the beta-carbon. The effect is greater in the more crowded *syn* case. This shift difference is present in any N-substituted piperidine derivative (and in many other systems). When the substituent is in the axial position as in **10.23**, it is *gauche* related to ring carbons 3 and 5, but when it is equatorial (**10.24**), the compression is relieved. At room temperature, the chemical shift was close to that of **10.24**, which indicated that the N-methyl group, while not fixed in position from the inversion processes, was nevertheless largely occupying the equatorial position.

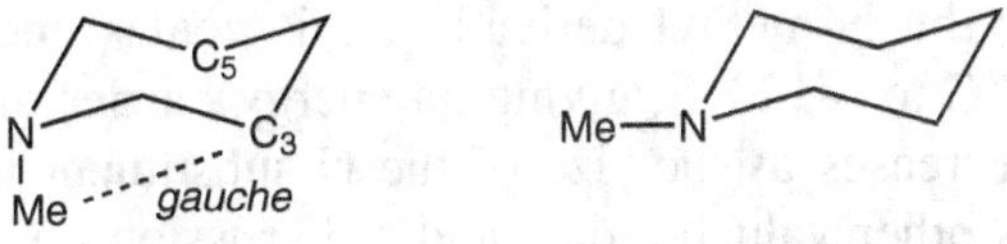

N C5 C3 Me *gauche* Me—N

10.23 **10.24**

As another example of the power of ^{13}C NMR in conformational analysis of heterocycles, portions of a study reported by Katritzky, Baker, and Brito-Palma[14] may be cited. Spectra were obtained at various temperatures for several derivatives of 1-oxa-3-aza-cyclohexane. This ring system gives well separated signals both in ^{13}C and ^{1}H NMR spectra because of the presence of the electron attracting (causing deshielding) O and N atoms. At room temperature, both ring and nitrogen inversion are proceeding rapidly, and signals for the individual conformers cannot be seen. Thus, at 20°C for the N-methyl derivative, the ring carbon signals (all single lines from decoupling of the protons) appeared at δ 87.6 (C-2), 68.2 (C-6), 53.3 (C-4), and 23.9 (C-5); the N-methyl signal was at δ 40.2. At −142°C, all signals were doubled as the equilibration was stopped, and the spectra of the individual molecules with axial and equatorial N-methyl substituents were observed. The axial N-methyl derivative (**10.25**) undergoes steric compression with the gamma-related ring carbon C-5, which is said to be associated with a 5.7 ppm upfield shift of that carbon (at δ 20.0, an effect already observed for N-methylpiperidine) relative to the signal for the equatorial conformer (**10.26**, δ 25.7); indeed, all carbons are shifted upfield by various amounts in this more crowded conformer, and the relative sizes of the signals for a particular ring carbon give a measure of the percentages of the conformers.

O Me N ⇌ O N Me

10.25 **10.26**

For the N-methyl derivative, the axial form is in slight excess. The C-5 signal in the 20°C spectrum falls between the values of the individual conformers, as would be expected for an equilibrating system, which is true for the other ring carbons as well. By observing the point at which coalescence of the two peaks occurs as the temperature is slowly raised, the energy barrier (the activation energy for the conformer interconversion) between the two nitrogen-inversion conformers can be calculated; for the N-methyl derivative, this coalescence temperature is about -110°C to -120°C, giving an energy barrier of 7.4 kcal/mol. This barrier decreases as the size of the N-substituent increases. This paper provides other valuable data and a discussion on the ring inversion and nitrogen inversion processes.

Another great aid in contemporary conformational analysis of heterocycles is the use of ab initio and density functional theory computational techniques, which can calculate energies of the conformers and other parameters. The 1-oxa-3-azacyclohexane system has been subjected to such study, with results comparing well with the experimental data.[15] The calculations confirmed experimental observations that Me, Et, and n-Pr groups on N were predominantly axial, whereas the larger groups i-Pr and t-Bu were predominantly equatorial. The effect of these substituents on ring bond angles, bond lengths, and dihedral angles in the ring were determined, and the general study emphasizes the importance of the computational approach in advanced studies of heterocyclic systems.

In this section, we have only made use of chemical shift effects in studying stereostructure of simple heterocycles. However, it needs to be pointed out that the determination of stereostructure in complex molecules is greatly aided by special NMR experiments, such as double irradiation techniques, especially using the nuclear Overhauser effect (NOE) to detect longer range nuclear interactions, as well as by various two-dimensional techniques. The study of these techniques is beyond the level of this introductory text.

10.5. CHIRALITY EFFECTS ON BIOLOGICAL PROPERTIES OF HETEROCYCLES

It has long been known that the individual enantiomers (said to be chiral) of a racemic mixture may have different biological properties even though, of course, their chemical reactivity is identical. Their biological difference arises when the drug must fit and bind in a receptor site of a complex enzyme. These sites can frequently discriminate between the enantiomers, because the sites can have a specific stereosequence of the groups where the drug must fit. The drug thalidomide provides a well-known and dramatic example of biological specificity of enantiomers.

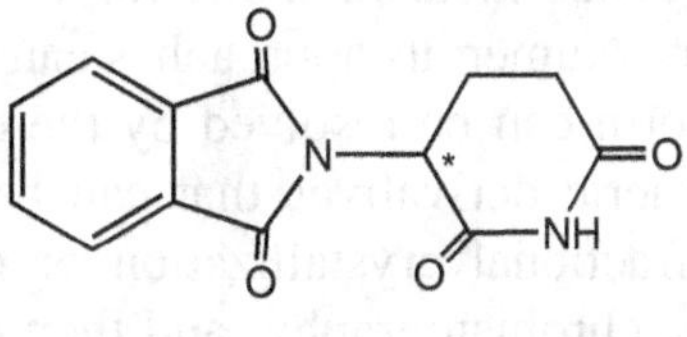

10.26, thalidomide

It had sedative, hypnotic, and anti-inflammatory properties and was introduced in 1957 in racemic form primarily as an aid in controlling the effects of pregnancy. However, it was recognized within a few years that it had the horrible side effect of being teratogenic, and thousands of babies were born with severe malformations. It was banned from commerce. However, it did have valuable beneficial medical effects in nonpregnancy cases, and research on it was continued. It was discovered that only the S(−) enantiomer was teratogenic, and the R(+) enantiomer had different and useful properties. However, an important factor in the use of thalidomide is that racemization of the separated enantiomers is facile under biological conditions, requiring great caution in therapeutic applications. Medical use of thalidomide is permitted again but only after thorough evaluation and passing through stringent requirements of authorization.

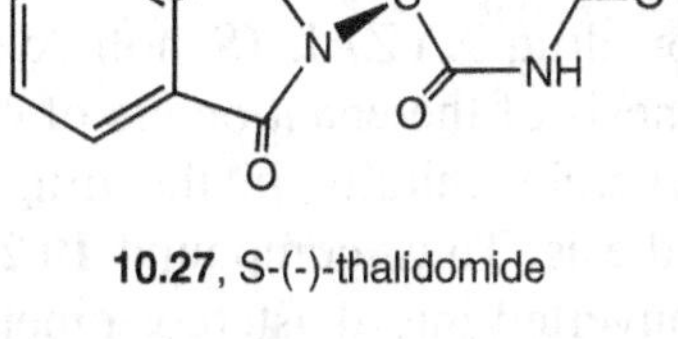

10.27, S-(-)-thalidomide

Starting in the 1980s, the pharmaceutical industry commenced extensive research into preparing the optically active versions of known and new drugs. Partly this was in the expectation that enhanced or even different biological properties would be found in the individual enantiomers. There was another incentive, however, in that it was possible to patent a single enantiomer of a drug previously patented as the racemic mixture, thus protecting the interests of the discoverer of the drug for another time period. In addition, the regulatory agencies require that the individual enantiomers of a racemic drug be prepared and tested before authorization is given to put a racemic drug on the market. As a result, the preparation of chiral drugs has become a major activity in the pharmaceutical industry. This has required the development of effective, economical methods for the procurement of the individual enantiomers. Numerous approaches can be taken. Drugs synthesized in racemic form can be resolved by the conventional method of forming diastereomeric derivatives that can be separated by physical means such as fractional crystallization or occasionally distillation, but especially by chromatography, and then caused to release the individual enantiomer. Direct separation of racemic mixtures can be accomplished by chromatography on chiral gas or high-performance liquid chromatography (HPLC) column packings. As an example,[16] the enantiomers of thalidomide can be separated cleanly by HPLC on the commercial chiral packing "Chiralpak AD-RH." In a different approach, chiral drugs can be synthesized directly by using starting materials available from optically active natural compounds (nature's "chiral pool"), as well as by new stereoselective processes in which chiral catalysts or enzymatic methods are employed to create preferentially one of the enantiomers. These and other aspects of creating chiral drugs are discussed in a 2009 review.[17] Many chiral drugs are now on the market or being developed; only a few will be described here as examples.

The antihistamine drug cetirizine, which is a piperazine derivative, is a racemic mixture, and has been used in the treatment of rhinitis and hay fever. It is available over the counter as the well-known ZYRTEC (McNeil-PPC, Inc). However the levo-rotatory R-enantiomer (**10.29**) was later found to be the more active form and is now on the market as the prescription drug XYZAL (Sanofi-Aventis U.S. LLC). An approach for the synthesis of the enantiomers of cetirizine is expressed by Scheme 10.11, wherein chiralty in the drug is introduced at an early stage in the synthesis. Thus, compound **10.28** was synthesized in racemic form then converted into diastereoisomeric salts with an optically active tartaric acid. To procure the (−)-enantiomer of **10.28**, the resolution was performed with (+)-tartaric acid; for the (+)-enantiomer, the resolution employed (−)-tartaric acid.[18] After separation of the salts by crystallization, the **10.28** enantiomers were released by treatment with base and each then converted to optically active cetirizine enantiomers or related compounds by use of the general reaction suggested by Scheme 10.11 for the synthesis of the (R)-isomer of cetirizine.

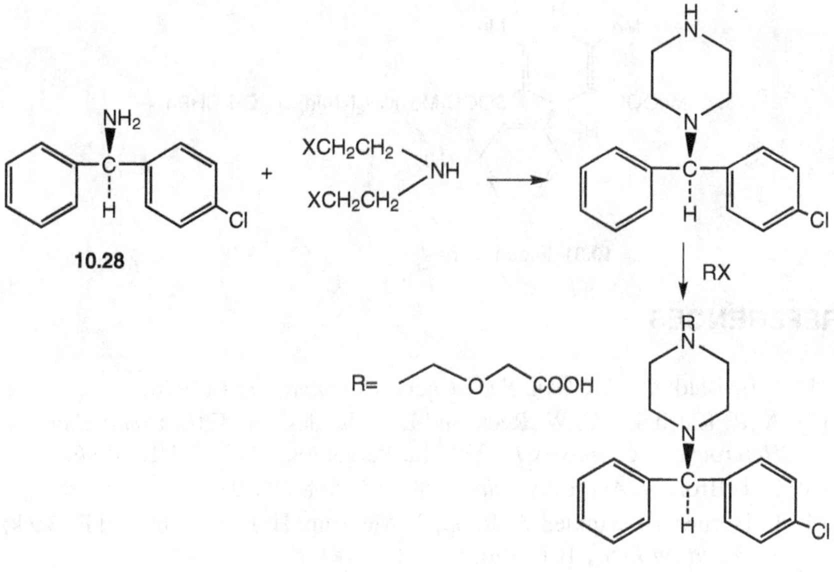

10.28

10.29, levocetirizine
(XYZAL)

Scheme 10.11

Another example of a chiral drug is escitalopram, which is an antidepressant acting as a selective serotonin reuptake inhibitor. It is optically active, having the S-configuration shown in structure **10.30**. It is a follow-up to the marketed racemic mixture of the same structure known as citalopram, and is more active than the racemic mixture.

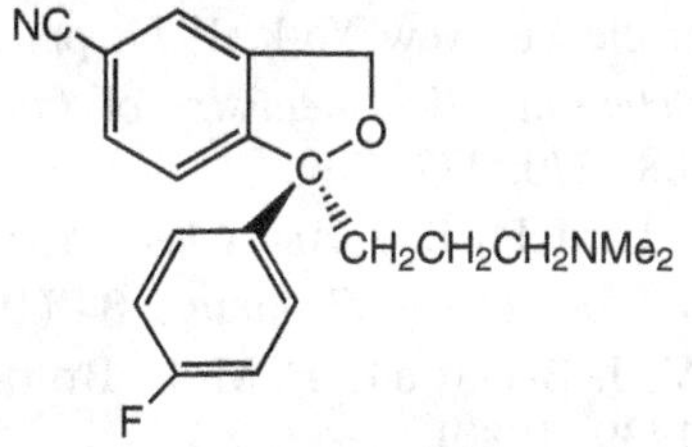

10.30, escitalopram

A final example is lercanidipine (Zanidip; Napp Pharmaceuticals), which is a calcium-channel blocker used for the treatment of high blood pressure. Its stereostructure is shown in **10.31**. It is the S-enantiomer; the R-enantiomer is inactive.

10.31, lercanidipine

REFERENCES

(1) J. E. Baldwin, *J. Chem. Soc., Chem. Commun.*, 734 (1976).

(2) A. R. Katritzky, C. W. Rees, and E. F. V. Scriven, Eds., *Comprehensive Heterocyclic Chemistry II*, Vol. 1a, Pergamon, Oxford, UK, 1996.

(3) S. J. Brois, *J. Am. Chem. Soc.*, **90**, 506, 508 (1968).

(4) L. Horner, H. Winkler, A. Rapp, A. Mentrup, H. Hoffmann, and P. Beck, *Tetrahedron Lett.*, 161 (1961).

(5) T. Katsuki and K. B. Sharpless, *J. Am. Chem. Soc.*, **102,** 5974 (1980); for an extensive review, see T. Katsuki and V. S. Martin, *Org. Reactions*, **48**, 1(1996).

(6) I. V. Alabugin, V. I. Timokhin, J. N. Abrams, M. Manoharan, R. Abrams, and I. Ghiviriga, *J. Am. Chem. Soc.*, **130**, 10,984 (2008).

(7) L. D. Quin, J. P. Gratz, and R. E. Montgomery, *Tetrahedron Lett.*, 2187 (1965).

(8) E. Juaristi and G. Cuevas, *The Anomeric Effect*, CRC Press, Boca Raton, FL, 1995.

(9) V. Prelog and P. Wieland, *Helv. Chim. Acta*, **27**, 1129 (1944).

(10) W. L. F. Armarego, *Stereochemistry of Heterocyclic Compounds*, Vol. 1, Wiley-Interscience, New York, 1977, pp. 366–367.

(11) L. D. Quin, *The Heterocyclic Chemistry of Phosphorus*, Wiley, New York, 1980, pp. 368–370, 377–382.

(12) E. L. Eliel and A. T. McPhail, *J. Am. Chem. Soc.*, **96**, 3021 (1974).

(13) H. Booth, *J. Chem. Soc., Chem. Commun.*, 34 (1979).

(14) A. R. Katritzky, V. J. Baker, and F. M. S. Brito-Palma, *J. Chem. Soc., Perkin Trans. 2*, 1739 (1980).

(15) M. Hurtado, J. G. Contreras, A. Matamala, O. Mó, and M. Yáñez, *New J. Chem.*, **32**, 2209 (2008).

(16) K. Sembongi, M. Tanaka, K. Sakurada, M. Kobayashi, S. Itagaki, T. Harano, and K. Iseki, *Biol. Pharm. Bull.*, **31**, 497 (2008).

(17) M. C. Nunez, M. E. Garcia-Rubino, A. Conejo-Garcia, O. Cruz-Lopez, M. Kimatrai, M. A. Gallo, A. Espinosa, and J. M. Campos, *Current Med. Chem.*, **16**, 2064 (2009).

(18) E. Cossement, G. Bodson, and J. Gobert, *U.S. Patent* 5,478,941 (Dec. 26, 1995).

REVIEW EXERCISES

10.1. When ethylene oxide is converted to ethylene glycol, a large excess of water is used. Write the major byproduct that is formed when an equimolar amount of water is used.

10.2. J. Ichikawa, R. Nadano, T. Mori, and Y. Wada, *Organic Synth.*, **83**, 111 (2006). Classify this ring closure per Baldwin's rules and indicate whether it is favored or disfavored.

F F HN Ts → F N Ts

10.3. V. Rauhala, K. Nättinen, K. Rissanen, A. M. P. Koskinen, *European J. Org. Chem.*, **2005**, 4119-4126 (2005). Classify both possible ring closures (path a and path b) per Baldwin's rules and indicate whether each is favored or disfavored.

BnO HO a b OH O OBn CH_3 → a OBn CH_3 O O BnO OH

↓ b

OBn CH_3 HO O BnO O

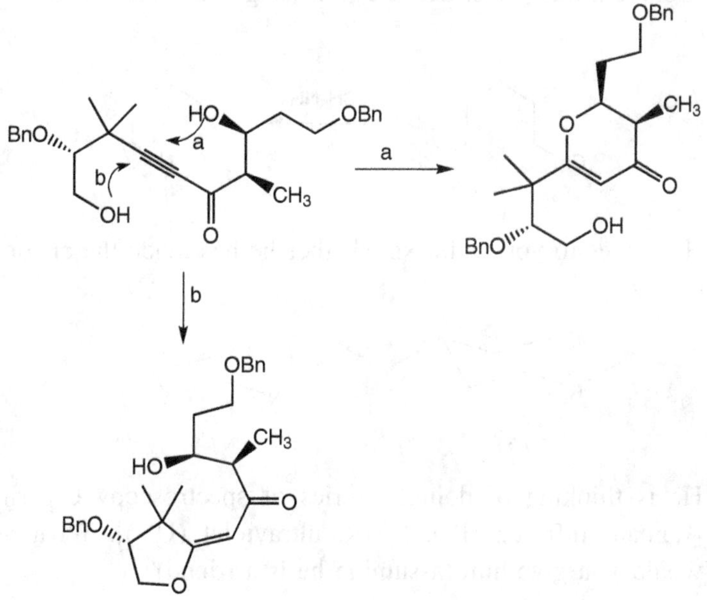

10.4. A. Cuppoletti, et al., *J. Phys. Org. Chem.*, **15**, 672-675 (2002). Predict the ring closure of the radical and classify per Baldwin's rules, indicating whether it is favored or disfavored.

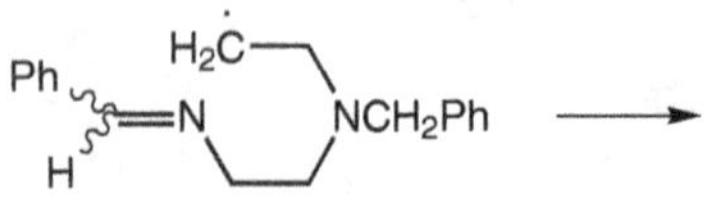

Ph H $H_2\dot{C}$ N NCH_2Ph →

10.5. For each pair, indicate whether each isomer can be prepared and isolated at room temperature.

a

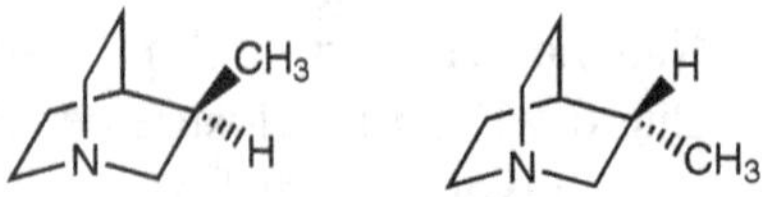

CH_3 H N H CH_3 N

b

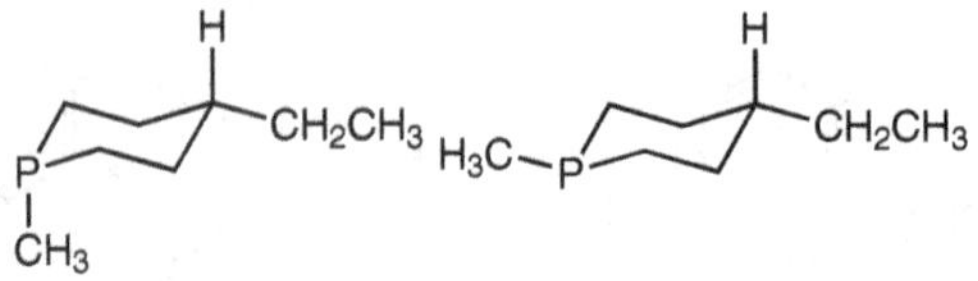

H CH_2CH_3 P CH_3 H H_3C–P CH_2CH_3

c

H CH_2CH_3 N CH_3 H H_3C–N CH_2CH_3

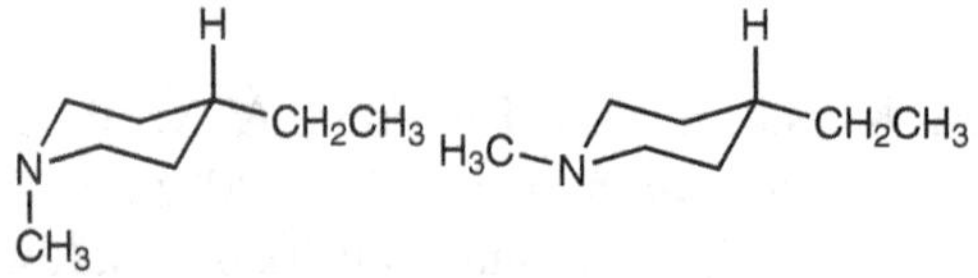

10.6. A student has performed the following synthesis.

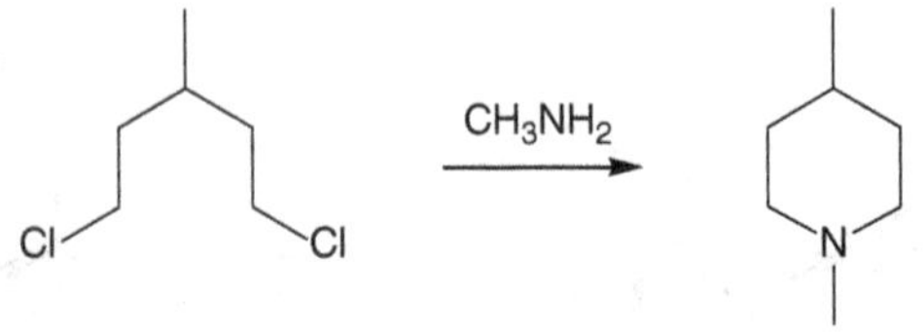

Cl Cl CH_3NH_2 → N

He comes to you and asks whether he has made the cis or trans product.

N trans N cis

He is thinking of doing a series of spectroscopy experiments [Raman, Infrared (IR), NMR, ultraviolet (UV)]. What advice would you give him (assuming he is a friend)?

10.7. Which of the following has a ring shape that can be described as butterfly like (bent and constantly inverting)?

a) Aziridine
b) Azetidine
c) Pyrrolidine
d) Piperidine
e) Azepane

CHAPTER 11

SYNTHETIC HETEROCYCLIC COMPOUNDS IN AGRICULTURAL AND OTHER APPLICATIONS

11.1. HETEROCYCLIC AGROCHEMICALS

11.1.1. General

Heterocyclic compounds play the same major role in crop and animal treatment as they do in medicine. We will consider here only the use of synthetic compounds on crops in the field; many agents have been developed that function as insecticides, fungicides, herbicides, and plant growth regulators (PGRs), and these agents are generally called pesticides. Naturally occurring heterocyclic compounds also find use in these areas but will not be considered here. As in pharmaceutical research, the industry synthesizes many thousands of compounds for testing in the previously mentioned areas; many are found to have some level of activity, but often not with enough to justify commercial development. The cost of production also may prevent the introduction of new agents to the market, as will mammalian toxicity considerations. Nevertheless, the field is large and of great importance in the vast area of agriculture. In this section, the discussion is organized around the heterocyclic ring systems that are found in the more important agents. Great structural diversity is found in the active agents, which possess high specificity for one of the modes of biological action. For a more detailed survey

Fundamentals of Heterocyclic Chemistry: Importance in Nature and in the Synthesis of Pharmaceuticals,
By Louis D. Quin and John A. Tyrell Copyright © 2010 John Wiley & Sons, Inc.

of heterocycles in agrochemical applications, a review by Crowley[1a] is available. This review was invaluable in the preparation of this section.

11.1.2. Results from the Formative Years of Pesticide Chemistry

The real value of synthetic agrochemicals only became apparent by the discovery in the 1940s of the powerful insecticides Dichlorodiphenyltrichloroethane [DDT; more properly 1,1-bis(p-chlorophenyl)-2,2,2-trichloroethane], and esters of organophosphorus acids such as parathion ($(EtO)_2P(S)OC_6H_4$-NO_2-p), among other compounds. Research on pesticides and PGRs became a major activity of many industrial firms, and a great variety of structures were found to have commercial value, heterocycles being prominent among them. This section will summarize some of the heterocyclic compounds that were commercialized during the early years of pesticide and PGR research. Viewing these compounds provides an excellent introduction to the wide range of heterocyclic ring systems found in the more active agents. The research effort has continued unabated, and far too many active structures have been discovered in more recent years to permit a survey in this book. However, section 11.1.3 will present some comments on newer areas of pesticide chemistry and modes of biological action.

11.1.2.1. Pyridine Derivatives. Some useful pyridine derivatives are shown in Figure 11.1. In the discussion to follow, it will be shown that these derivatives exhibit a variety of types of biochemical activity.

Picloram is a herbicide that selectively kills broadleaf weeds. Its synthesis is shown in scheme 11.1. Picloram has auxin-like (growth-promoting) properties; it acts by increasing plant growth so rapidly as to use up the normal nutrients and kill the plant.

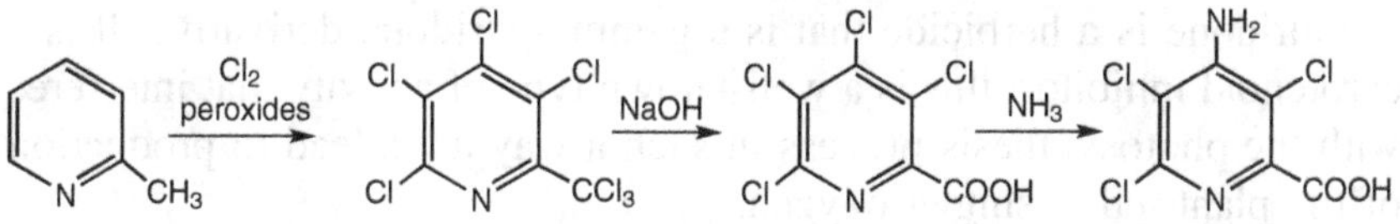

Scheme 11.1

Chlorpyrifos belongs to the family of organophosphorus insecticides, all of which function by inhibition of the enzyme cholinesterase. It is synthesized by the reactions of Scheme 11.2.

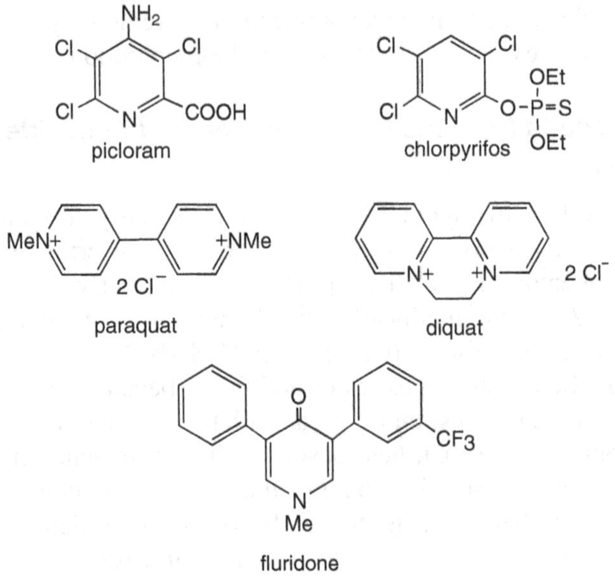

Figure 11.1. Some pyridine agrochemicals.

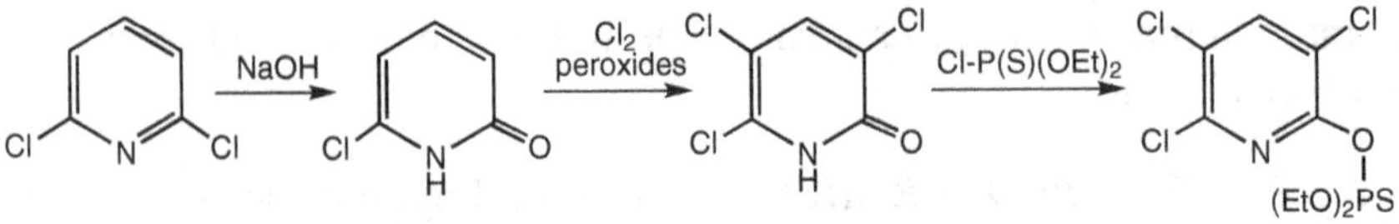

Scheme 11.2

Paraquat and diquat are both quaternary salts based on bipyridyls. Pyridine is the starting material for both compounds. They are non-selective, rapid action herbicides that act on all green plants through interference with the photosynthetic electron-transport system.

Fluridone is a herbicide that is a gamma-pyridone derivative. It is a carotenoid inhibitor; this is a well-known type of activity that interferes with the photosynthesis process in such a way as to lead to production of the plant-lethal singlet oxygen.

11.1.2.2. Pyrimidine Derivatives. Some prominent pyrimidine-based agrochemicals are shown in Figure 11.2.

The synthesis of these compounds employs the conventional cyclization methods of pyrimidine chemistry described in Chapter 6. For

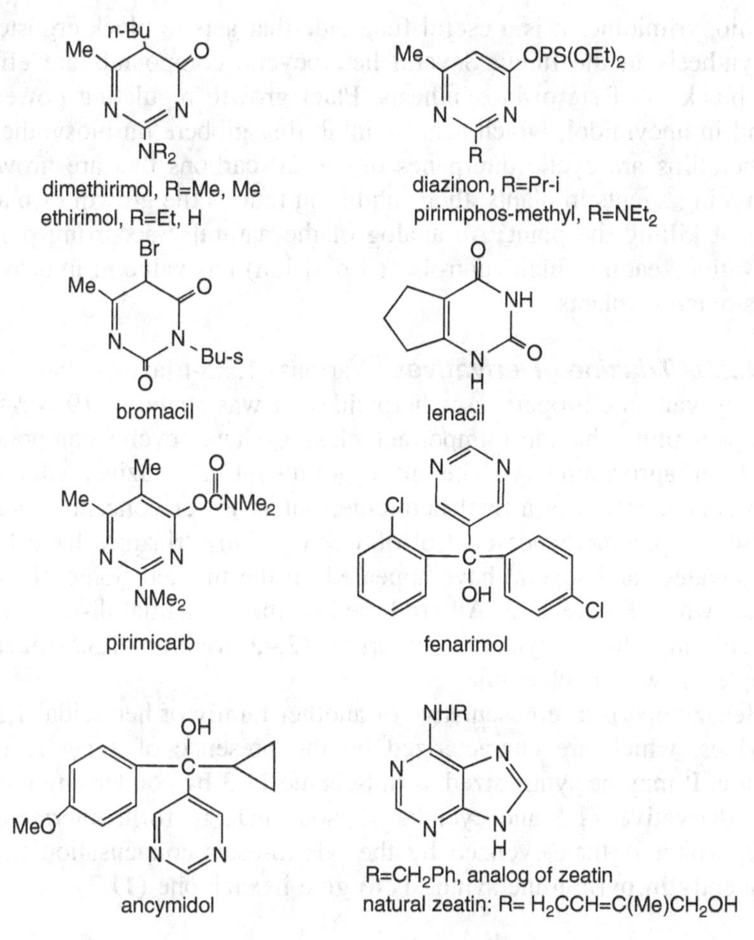

Figure 11.2. Pyrimidine derivatives as agrochemicals.

example, dimethirimol is prepared by reaction of the n-butyl derivative of ethyl acetoacetate with N,N-dimethylguanidine. This agent and ethirimol are fungicides, which seem to act through interference with the metabolism of adenine in the plant. Bromacil and lenacil are analogs of uracil and are made from cyclizations around urea; both are used as total herbicides. Diazinon, pirimiphos, and pirimicarb are insecticides that react with the enzyme cholinesterase that is involved in the nervous system. Their immediate precursors are 4-carbonyl derivatives of the pyrimidines, which are dominant forms in keto-enol tautomerism; the acyl groups are added by attack on oxygen of the enolic form. Fenarimol is prepared by condensation of dichlorobenzophenone and

5-lithiopyrimidine. It is a useful fungicide that acts to block ergosterol biosynthesis in the fungi. Several heterocyclic compounds are effective blockers of sterol biosynthesis. Plant growth regulating power is found in ancymidol, which acts by inhibiting gibberellin biosynthesis. Gibberellins are cyclic diterpenes of 19–20 carbons that are growth-promoting agents in plants; their inhibition retards the growth of plants without killing the plant. An analog of the naturally occurring purine derivative zeatin (which controls cell division) has value in improving the storage of plants.

11.1.2.3. Triazine Derivatives. Various 1,3,5-triazines have extremely valuable properties as herbicides. It was stated in 1984[1a] that they constitute the most important class of heterocyclic compounds in all of agrochemistry. The most prominent is atrazine, which at high concentration is a total herbicide, but at lower concentrations is useful for preemergence control of weeds. Many triazines have been investigated, and several have appeared on the market. Some of these are shown in Figure 11.3. All are based on the sequential displacement of chlorine from cyanuric chloride (2,4,6-trichloro-1,3,5-triazine, Chapter 6) with nucleophiles.

Hexazinone is a representative of another family of herbicidal 1,3,5-triazines, which are characterized by the presence of a cyclic urea feature. It may be synthesized as in Scheme 11.3 by condensing guanidine derivative **11.1** and cyclohexyl isocyanate to form intermediate **11.2**, which is then cyclized by the NH-to-ester condensation found frequently in pyrimidine synthesis to give hexazinone (**11.3**).

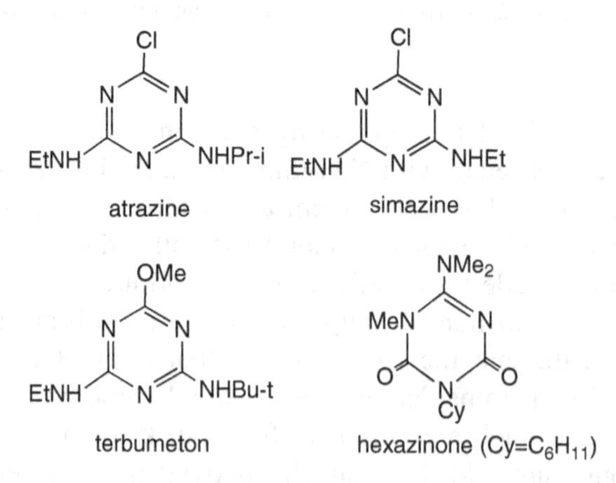

Figure 11.3. 1,3,5-Triazines as herbicides.

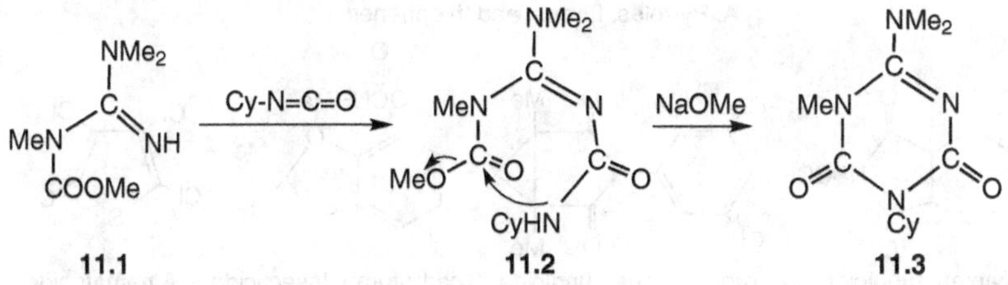

Scheme 11.3

Keto derivatives of 1,2,4-triazines are also useful herbicides. An example is metamitron, which is formed as in Scheme 11.4 by intramolecular condensation of a hydrazine group with an ester.

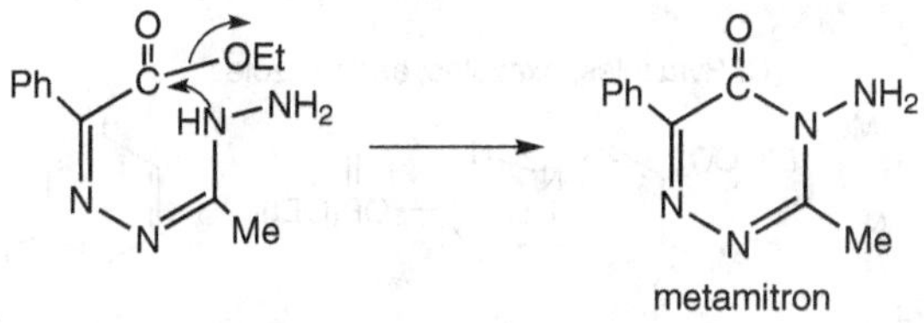

metamitron

Scheme 11.4

11.1.2.4. Derivatives of 5-Membered Ring Systems. Many derivatives of various 5-membered rings have been introduced successfully as agrochemicals. This is well illustrated by the collection of compounds shown in Figure 11.4. For the most part, these agents have been synthesized by the conventional methods described elsewhere in this book and demonstrate the value of a fundamental knowledge of heterocyclic chemistry in an advanced applications area such as agrochemistry. The same statement would also be true in the field of synthetic medicinal compounds. In sections 11.1.2.1 through 11.1.2.3, some of the syntheses of 6-membered heterocycles valuable in agrochemistry are reviewed, to make the point of the use of conventional methods in practical chemistry. The construction of the 5-membered rings need not be reviewed here, but many can be understood on the basis of material already presented in this book.

11.1.3. Examples of the Results from More Recent Research

Only a few illustrative structures will be shown here, defining some trends in research of the 2000s.

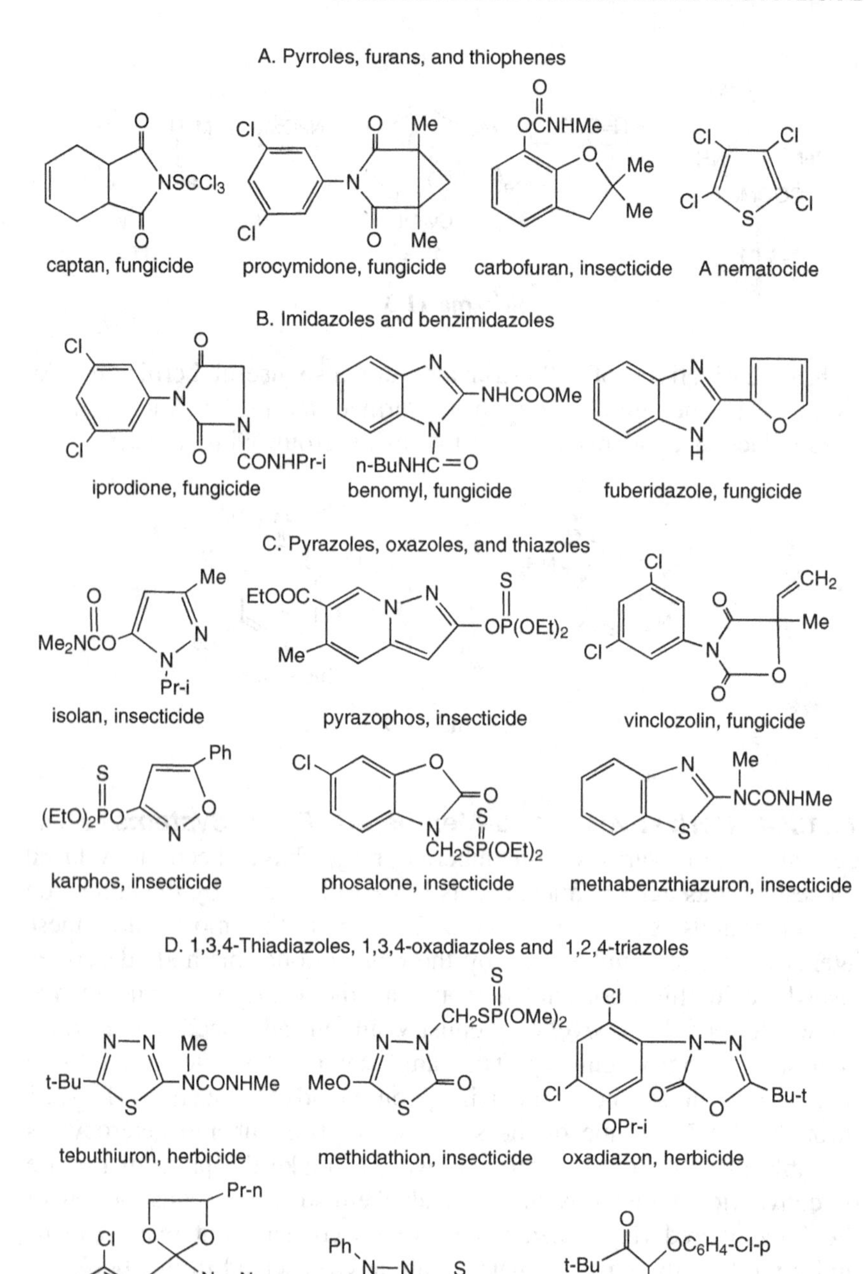

Figure 11.4. Derivatives of 5-membered rings.

11.1.3.1. Triazolopyrimidines. Flumetsulam was the first of a new family of herbicides containing the triazolo[5,1-a]pyrimidine ring system.[2] Important in this compound also was the presence of a sulfonamido group. Research has continued in the triazolopyrimidine area; a report published in 2009[3] describes structure–activity relations in this family and the course of research that led to the discovery of the new highly valuable herbicide penoxsulam.[3] Here, the ring system has [1,5-c] fusion and the reversed structure of the sulfonamide group. It is active against grass and broadleaf weeds. Triazolopyrimidines exhibit their herbicidal activity by inhibiting the enzyme acetolactate synthase.

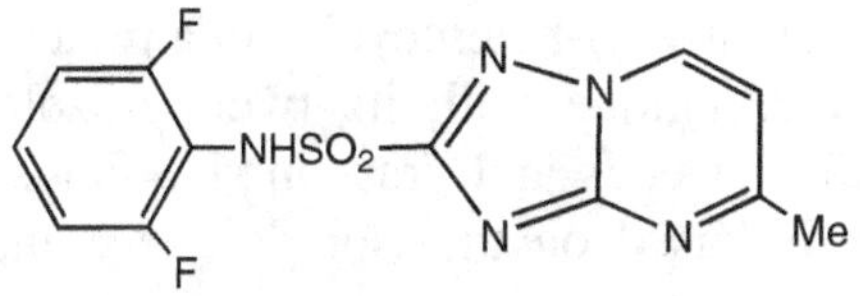

flumetsulam

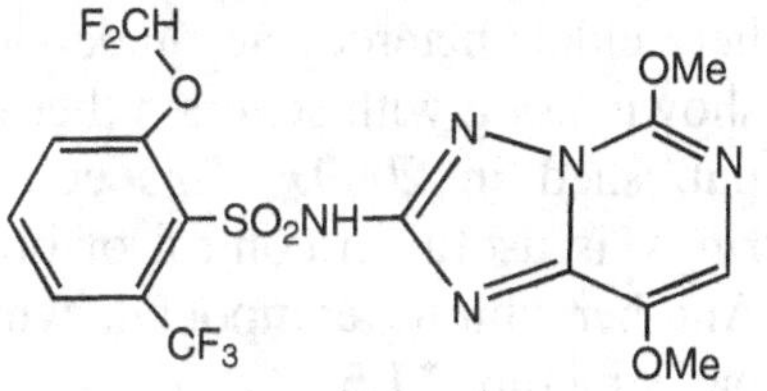

penoxsulam

11.1.3.2. Pinoxaden. This herbicide is active against grass weed species in the growing of grain cereal crops, especially rice. It is an inhibitor of acetylcoenzyme A carboxylase. Its structure evolved from considerations of the pyrazolidine-3,5-diones (and their enol derivatives), many of which have valuable herbicidal activity (e.g., structure **11.4**, designated CGA 271312 by Ciba-Geigy). The development of pinoxaden is described in a recent review article.[4]

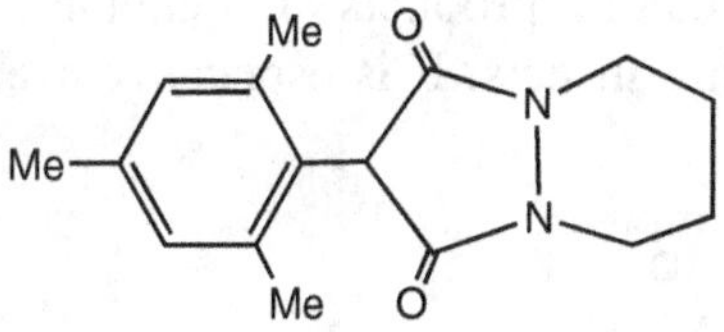

11.4

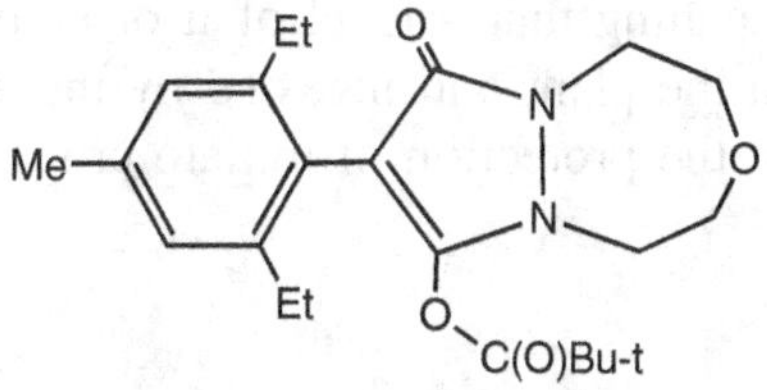

pinoxaden

The key step in the construction of the pyrazolo[1,2-d]-1, 4, 5-oxadiazepane ring is shown in Scheme 11.5.

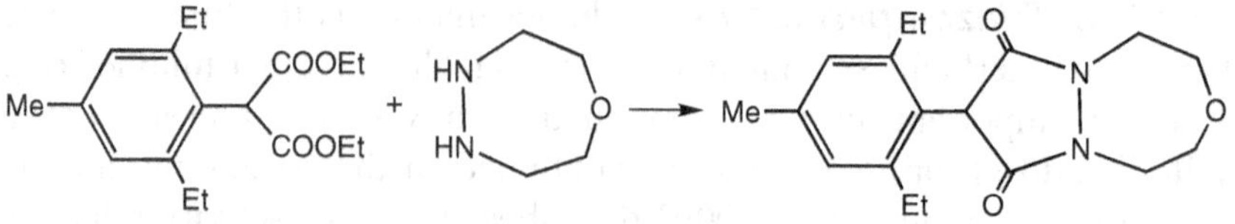

Scheme 11.5

11.1.3.3. Alkyne-Containing Heterocycles. Many heterocycles with alkyne groups have potent pesticidal activity and constitute another broad family receiving attention. As herbicides, they serve as inhibitors of the enzyme protoporphyrinogen-IX oxidase, which catalyzes the last step in the biosynthesis of chlorophyll. A typical herbicidal heterocycle under development is pyraclonil, which is shown along with several other useful alkynyl-heterocycles in a review published in 2009.[5] This compound, unique with its two pyrazole rings, is useful in control of broadleaf weeds and grass in rice fields. Another unique compound, which is based on the isoindole system, has structure **11.5**.

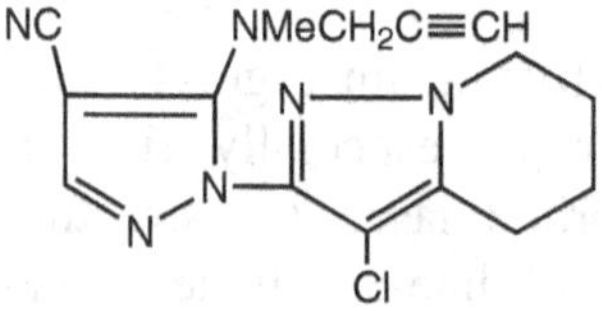

pyraclonil

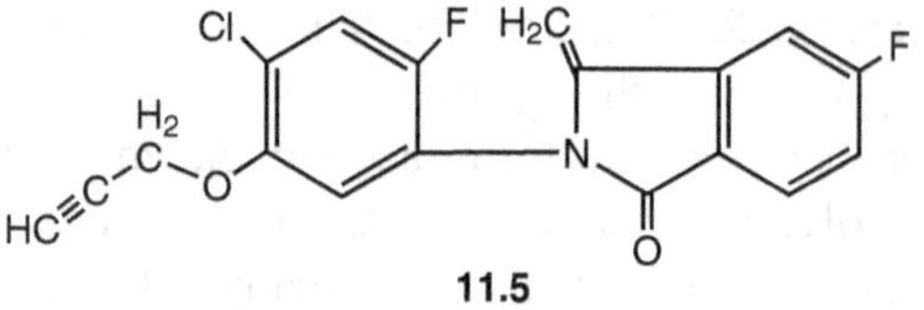

11.5

11.1.3.4. Thiamethoxam. This new insecticide is classed as a member of the important neonicotinoid family, which act as agonists of the nicotinic acetylcholine receptor. Thiamethoxam has systemic activity, meaning that a level of it or active metabolic products[6] is maintained in the plant and ingested by the attacking insects. It is especially used in the protection of tomato crops.

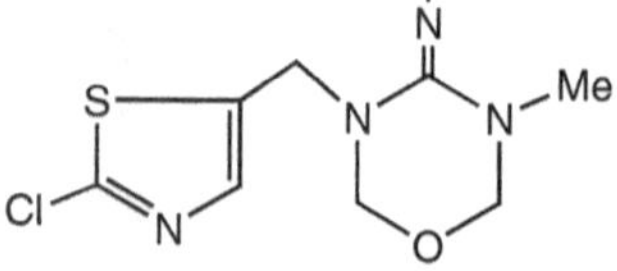

thiamethoxam

11.1.3.5. Chlorantraniliprole. Diamide insecticides are another class of recently introduced crop protection agents, which behave as activators of ryanodine receptors in the insect. This leads to uncontrolled calcium release in muscles. Chlorantraniliprole is a member of this family[7] and is in commercial use for protection from various pests.

chlorantraniliprole

11.1.3.6. Triketones with Heterocyclic Substituents. Triketones represent a well-studied, but still developing, family of herbicides. Some with pyridyl substituents are among the most active. Much of the research in this area has been reviewed.[8] The triketones are inhibitors of the plant enzyme 4-hydroxyphenylpyruvate dioxygenase (HPPD), which plays a key role in the biosynthesis of plastoquinone and tocopherol. Compound **11.6**, which is known as nicotinoyl syncarpic acid, is shown as a typical structure of this type. Its potent herbicidal activity led to synthetic work that has yielded many related structures in an effort to improve selectivity in the herbicidal action.

11.6, nicotinoyl syncarpic acid

11.1.3.7. Fipronil. This insecticide, which is a pyrazole derivative, is an effective agent for the elimination of various pests, such as wasps, bees, cockroaches, fleas, etc. It acts by disrupting the central nervous system of the insects, specifically by blocking chloride ion passage in the system. It is another discovery of the biological effectiveness of fluorine-containing substituents on a heterocyclic ring.

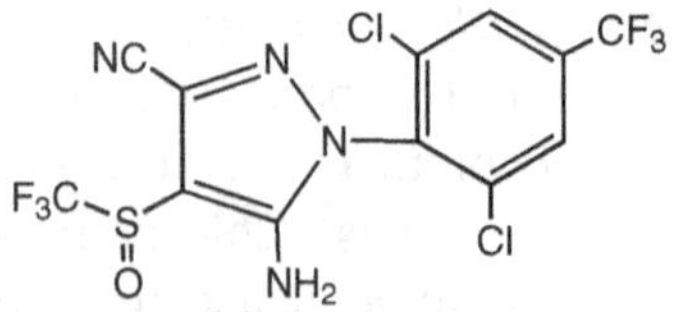

fipronil

11.2. APPLICATIONS OF HETEROCYCLIC COMPOUNDS IN COMMERCIAL FIELDS

Heterocyclic compounds are of great importance in many different fields of commerce. They represent specialized, well-developed areas of technology, and only a brief comment on such areas can be provided here. Much more detailed presentations are given in *Comprehensive Heterocyclic Chemistry*[1] and in the text *Heterocycles in Life and Society*.[9]

An extremely important application of heterocyclic compounds is in the field of dyes and pigments. Extended conjugation is an important ingredient for a compound to be colored, and heterocyclic systems, usually multicyclic, in great numbers have been constructed around this principle. The field is enormous, as is demonstrated in the review by D. R. Waring in *Comprehensive Heterocyclic Chemistry*.[1b] Industrial organic chemistry can trace its beginnings back to the (accidental) discovery of mauveine (**11.7**) in 1856 by W. H. Perkin; it was the first organic compound to be prepared synthetically at the industrial scale. Another heterocyclic compound, indigo (**11.8**), was derived from natural sources and was used for centuries before it was synthesized in 1883 and later made commercially. These two early compounds display the extended conjugation so important in the development of new dye and pigment chemicals.

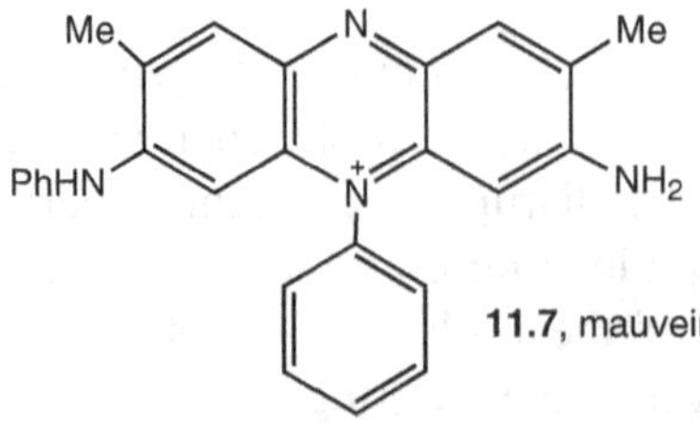

11.7, mauveine

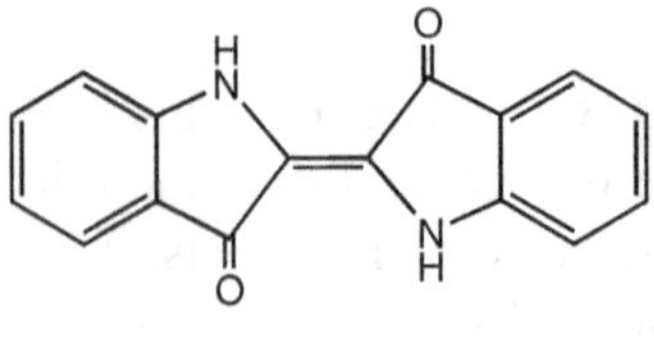

11.8, indigo

Many types of dyes are available, which require the presence of the proper functional groups to cause adherence to fabrics and other materials, and many techniques for the dying process are in use today.

Technology in the area of photography is highly developed, making use of heterocyclic compounds in various ways in the several steps of the process. Discussions will be found in *Comprehensive Heterocyclic Chemistry*.[1c,d]

Heterocyclic compounds can participate in polymer technology in several ways. They can be pendants on a polymer chain, as might be formed from the polymerization of vinyl monomers with heterocyclic substituents. There are processes where the polymer is formed by closing heterocyclic rings. Finally, heterocyclic groups can be added to previously formed polymers. These processes are described by S. M. Heilmann and J. K. Rasmussen in *Comprehensive Heterocyclic Chemistry*.[1e]

Hindered heterocyclic amines are used as light stabilizers in plastic and coating formulations, protecting against degradation by ultraviolet radiation. These agents are known as hindered amine light stabilizers (HALSs) and are commonly derivatives of 2,2,6,6-tetramethylpiperidine.[10] An example of a HALS agent is Tinuvin 770 (BASF), which is a diester of sebacic acid and 4-hydroxy-2,2,6,6-tetramethylpiperidine. It is thought to act through the formation of a piperidinoxyl radical.[11]

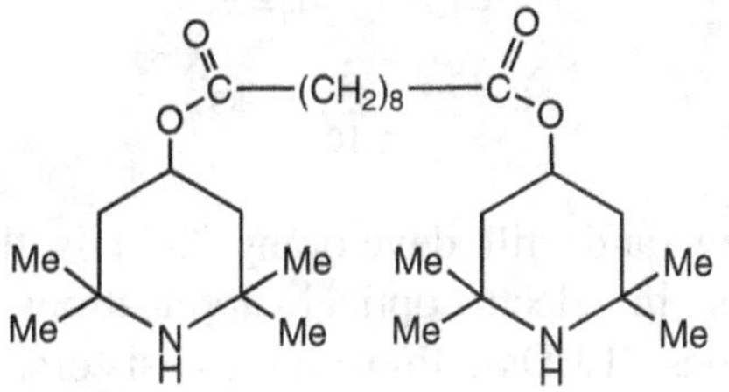

O C —(CH2)8— C O
Me Me Me Me
Me N H Me Me N H Me

tinuvin 770

A thriving and highly important field is the construction of coordination complexes from metallic species and heterocycles. These complexes can be useful as reaction catalysts and have other uses as well. To illustrate the catalyst area (which is large), the zirconium complex formed from the anion of indenylindoyl anion (**11.9**), and $ZrCl_4$ is offered as an example. The complex has the formula Zr(**11.9**)$_2$$Cl_2$ and is an excellent catalyst for the polymerization of olefins.[12]

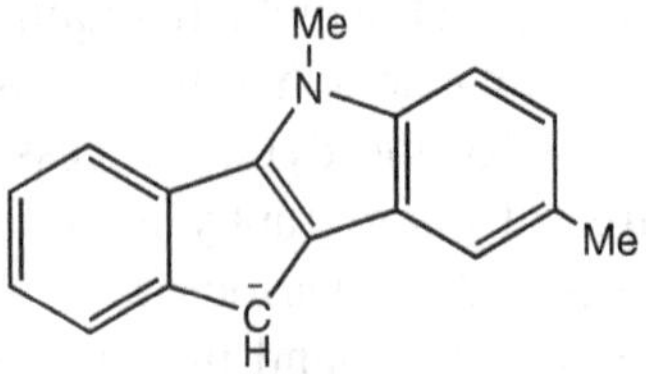

11.9

Also, we have noted in Chapter 10 that heterocycles with chirality can form complexes that are useful catalysts for asymmetric synthesis. This is a field of great contemporary interest.

Another valuable property is the selective binding of certain metallic species.[13] An example of this type of ligand is the 1,10-phenanthroline derivative **11.10** (PDALC), which selectively binds to larger metallic cations such as Ca^{++} and Pb^{++}. The crystallized coordination complex formed from calcium perchlorate has the formula $[Ca(PDALC)_2](ClO_4)_2 \cdot H_2O$. Certain heterocyclic ligands have special value in selective complexation because, as in the case of PDALC, the backbone containing the ligating nitrogen atoms can be rigid and offer a cavity of fixed geometry to receive the metal.

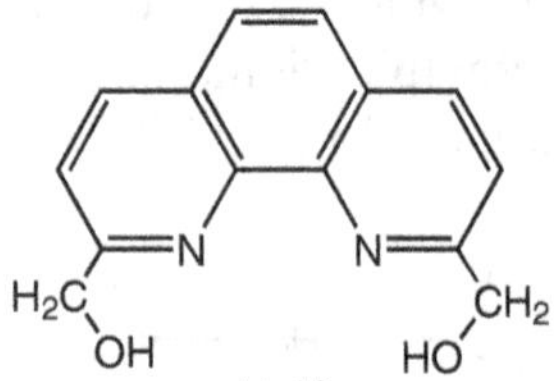

11.10

A relatively new and still developing field is the use of heterocyclic compounds in electro-optical applications, which includes light-emitting diodes (LEDs), thin-film transistors, and photovoltaic cells. To possess these properties, molecules must have extended conjugated unsaturation. This lowers the highest occupied molecular orbital (HOMO)-lowest unoccupied molecular orbital (LUMO) energy gap and causes light absorption at long wavelengths. One type of useful structure has several heterocyclic rings such as pyrrole or thiophene joined in a linear fashion. The phosphole ring system is a new participant in this type of array. This is illustrated by compound **11.11**, in which two thiophene rings are attached to a central phosphole ring (as the sulfide).

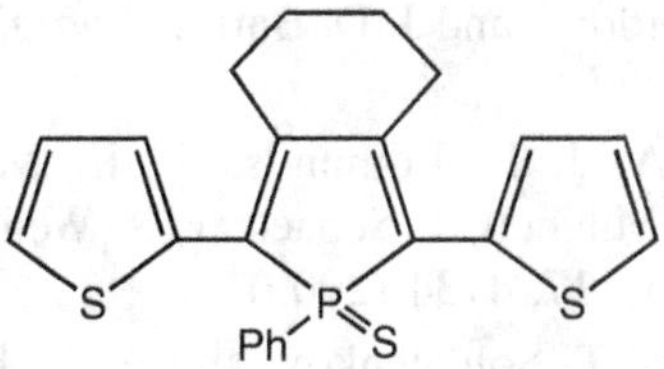

11.11

This compound has LED properties; when deposited as a thin film between a bilayer cathode and anode, yellow light was emitted by application of a low voltage.[14] Other related structures are being examined for similar electro-optical activity.

Another new application of heterocyclic compounds is in the field of ionic liquids. These compounds generally are quaternary salts of certain heterocyclic bases, and they are finding use as high-boiling polar solvents for extractions or as reaction media.[15] Common among the ionic liquids known so far are salts of imidazole, which are shown as follows.

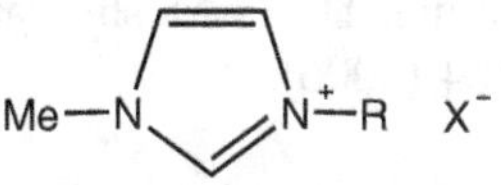

REFERENCES

(1) A. R. Katritzky and C. W. Rees, Eds., *Comprehensive Heterocyclic Chemistry*, Vol. **1**, Pergamon, Oxford, UK, 1984; (a) P. J. Crowley, Chapter 1.07; (b) D. R. Waring, Chapter 1.12; (c) J. Bailey and B. A. Clark, Chapter 1.14; (d) J. Stevens, Chapter 1.13; (e) S. M. Heilmann and J. K. Rasmussen, Chapter 1.11.

(2) W. A. Kleschick, B. C. Gerwick, C. M. Carson, W. T. Monte, and S. W. J. Snider, *J. Agric. Food Chem.*, **40**, 1083 (1992).

(3) T. C. Johnson, T. P. Martin, R. K. Mann, and M. A. Pobanz, *Bioorg. Med. Chem.*, **17**, 4230 (2009).

(4) M. Muehlebach, M. Boeger, F. Cederbaum, D. Cornes, A. A. Friedmann, J. Glock, T. Niderman, A. Stoller, and T. Wagner, *Bioorg. Med. Chem.*, **17**, 4241 (2009).

(5) C. Lamberth, *Bioorg. Med. Chem.*, **17**, 4047 (2009).

(6) R. Karmakar, R. Bhattacharya, and G. Kulshrestha, *J. Agric. Food Chem.*, **57**, 6360 (2009).

(7) G. P. Lahm, D. Cordova, and J. D. Barry, *Bioorg. Med. Chem.*, **17**, 4127 (2009).

(8) R. Beaudegnies, A. J. F. Edmunds, T. E. M. Fraser, R. G. Hall, T. R. Hawkes, G. Mitchell, J. Schaetzer, S. Wendeborn, and J. Wibley, *Bioorg. Med. Chem.*, **17**, 4134 (2009).

(9) A. F. Pozharskii, A. T. Soldatenkov, and A. R. Katritzky, *Heterocycles in Life and Society: An Introduction to Heterocyclic Chemistry and Biochemistry and the Role of Heterocycles in Science, Technology, Medicine and Agriculture*, Wiley, New York, 1997.

(10) H. Jia, H. Wang, and W. Chen, *Radiation Phys. Chem.*, **B76**, 1179 (2007).

(11) C. Saron, M. I. Felisberti, F. Zulli, and M. Giordano, *J. Braz. Chem. Soc.*, **18**, 900 (2007).

(12) S. Nagy, B. P. Etherton, R. Krishnamurti, and J. A. Tyrell, *U.S. Patent* 6,376,629 (April 23, 2002).

(13) R. T. Gephart III, N. J. Williams, J. H. Reibenspies, A. S. De Dousa, and R. D. Hancock, *Inorg. Chem.*, **47**, 10342 (2008).

(14) C. Fave, T.-Y. Cho, M. Hissler, C. W. Chen, T.-Y. Luh, C.-C. Wu, and R. Réau, *J. Am. Chem. Soc.*, **125**, 9254 (2003).

(15) R. P. Singh, R. D. Verma, D. T. Meshri, and J. M. Shreeve, *Angew. Chem. Int. Ed.*, **45**, 3584 (2006).

APPENDIX

UNIFIED AROMATICITY INDICES (I_A) OF BIRD

[Taken from C. W. Bird, *Tetrahedron*, 48, 335 (1992)]

Compound	I_A	Compound	I_A
1,3-Oxazole	47	Pyrimidine	79
1,2,4-Oxadiazole	48	Pyrrole	85
1,2-Oxazole	52	Pyridine	86
Furan	53	1,2,4-Triazine	86
1,2,5-Oxadiazole	53	1,2,4-Thiadiazole	89
Tellurophene	59	1H-Tetrazole	89
1,3,4-Oxadiazole	62	Pyrazine	89
1,2,3-Thiadiazole	67	Pyrazole	90
Selenophene	73	1H-1,2,3-Triazole	90
1,2,3-Triazine	77	1,2-Thiazole	91
Imidazole	79	1,2,4,5-Tetrazine	98
Thiazole	79	Benzene	100
Pyridazine	79	1,3,5-Triazine	100
1,3,4-Thiadiazole	80	1H-1,2,4-Triazole	100
Thiophene	81.5	Pentazole	109

Fundamentals of Heterocyclic Chemistry: Importance in Nature and in the Synthesis of Pharmaceuticals,
By Louis D. Quin and John A. Tyrell Copyright © 2010 John Wiley & Sons, Inc.

INDEX

Fundamentals of Heterocyclic Chemistry: Importance in Nature and in the Synthesis of Pharmaceuticals,
By Louis D. Quin and John A. Tyrell Copyright © 2010 John Wiley & Sons, Inc.